5th International Conference on Advanced Engineering and Technology (ICAET 2018)

IOP Conference Series: Materials Science and Engineering
Volume 507

Incheon, South Korea
14 – 16 December 2018

ISBN: 978-1-5108-8630-8
ISSN: 1757-8981

Printed from e-media with permission by:

Curran Associates, Inc.
57 Morehouse Lane
Red Hook, NY 12571

Some format issues inherent in the e-media version may also appear in this print version.

Printed by Curran Associates, Inc. (2019)

For permission requests, please contact the Institute of Physics at the address below.

Institute of Physics
Dirac House, Temple Back
Bristol BS1 6BE UK

Phone: 44 1 17 929 7481
Fax: 44 1 17 920 0979

techtracking@iop.org

Additional copies of this publication are available from:

Curran Associates, Inc.
57 Morehouse Lane
Red Hook, NY 12571 USA
Phone: 845-758-0400
Fax: 845-758-2633
Email: curran@proceedings.com
Web: www.proceedings.com

Table of contents

Volume 507

2018 5th International Conference on Advanced Engineering and Technology (5th ICAET)

14–16 December 2018, Incheon, South Korea

Accepted papers received: 7 March 2019
Published online: 24 April 2019

Preface

Papers

IOP Conf. Series: Materials Science and Engineering **507** (2019) 011001 doi:10.1088/1757-899X/507/1/011001

Preface

The 5th ICAET 2018 is the 2018 5th International Conference on Advanced Engineering and Technology (5th ICAET) took place in Incheon, South Korea, on December 14-16, 2018.

The conference program covered invited, oral, and poster presentations from scientists working in similar areas to establish platforms for collaborative research projects in this field. This conference will bring together leaders from industry and academia to exchange and share their experiences, present research results, explore collaborations and to spark new ideas, with the aim of developing new projects and exploiting new technology in this field.

The committee of ICAET expresses their sincere thanks to all authors for their high-quality research papers and careful presentations. All reviewers are also thanked for their careful comments and advices. Thanks are finally given to IOP Publication as well for producing this volume.

The Organizing Committee of ICAET 2018

Committee Chair

Prof. Dongyan SHI

College of Mechanical and Electrical Engineering

Harbin Engineering University

IOP Conf. Series: Materials Science and Engineering **507** (2019) 011001 doi:10.1088/1757-899X/507/1/011001

Conference Photograph

IOP Conf. Series: Materials Science and Engineering **507** (2019) 011001 doi:10.1088/1757-899X/507/1/011001

Sponsors

Incheon Disaster Prevention Research Center (IDPRC)	

Recently, various efforts to prevent and prepare are vitally needed for prevention of Disasters and calamities. So we understand the necessity for technology of disaster and we built up the Incheon Disaster Prevention Research Center (IDPRC) in Incheon National University (INU) in 1997.

Accordingly, Incheon Disaster Prevention Research Center (IDPRC) in Incheon National University has progressed of research on the prevention of disasters and calamities through the various seminars, conference and lectures. These research could be conducted cooperation with Incheon National University (INU) in various fields structure, soil, hydraulic and environment.

Incheon Disaster Prevention Research Center (IDPRC) will try to be a leader in the disaster of industry through the various research activities and global conference.

Incheon National University	

Incheon National University established a global campus by integrating two city council funded colleges into one in March 2010 and became a national university operated by the legal entity. With the start of attracting the branch of Lawrence Berkeley Research Center that is a world famous US national policy research center for education and research, Incheon National University will attract St. Petersburg University in Russia, Polymers University in Britain, and Kent University in Belgium and other foreign universities and research centers. These efforts will make INU spring to a world-wide competitive university.

As a local development leading university, Incheon National University established INU VISON 2020 and improved its competition actively through investment in selection and concentration. In compliance with the geographic feature and areal specialty, it will concentrate on the international trade, goods distribution, applied technology convergence, BNT-based life science, urban science and the sinology- based local humanity fields.

IOP Conf. Series: Materials Science and Engineering **507** (2019) 011001 doi:10.1088/1757-899X/507/1/011001

Lists of Committees

Committee name	Organizer name	Organization
Honorary Chair	Dongyan SHI	Harbin Engineering University
Conference General Chairs	Jongwan. Hu	Incheon National University
Honorary Chair	Jongwon. Jung	Chungbuk University
Technical Committees	I. Mansouri	Birjand University of Technology
	J. W. Hu	Incheon National University
	J. K. Ahn	Incheon National University

IOP Conf. Series: Materials Science and Engineering **507** (2019) 011002 doi:10.1088/1757-899X/507/1/011002

Peer review statement

All papers published in this volume of *IOP Conference Series: Materials Science and Engineering* have been peer reviewed through processes administered by the proceedings Editors. Reviews were conducted by expert referees to the professional and scientific standards expected of a proceedings journal published by IOP Publishing.

Comparison of cut-resistance performance of gloves made from virgin and recycled Para-Aramid fibres

M H Malik[1], N Akhtar[2], A Bakkar[3] and A Fraz[4]

[1]Professor, School of Textile and Design, University of Management and Technology, Lahore, Pakistan
[2]Assistant Professor, School of Textile and Design, University of Management and Technology, Lahore, Pakistan
[3]National Textile University, Faisalabad, Pakistan
[4]Assistant Professor, School of Textile and Design, University of Management and Technology, Lahore, Pakistan

Abstract. Application areas of high performance fibres are increasing day by day due to their superior physical, thermal and chemical properties. Para-aramid is one of these high-tech fibres which is extensively used for making technical textiles like cut-resistant gloves, cut-protective seat covers, and puncture-resistant fabrics, etc. As the use of this fibre has increased manifold, an attempt has been made to assess the performance of recycled fibres in comparison with the virgin fibres, by opening the para-aramid knitted fabric rags into fibrous state and convert the reclaimed fibres into yarn for making cut-resistant cost effective gloves. Four yarns, each of 70 tex, were spun on ~~a~~ DREF 3000 friction spinning machine. Two yarns were made from 100% virgin para-aramid fibres, the first one was the core-wrapped yarn consisting of a 50 micrometer steel wire in its core and the second one without it. The other two yarns were spun from a blend of 50% virgin and 50% recycled para-aramid fibres, one yarn with a 50 micrometer steel core while the other without it. The test results on the manufactured gloves from these yarns showed that the cut-resistance performance of the gloves made from 50:50 blend of virgin and recycled para-aramid fibres was comparable with the gloves made from virgin para-aramid fibres.

1. Introduction

Injuries at workplace can be caused by accidents, a growing number of which are due to improper or insufficient protection for workers. Industrial injuries can be caused by many reasons including mechanical failure, electronic failure, hazardous environmental conditions, faulty design of equipment, human failure, defective material, and poor working conditions [1]. Researchers have found that hands and fingers were most likely to get injured among body parts [2, 3]. Hands of workers are exposed to different types of dangers in various occupational fields like gardening, textiles, steel manufacturing, and glass furnaces, etc. Such dangers include burning, cutting, and rubbing, etc. of body parts. Majority of these hand injuries include puncture and cutting type lacerations [4]. Such injuries can be avoided if protective gloves are used [5]. Cut resistant protective gloves are used to protect the wearer's hands from cuts while working with sharp tools or edges in workplaces such as edible meat processing units, glass producing and processing works, and metal sheet processing plants, etc. [6]. Such gloves should have good grip ability and dexterity in addition to desired properties of tear strength and cut as well as abrasion resistance [7, 8]. The resistance to cutting is measured by cut resistance performance which has been classified into 5 categories by International Standards Organization [9]. Cut resistant gloves are made from different types of materials which are selected in different combinations to achieve certain level of cut resistance. Material selection cannot be done merely on the basis of tensile properties of materials which essentially show material's resilience towards pulling in lateral direction [10]. High strength materials, showing excellent properties against tensile loads, showed reduced strengths under traverse loads when loaders with sharpened contact edges were used [11]. Gloves made from latex give the lowest level of cut resistance whereas gloves made from metal mesh provide the best cut-resistant performance, but such gloves of metal mesh become heavy, uncomfortable, and possess poor dexterity. Therefore, attempts are made to make lighter and more comfortable gloves with better cut resistance performance. Recently, this objective has been achieved by using high performance materials such as para-aramids whose high strength to weight ratios resulted into strong yet light and comfortable gloves [12]. Para-aramids are about five times stronger than steel and gloves made from para-aramids give exceptional cut-resistance [12]. Para-aramids can also be used with reinforcement by steel wire or glass filaments. E-glass reinforced para-aramid gloves were found to be effective for cut protection under EN308:2003 [13].

Cut resistance is the material's ability of resisting damage or failure when an attempt to cut it is made by a moving sharp-edged object. Cutting, which involves sliding movement of a sharp edge against the subject material, is affected by friction between the two. Greater the work required to deform the material in transverse compression, higher will be the energy dissipated showing better cut resistance performance of the material [14]. Any increase in the coefficient of friction of the yarn changes the cut resistance of the material, depending on the thickness and the microstructure of the material to be cut [15]. Cut resistance can be improved by increasing areal density, or by using high-performance materials and composite yarns made with varying combination of materials [16, 17]. These methods can be used separately, or in combination, to improve cut resistance of textile materials.

Cut resistance performance of gloves and clothing can be measured by using two testing methods- EN388 applicable for European Union whereas ANSI/ISEA 105-2016 for American markets [18]. ASTM F1790 is based on the ANSI/ISEA procedure which uses TDM 100 machine in which straight edges with three different loads are slid against the fabric specimen to cut the specimen within 2-inch distance of movement, and then load for cutting at standard reference length of 25.4 mm (1 inch) is calculated by regression [19]. In EN388:2003 blade cut resistance method based on Coup Test, a circular blade with fixed 5N load is used to cut the glove specimen by moving back and forth on it. The number of cycles of blade to cut the fabric are noted; the higher the number of cycles, greater the cutting resistance. EN388 was revised in 2016 by combining both "Coup Test" and the "TDM-100 Test". The objective was to increase the accuracy of the testing instrument. In the new test a maximum limit of 60 cycles is fixed for coup test. If specimen is cut within this limit, TDM-100 test is skipped and result of Coup test is given mentioning skipping of the other test. On the other hand, if specimen does not cut within 60 cycles, then TDM-100 test becomes necessary and results are given accordingly. A two-year time allowance (up to 2018) was given to glove manufacturers selling in Europe to comply with the new EN388:2016 standard. The standard was revised to accommodate for the error induced by fabrics made of high performance yarns containing glass and steel which resulted into dulling of cutting edge.

Weight of the material, and different coatings applied to the outer surface, can also affect the performance characteristics. Lighter weight gloves are found to be more flexible, having better dexterity and result in less hand fatigue, whereas the heavier gloves generally provide better cut protection. Different materials have been used to improve cut resistance of the gloves. Nowadays, cut resistant gloves are made from different types of high performance composite yarns with various core and sheath combinations which give different levels of cut resistance performance. This study was done to evaluate the cut resistance performance of gloves made from virgin and recycled para-aramid fibres, and core-spun yarns made from stainless steel monofilament wire core. In addition, a comparison of gloves made from virgin and recycled para-aramid fibres was also made to evaluate the suitability of using recycled materials in order to address environmental aspects of using these materials.

The tensile properties and regularity of spun-yarns are influenced by length, strength, fineness and length uniformity of the constituent fibres [20-23]. To obtain the consistent quality of spun yarns, it is essential to have the minimum variation in the properties of the constituent fibres [24-26]. Two drawing frame passages are given to improve the regularity and uniformity of the material.

The production and use of plastics leaves bad impact on environment firstly because these are not easily biodegradable, and secondly because they consume about 8% of world oil and gas during production [27]. These impacts can be reduced by recycling of materials. Further, use of recycled materials also results into reduction in carbon dioxide emissions due to less need for new materials. Recycled materials have been in use for decades to cut down the costs in addition to protection of environment. Literature on use of recycled polymers for making cut-resistant gloves is scarce. In such a study use of recycled PPTA fibres with and without steel core gave better cut-resistant properties than virgin PPTA fibres probably due to higher coefficient of friction because of fibrillation during recycling process [28]. The present study was aimed at studying the possibilities of cost reduction by using blends of virgin and recycled para-aramid fibers and their effect on cut-resistance properties.

2. Materials and methods

In this study four types of yarns were made- yarns of virgin para-aramid (poly-p-phenylene terephthalamide or PPTA) fibres with and without steel core, and yarns of recycled para-aramid fibres of the same type mixed with virgin para-aramid fibres in the 50:50 ratios, with and without steel core. The core of stainless steel monofilament of 50 micrometer (micron) diameter was used for the study. All the yarns were spun on DREF 3000 friction spinning system. The recycled fibres for this study had been reclaimed by gently opening the knitted fabric rags using a textile recycling machine.

2.1. Raw material properties

The linear density, breaking strength, elongation at break and tenacity of stainless steel wire of thickness 50 microns were 139 deniers, 150 cN, 18.1% and 9.72 cN/tex respectively. The length, fineness, strength and elongation of virgin and recycled PPTA fibres were measured in the standard atmosphere in accordance with ISO 139:2005 following the test standards ISO 6989:1981, ISO 1973:1995 and ISO 5079:1995 respectively [29-32]. Physical properties of the fibres are summarized in table 1.

Table 1. Physical properties of para-aramid fibres.

Properties	Virgin PPTA Fibre	Recycled PPTA Fibre
Mean fibre length (mm)	46.05	35.50
Maximum fibre length (mm)	47.00	46.00
Minimum fibre length (mm)	43.00	16.00
CV %	2.34	27.05
Fibre fineness (tex)	0.25	0.25
Mean breaking force (cN)	35.74	21.94
Maximum breaking force (cN)	39.73	30.94
Minimum breaking force (cN)	31.91	15.22
CV %	6.69	20.03
Elongation at break (%)	17.68	11.96
Mean tenacity (cN/tex)	144.3	90.3
Maximum tenacity (cN/tex)	158.9	123.7
Minimum tenacity (cN/tex)	127.6	60.9
CV %	6.84	19.56

IOP Conf. Series: Materials Science and Engineering **507** (2019) 012001 doi:10.1088/1757-899X/507/1/012001

2.2. Pre-spinning processing

Virgin and recycled PPTA were processed separately in the blow room, carding section and drawing section of a textile spinning mill. The virgin PPTA was opened manually and fed to the blow-room line, which consisted of a bale-breaker, a transport fan, a fine opener, a metal detector, a transport fan and a chute feed system for feeding the smaller opened tufts in the form of a batt to the carding machines.

However, the recycled PPTA was first sprayed with antistatic oil in the ratio of 1.5% by weight of input material, and then thoroughly mixed by hand before feeding to the blow-room line. The virgin PPTA batt was carded at a rate of 24.25 kg/h, and sliver of 4.25 g/m (60 grain/yard) linear density was produced at a delivery rate of 100 m/min. On the other hand, recycled PPTA batt was carded at a rate of 16.95 kg/h and sliver of 4.25 g/m (60 grains/yard) linear density was produced at the delivery rate of 70 m/min. The carded slivers of virgin PPTA fibres were fed to the breaker drawing machine with a doubling of six slivers to obtain the breaker drawn sliver of 3.56 g/m (55 grain/yard) at the delivery speed of 335 m/min. The drawn slivers of virgin PPTA fibres were then fed to the finisher drawing machine with a doubling of six slivers to obtain the finisher drawn sliver of 2.91 g/m (45 grains/yard) at the delivery speed of 350 m/min. The blending of virgin and recycled PPTA fibres in 50:50 ratios was carried out on the breaker drawing machine having doubling of six slivers. The drawn sliver of linear density 3.56 g/m (55 grains/yard) was delivered at a speed of 335 m/min. A second passage of finisher drawing with doubling of six slivers was used to improve the blending and regularity within the material. The sliver of linear density 2.91 g/m (45 grains/yard) was drawn from finisher frame at a speed of 350 m/min.

2.3. Yarn spinning

Yarns from virgin PPTA and blend of virgin and recycled PPTA drawn slivers were spun with and without stainless steel monofilament core on DREF 3000 friction spinning machine. The linear density of each yarn was 70 tex (Ne 8.4). The first two types of yarns were made by using three finisher drawn slivers of virgin PPTA fibres as sheath, while the remaining two types of yarns were made by using three finisher drawn slivers of virgin and recycled PPTA fibres in 50:50 ratios. Each set of yarn types were further divided into two types based on existence or absence of monofilament stainless steel cores in addition to one sliver of same material as in sheath. The physical properties of the spun yarn were measured as per ISO 2062:2009 and ASTM D 3108-01 test methods and are given in table 2 [33, 34].

Table 2. Physical properties of yarns.

	Physical Properties				
Material Composition	**Breaking Strength (cN)**	**CV (%)**	**Elongation (%)**	**Tenacity (cN/tex)**	**Coefficient of Friction**
VPA[a]	998	8.69	6.06	14.26	0.233
RPA/VPA[b]	641	24.26	5.92	9.16	0.263
VPA+SC[c]	524	4.22	10.98	7.49	0.237
RPA/VPA+SC	417	12.76	5.85	5.96	0.277

[a] VPA = Virgin para-aramid
[b] RPA = Recycled para-aramid fibre
[c] SC = Steel core

2.4. Gloves manufacturing

Five pieces of gloves were made from each composite yarn on a 7-guage Shima Seiki gloves knitting machine. In each case the yarn feed rate and courses per unit length were kept constant to achieve the areal density of 285 g/m^2. Each glove had the size of 240mm x 100mm.

The methodology of the study can be expressed pictorially as below in 'figure 1'.

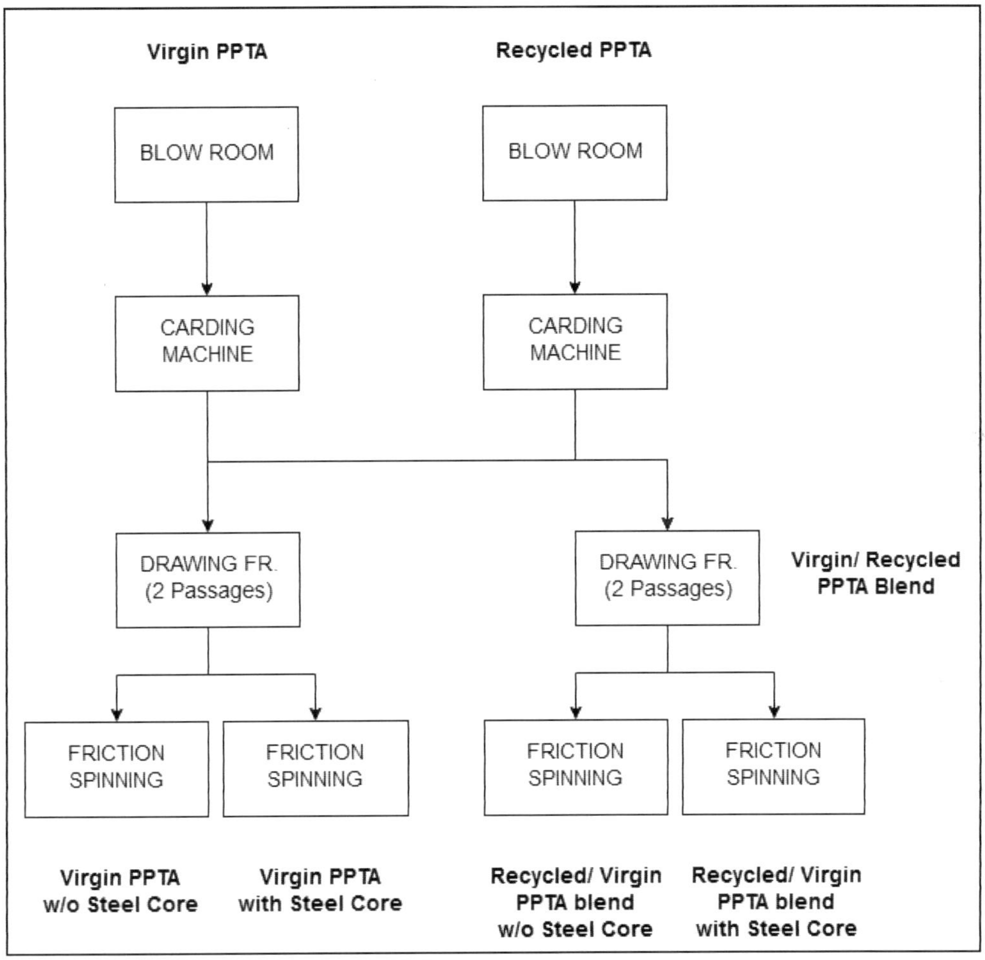

Figure 1. Flowchart of the study

3. Results and discussion

3.1. Physical properties of fibres and yarns

Physical properties of virgin and recycled materials were measured and compared for both fibre and yarn states. The physical properties of recycled fibres were poor as compared to virgin fibres (table 1), although maximum care had been taken during opening of knitted fabric rags on a textile recycling machine to avoid fibre breakage and loss of strength. In order to avoid fibre damage due to fibre lapping to rollers, anti-static oil was sprayed and atmosphere of the recycling department was controlled. Para-aramid fibres exhibited good fibre-to-fibre and fibre-to metal resistance. Nevertheless, the recycling process created fibrillation on the surface of reclaimed fibres (figure 2), compared to clear surface of virgin PPTA fibres shown in figure 3. The linear density remained unchanged in both cases.

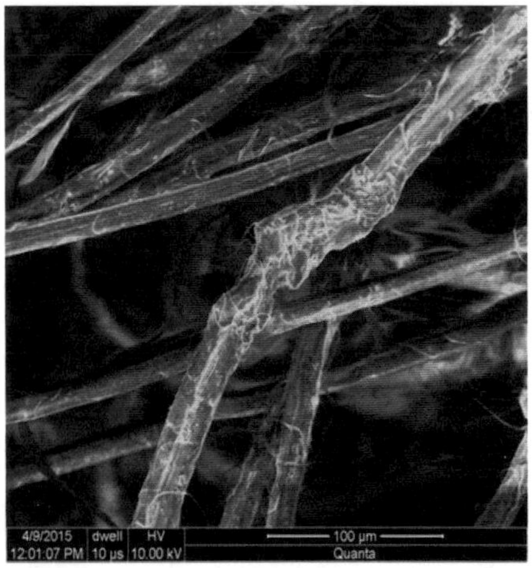

Figure 2. SEM image of recycled para-aramid fibre.

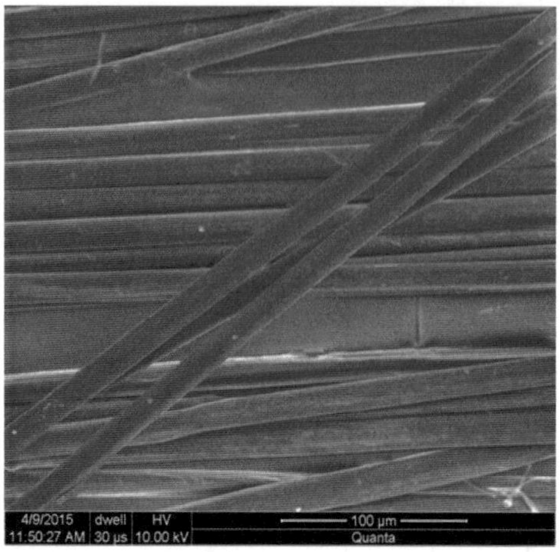

Figure 3. SEM image of virgin para-aramid fibre.

Tensile properties of yarns made from blend of recycled and virgin PPTA fibres in 50:50 ratios were considerably poor as compared to yarns made from virgin PPTA fibres due to inferior properties of recycled PPTA fibres. That showed the direct linear relationship between tensile properties of yarns and fibre length and tensile properties of their constituent fibres.

When a steel wire of 50 microns was used as a core of 70 tex yarns spun from a blend of recycled/virgin and virgin fibres, the tensile properties of composite yarns deteriorated as compared to the yarns of the same composition but without steel core due to low tenacity (9.72 cN) of steel wire in comparison with high tenacities of virgin and recycled PPTA fibres. Another factor for this decrease was use of 22% w/w of steel wire in the composite yarn. The coefficient of friction of yarns made from partly recycled fibres was higher due to presence of fibrils on the surface of these fibres. The protruding fibrils offered higher friction to metal during testing (table 2).

3.2. Glove cut resistance

In this research work, EN 388:2003 test method was used to measure the cut resistance performance of gloves as it was still valid and condition of the cutting blade was taken special care of. Another care was to ensure proper cutting of gloves fabric having yarns with stainless steel cores. In this method a circular blade with fixed load of 500g/5N was moved to and fro to cut the material. The number of cycles to cut the material is counted. The sharpness of the blade is checked before and after every test by using a standard reference fabric of cotton. The test is repeated five times to get the highest value of cut resistance index "I" which is calculated using equation (1).

$$I = \frac{(C + T)}{C} \tag{1}$$

Where "C" is the cut resistance of reference fabric and "T" is the cut resistance of testing specimen. Higher index values show better cut resistance performance of the material. Cut performance level ranging from 0 to 5 is then calculated based on average cut resistance index value as shown in table 3.

Table 3. Cut Performance Level

Cut Resistance Index	Cut Performance Level
20.0	5
10.0	4
5.0	3
2.5	2
1.2	1
<1.2	0

Higher cut performance level shows more resistance to cutting, which is desirable for protective gloves.

Further studies using updated testing method are expected by researchers in future. Mean values of cut indices and performance level of each sample are given in table 4.

Table 4. Cut performance of gloves.

Material Composition	Cut Index (lengthwise)	CV (%)	Cut Performance Level	Cut Index (Widthwise)	CV (%)	Cut Performance Level
VPA	2.78	2.27	2	3.53	6.28	2
RPA/VPA	3.07	2.18	2	4.02	3.73	2
VPA+SC	2.82	2.62	2	3.79	7.94	2
RPA/VPA+SC	3.47	5.28	2	5.90	7.79	3

The mean cut index values along the length and width of glove samples were in the range of 2.78-3.47 and 3.53 – 5.90 respectively. The cut performance of all samples in length and width directions was of level 2 except one, where the glove made from recycled/virgin fibres with steel wire core showed a mean cut index of 5.9 along the width corresponding to cut performance level 3.

High level of coefficient of variation of cut indices in both directions for yarns using stainless steel cores, perhaps, showed the insignificance of using that core, may be due to slippage between core and wrapping fibres.

In general, the performance of all glove samples made from virgin or recycled/virgin fibres either with or without steel core in yarns fall in the cut performance level 2, which shows that cost effective cut-resistance gloves of same performance level can successfully be made from recycled para-aramid fibres instead of expensive virgin para-aramid fibres. This result is also not against a recent study conducted by Awais et al. [28]. Moreover, the usage of steel core could not result into any significant improvement in blade cut resistance of gloves. Higher cut performance levels can be achieved either by increasing the weight of gloves using coarser yarns or placing glass yarn in the core of yarn instead of steel wire.

4. Conclusion

Cost effective cut-resistant gloves can be manufactured by using friction-spun yarns consisting of recycled para-aramid fibres. However, recycling process should be gentle to avoid fibre length breakage which reduces the tensile properties of the yarns and may cause lower efficiency in the knitting process. The recycled process created fibrils on the surface of fibres which seem to be favorable to the cutting resistance of the knitted gloves. The movement of the fibrils with the moving blade dissipates a part of the cutting energy in friction resulting in enhanced cut index. Further the mobility of fibres and yarn improves the cut performance level of knitted structures. The use of steel core in making the yarns could not show justification due to high levels of variation within readings probably due to slippage between core and sheath materials. In fact, the use of monofilament core becomes questionable as a result of this study.

IOP Conf. Series: Materials Science and Engineering **507** (2019) 012001 doi:10.1088/1757-899X/507/1/012001

References

[1] Sanders M S and McCormick E J 1993 *McGraw-Hill.*
[2] Khan M Y 2000 *Pakistan J. Med. Sci.* **16** 123-7.
[3] Khan M M A, Halim Z I and Iqbal M 2006 *Int. J. Occup. Saf. Ergon.* **12(3)** 327-41.
[4] Dolez P I, Gauvin C, Lara J and Vu-Khanh T 2010 *Int. J. Occup. Saf. Ergon.* **16(2)** 169-83.
[5] Dolez P I 2010 *Int. J. Occupational Safety and Ergonomics.* **16 (2)** 169-83.
[6] The Superior Book of Cut Protection available at https://www.superiorglove.com/blog/wp-content/uploads/Superior-cut-book-2017-email.pdf accessed on January 20, 2019
[7] Mellstrom G A and Boman A S 2005 *CRC Press.*
[8] Bhardwaj A, Kaushik J and Jajpura L 2016 *Int. J. Enhanced Res. in Sci., Tech. and Eng.* **5(11)** 16-22
[9] International Standards Organization ISO 13997:1999 *Standard test method for determination of resistance to cutting by sharp objects.*
[10] Shin H S, Erlich D C, Simons J W and Shockey D A 2006 *Tex. Res. J.* **76(8)** 607-13.
[11] Shin H S, Erlich D C and Shockey D A 2017 *Failure Behavior of High Strength Yarns under Transverse Loads* Proc. of the Asian Pac. Conf. on Fract. and Strength and Int. Conf. on Adv. Technol. in Exp. Mech. (Tokyo) 701-5.
[12] Ertekin M and Kirtay H E 2016 *J. Tex. Inst.* **107(10)** 1276-83.
[13] Bhardwaj A, Kaushik J, Jajpura L and Kumar N 2017 *Int. J. Res. In App. Sci. and Eng. Tech.* **5(10)** 291-96.
[14] Kothari V K, Das A and Sreedevi R 2007 *Ind. J. Fibre and Tex. Res.* **32(3)** 306-11.
[15] Thi B V, Vu-Khanh T and Lara J 2005 *J. Thermoplastic Compos. Mater.* **18(1)** 23-35.
[16] Boman A, Estlander T, Wahlberg J E and Maibach H I 2004 2ed *Taylor and Francis*
[17] Hearle J 2001 *High Perform. Fibers (Taylor & Francis)*
[18] European Union EN 388:2016 *Standard Test Method for Gloves Giving Prot. from Mech. Risks*
[19] American Society of Testing and Materials ASTM F1790-14:2014 *Standard Test Method for Measuring Cut Resistance of Mater. Used in Prot. Clothing.*
[20] Lyingar R L N and Gupta A K 1974 *Tex. Res. J.* **7** 489-92.
[21] Neelakantan P and Subramanian T A 1976 *Tex. Res. J.* **1** 822-7.
[22] Pan N, Hua T and Qiu Y 2001 *Tex. Res. J.* **11** 960-4.
[23] Ramey H H, Lawson R and Worley S 1977 *Tex. Res. J.* **10** 685-91.
[24] Patel G S and Patil N B 1975 *Tex. Res. J.* **2** 168-72.
[25] Pillay K P R 1964 *Tex. Res. J.* **8** 663-74.
[26] Zhu R and Erthridge M D 1997 *Tex. Res. J.* **9** 694-8.
[27] Hopewell J, Dvorak R and Kosior E 2009 Plil. *Trans. R. Soc. B* **364** 2115-26.
[28] Awais M, Tausif M, Ahmad F, Jabbar A and Ahmad S 2014 *J. Tex. Ins.* **106(4)** 354-58
[29] International Standards Organization ISO 139:2005 *Standard test method for standard atmospheres for conditioning and testing*
[30] International Standards Organization ISO 6989:1981 *Standard test method for determination of length and length distribution of staple fibres (by measurement of single fibres)*
[31] International Standards Organization ISO 1973:1995 *Standard test method for determination of linear density- gravimetric method and vibroscopic method*
[32] International Standards Organization ISO 5079:1995 *Standard test method for determination of breaking force and elongation at break of individual fibres*
[33] International Standards Organization ISO 2062:2009 *Standard test method for yarns from packages- determination of single-end breaking force and elongation at break using constant rate of extension (CRE) tester*
[34] American Society of Testing and Materials ASTM D 3108-01:2001 *Standard test method for coefficient of friction, yarn to solid material*

IOP Conf. Series: Materials Science and Engineering **507** (2019) 012002 doi:10.1088/1757-899X/507/1/012002

Comparative Investigations on Internal Force Distribution Characteristics on Concrete Segments in Qiongzhou Strait Sub-Sea Tunnel

Zhiyuan Wang[1], Shouju Li[2], Kai Guo[2]

[1]College of Ocean and Civil Engineering, Dalian Ocean University, Dalian, 116023, China
[2]State Key Laboratory of Structural Analysis for Industrial Equipment, Dalian University of Technology, Dalian, 116024, China

Abstract. In the design of shield tunnel concrete segments, the calculation of internal forces on segment sections is an important part. In order to study the influence of combining calculation of water and soil pressure and separate calculation of water and soil pressure towards internal forces on segment sections, taking the Qiongzhou Strait sub-sea tunnel as the background, ANSYS software is used to establish the equivalent stiffness circle model, hinged model and beam spring model, and the internal forces on segment are calculated by two methods of combining calculation of water and soil pressure and separate calculation of water and soil pressure. The results show that the difference of the bending moment between the two methods is great while the difference of the axial force is small, the maximum bending moment of combining calculation of water and soil pressure is 2.2~2.9 times larger than that of separate calculation of water and soil pressure, and the axial force is 11%~16% larger than separate calculation of water and soil pressure.

1. Introduction

With the economic development and increasement of people living level, the demands for transportation infrastructures increase daily. Metro tunnel and underwater tunnel have been widely constructed by means of shield machine in worldwide because shield tunneling has some advantages of safety operations, lower effects on environments, high applicable for soil layers and higher operating speed. The concrete segments are used to support loading from soil and water pressure. There some segments in a lining ring. The segments are linked with bolts in a lining ring. The joints between concrete segments are weak parts in burdening soil and water pressure. So, to investigate the influence of joint mechanical characteristics on internal force distribution is very important for design of concrete segment section. Cao analyzed tunnel lining structures using multi-scale modeling method [1]. Arnau performed three-dimensional structural response of segmental tunnel linings [2]. Chen simulated crack problems in segments of shield tunnel using finite element method [3]. Do investigated segmental tunnel lining behavior using numerical method [4]. Tan studied engineering geological conditions and railway tunnel scheme across Qiongzhou strait [5]. Taking Qiongzhou Strait Submarine Tunnel as an example, this paper studies the internal force distribution characteristics of concrete segments under two loading conditions.

Taking underwater tunnel in Qiongzhou strait as an example, the two soil and water pressure loading models are proposed to investigate influence of segment joint mechanical characteristics on internal force distributions on concrete segments. Qiongzhou strait is located in between Leizhou peninsula and Hainan island, China. The minimum straight distance is 18km. The out-diameter of designed undersea tunnel is 11.1m. The height of concrete segments is 0.6m. The deposited depth of tunnel in soil layers is 50m, and depth of water is 80m. The concrete strength grade is C60. The elastic modulus of concrete is 40GPa. The Poisson's ratio of concrete is 0.2. The elastic modulus of soil layers is chosen as 40MPa. The Poisson's ratio of soil layers is chosen as 0.4. Two soil and water pressure loading models are considered to investigate internal force distribution characteristics on concrete segments. The first soil and water pressure loading model is called as combined calculation model of soil and water pressure. In this loading model, the soil and water pressure acting concrete segments is assumed that soil and water pressure is mixed together, the density of soil layer is saturated density. In the first soil and water pressure loading model, it is assumed that the water in soil layers is not free gravity water, and water will be constrained by soil skeleton and cannot flow in soil skeleton. So, soil and water including in soil skeleton is regarded as indivisible whole. This pressure loading model is applicable for poor permeability clay.

The second soil and water pressure loading model is called as separately calculation model of soil and water pressure. In this loading model, the soil and water pressures acting concrete segments are assumed to independently act on concrete segments. The density of soil layer is buoyant density. The water pressure loading for both loading models acts on normal direction of concrete segment surfaces. In the second soil and water pressure loading model, it is assumed that the water in soil layers is free gravity water, and water can freely flow through soil skeleton. This pressure loading model is applicable for sandy and silt soils. In the sandy and silt soils, there are more pores and poor cohesion between granules.

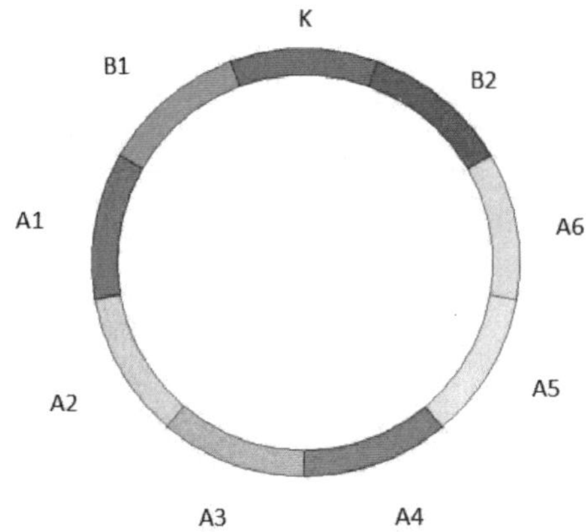

Figure 1. Arrangement of segments in a lining ring.

As shown in figure 1, one lining ring of concrete segments consists of 9 segments and 9 joints. The two segments are connected by 2 bolts in transverse section direction of tunnel. The segments are simplified as beam element (*BEAM3*) in ANSYS software. The soil layers are simulated by *PLANE42* element. The relations between concrete segments and soils are simulated by contacting elements.

2. Unchangeable stiffness beam model for concrete segment ring

To ignore the influence of joints on stiffness of segment linings, the concrete segment lining is simplified as homogenous circle ring with unchangeable stiffness beam. The bending moment distributions on concrete segment beam for two soil and water loading models are shown in figure 2. It can be observed from figure 4 that the maximum bending moment on concrete segment for combined calculation model of soil and water pressure is greater 2.28 times than separately calculation model of soil and water pressure. The maximum positive bending moments for two soil and water pressure computing models are located on roof and floor of concrete segment lining. The maximum negative bending moments for two soil and water pressure computing models are located on strait wall of concrete segment lining. The maximum negative bending moment is equal to maximum positive bending moment in absolute value.

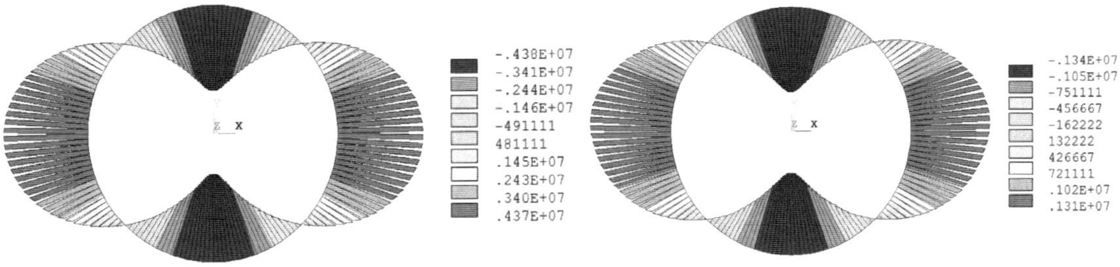

(a) Combined soil and water pressure loading model (b) Separately soil and water pressure loading model

Figure 2. Bending moment distributions on concrete segments using unchangeable stiffness beam model (unit:N·m/m).

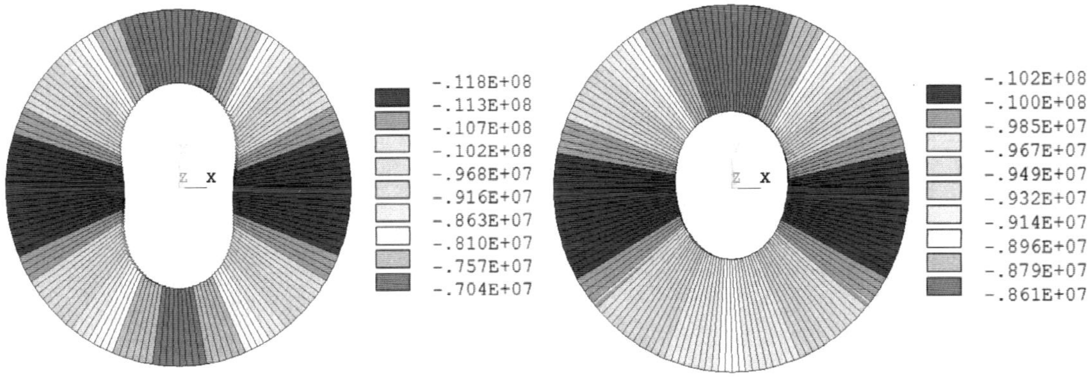

(a) Combined soil and water pressure loading model (b) Separately soil and water pressure loading model

Figure 3. Axial force distribution on concrete segments using unchangeable stiffness beam model (unit:N/m).

It can be observed from figure 3 that the maximum axial force on concrete segment for combined calculation model of soil and water pressure is greater 1.15 times than separately calculation model of soil and water pressure. The shapes of axial force on concrete segment for two soil and water pressure computing models are same. The maximum axial forces for two soil and water pressure computing models are located on straight wall of concrete segment lining.

3. Beams with hinges model for concrete segment ring

To ignore the bending resistance of joints on concrete segment linings, the joints of concrete segment lining are simplified as hinge joints without burdening bending moments.

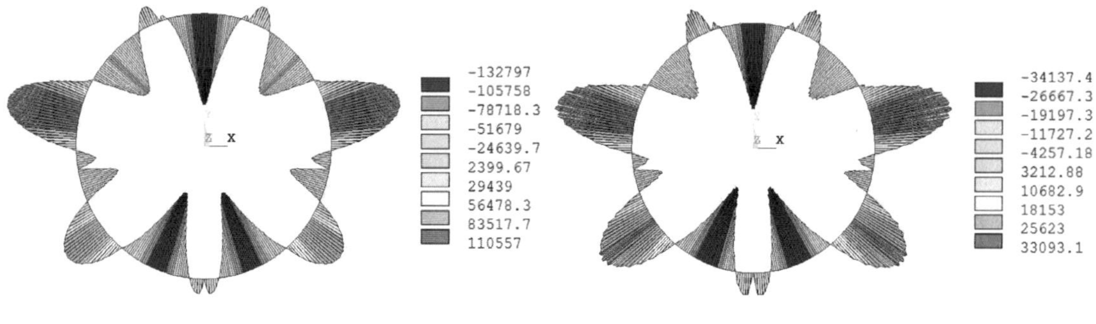

(a) Combined soil and water pressure loading model (b) Separately soil and water pressure loading model

Figure 4. Bending moment distribution on concrete segments using hinge model(unit:N·m/m).

It can be observed from figure 4 that the maximum bending moment on concrete segment for combined calculation model of soil and water pressure is greater 2.89 times than separately calculation model of soil and water pressure.

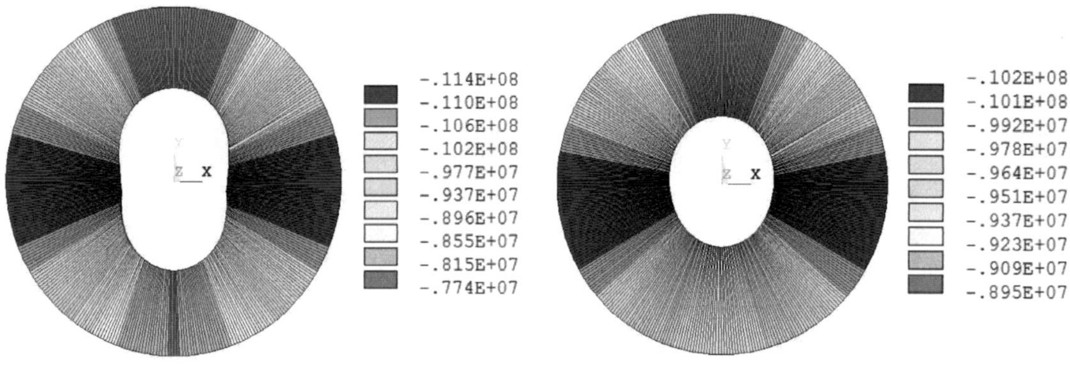

(a) Combined soil and water pressure loading model (b) Separately soil and water pressure loading model

Figure 5. Axial force distribution on concrete segments using hinge model (unit:N/m).

It can be observed from figure 5 that the maximum axial force on concrete segment for combined calculation model of soil and water pressure is greater 11.4% times than separately calculation model of soil and water pressure.

4. Beams with rotating stiffness spring model for concrete segment ring

In fact, the homogenous circle ring model with unchangeable stiffness beam overvalues bending resistance of joints, and hinged model ignores bending resistance of joints. So, the reasonable computing model for simulating joints of concrete segments is to take joints of concrete segments as rotating springs with rotating stiffness. Based on the section size of tunnel and connecting bolt property and experimental data [6], the rotating stiffness of rotating springs is determined as k_θ=70000kN·m/rad.

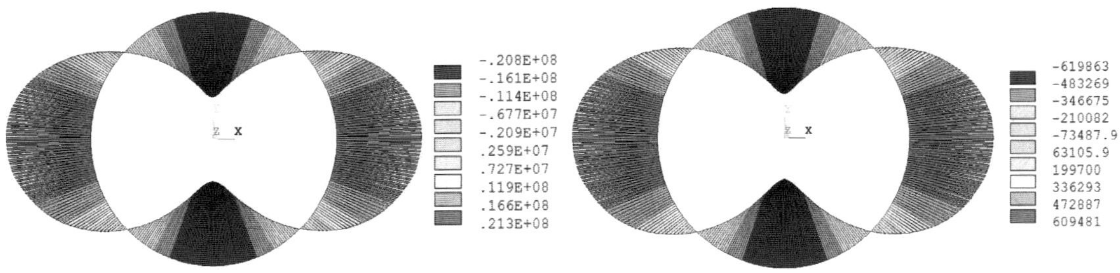

(a) Combined soil and water pressure loading model (b) Separately soil and water pressure loading model

Figure 6. Bending moment distribution on concrete segments using rotating stiffness spring model (unit: N·m/m).

It can be observed from figure 6 that the maximum bending moment on concrete segments for combined calculation model of soil and water pressure is greater 2.5 times than separately calculation model of soil and water pressure.

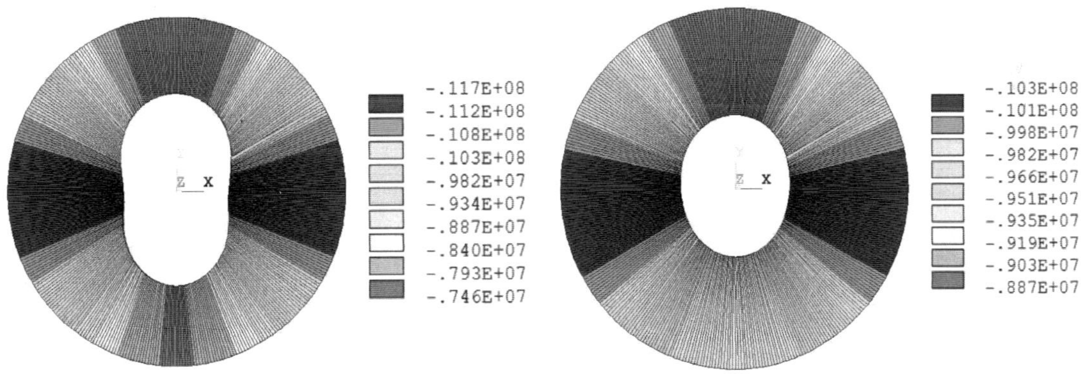

(a) Combined soil and water pressure loading model (b) Separately soil and water pressure loading model

Figure 7. Axial force distribution on concrete segments using rotating spring model (unit:N/m).

It can be observed from figure 7 that the maximum axial force on concrete segments for combined calculation model of soil and water pressure is greater 13.7% times than separately calculation model of soil and water pressure.

5. Discussion

From the above results, it can be seen that under the two loading conditions, there is no difference in the internal force distribution style in the segments, but there is a significant difference in the numerical value. It can be seen that under different soil conditions, the loads formed by the combination of soil and water are different, which has important reference value for construction design.

6. Conclusions

Shape of bending moments on concrete segments are basically same for combined model of soil and water pressure and separately model of soil and water pressure. The maximum bending moments on concrete segments for combined model of soil and water pressure are greater 2.28, 2.89 and 2.5 times than for separately model of soil and water pressure when using unchangeable stiffness model, range model and rotating spring model for concrete segment joints.

The axial forces on concrete segments are compressive states for both combined model of soil and water pressure and separately model of soil and water pressure. The maximum axial forces on concrete segments for combined model of soil and water pressure are greater 15.5%, 11.3% and 13.7% times than for separately model of soil and water pressure when using unchangeable stiffness model, range model and rotating spring model for concrete segment joints.

Acknowledgment
This work was supported by the National Program on Key Basic Research Project of China (973 Program) (2015CB057804) and the National Natural Science Foundation of China (11572079, 51209028) and Open Foundation of State Key Laboratory of Structure Analysis for Industrial Equipment (S14206) and the Liaoning BaiQianWan Talents Program 2016(2017-94) and the Program for Liaoning Excellent Talents in University (LJQ2014080).

References
[1] Cao Y, Wang P and Jin X 2012 *Tunneling and Underground Space Technol.* **28(1)** 124-34.
[2] Arnau O and Molins C 2012 *Eng. Struct.* **44** 210-21.
[3] Chen J S and Mo H H 2009 *Tunneling and Underground Space Technol.* **24** 91-102.
[4] Do N A, Dias D and Oreste P 2013 *2D Tunneling and Underground Space Technol.* **37(6)** 115-27.
[5] Tan Z, He W and Wang M 2018 *Tunnel Construction* **38(1)** 1.
[6] Koizumi C 2012 *Segment design for shield tunnel (Beijing, China Architecture & Building Press)*

IOP Conf. Series: Materials Science and Engineering **507** (2019) 012003 doi:10.1088/1757-899X/507/1/012003

Recent Advances in Bridge Engineering – Application of Steel Sheet Piles as Durable Structural Elements in Integral Bridges

P G Kossakowski[1]

[1]Department of Strength of Materials, Concrete and Bridge Structures, Faculty of Civil Engineering and Architecture, Kielce University of Technology, Al. Tysiąclecia Państwa Polskiego 7, 25-314 Kielce, Poland

Abstract. This paper discusses some of the problems related to the design and construction of integral and semi-integral bridges and the potential use of steel sheet piles as their permanent structural elements. The study focused on analyzing a recent application of this approach to support two overpasses in the TS-1 tunnel project completed in Kielce, Poland. The findings concerning the load-carrying capacity of steel sheet piles acting as supports indicate that this solution has a very high potential and it can be used on a much larger scale, not only in integral bridges but also in buildings and other non-building structures.

1. Introduction

Integral bridges are one of the most dynamically developing civil engineering structures, continually improved over time. Of interest is the fact that design concepts for integral bridges may involve the use of the latest structural systems and materials, e.g. high-strength steels, concretes or composites.

Structural durability has been an important factor determining the direction of research in this area for over a century. Actually, there have been two major trends: one focusing on the reduction in construction costs and the other on the reduction in maintenance costs. When a bridge has a service life of about 100 years [1], these problems become particularly crucial. Inspections of old bridges, older than 100 years, reveal that their condition is frequently poor, i.e. their load-carrying capacity and fatigue strength are no longer sufficient [2].

The research findings show that the phenomena responsible for the structural failure and, consequently, a shorter service life of a bridge are first observed at structural discontinuities. An example of these is the expansion joint between the bridge deck and the embankments. Some approaches to solving this problem assume that no expansion joint can be used; this means that the bridge needs to be designed and constructed as a continuous structure. Such bridges are called integral bridges, and their main feature is that there are no movement joints between the spans and the abutments in order to prevent relative horizontal displacements. Although integral bridges are usually designed as single-span structures, a multi-span continuous option is also possible. Most structures of this type are made of mass concrete.

The concept of integral bridges is not new in bridge engineering; the first studies were conducted in Great Britain and the US as early as in the 1930s. The first integral bridge was the Teens Run Bridge built in 1938 near Eureka in Gallia County, Ohio, US. Its five continuous spans made of reinforced concrete are supported by capped pile piers and abutments [3]. The idea of integral bridges has spread across the world. In Asia, for instance, the first integral bridges were constructed around the turn of the 21st century (Japan 1996, South Korea 2002). However, the country where integral bridges have become a common solution is Great Britain [4]. The recently completed integral bridges include the

Unstrut Valley Bridge in Germany, the bridges of the A1 motorway in Romania as well as the Dankuni-Palsit Flyover and the Kalkaji Flyover in India.

Many research centres and management companies, for example, those described in [5-13] are involved in improving the concept of integral bridges, because there are still many problems to be solved, such as problems related to the assessment of the ultimate limit state. The context of failure prediction based on damage mechanics (e.g. [14-16]) is important in this case. One of the areas of development of bridge structures is the introduction of elements that have had different applications, not even as structural members. An interesting and innovative solution is the use of steel sheet piles to build bridge abutment walls, which is the subject of this paper. There are very few bridges where steel sheet piles are used as integral structural elements; the examples in Poland include the abutments of the railway bridges in Lewin Brzeski and Swarzędz and the road bridges in Rzeszów and Kielce. In the Kielce project, described in this paper, the overpasses above the TS1 tunnel are some of the Polish biggest bridges where steel sheet piles were used as permanent structural elements. Obtaining a required load-carrying capacity of sheet piles as supports, and consequently, constructing a bridge with a considerable span length are significant achievements in the area of integral bridges.

2. Application of steel sheet piles as integral elements in buildings and non-building structures

Steel sheet piles are commonly used in the construction industry. In bridge engineering, their applications are numerous. Generally, they are employed as temporary walls to support excavations in various civil engineering projects. They are also suitable as elements protecting bridge piers and pier foundations from water action. For some time now, steel sheet piles have been introduced as permanent structural elements. In certain types of structures, steel sheet piles have become common as the main or self-supporting structural elements. However, globally, the number of projects with structures employing steel sheet piles on a regular basis is not high. In most cases, steel sheet piles are used as temporary elements supporting excavations.

Obviously, steel sheet piles have many benefits, which is why researchers and engineers are considering how to extend their applications. Steel sheet piles can be used in a number of complex civil engineering structures to solve specific problems and overcome certain limitations. A frequent problem encountered during construction or reconstruction of bridges located over operating rail lines or during maintenance, repair or removal of rail tracks is that the time limit allocated for such work is very short. Under such circumstances, it is not technologically feasible to use traditional reinforced concrete abutments; it is essential to look for alternative solutions offering, for example, fast installation of elements. One option is steel sheet piles, where the assembly time is indeed very short. The financial aspects of bridge projects based on sheet pile profiles are also very important, sometimes, more important than the engineering aspects. Another significant feature of this solution is that sheet piles have a double role when used as permanent structural elements: they protect excavation from collapsing during and after construction and they contribute to lower total costs because there is no need for their removal.

The benefits of steel sheet piles were noticed a long time ago. Initially, they were applied as self-supporting and main structural elements. Over the recent years, however, the solutions implemented globally have included tunnels, bridge abutments and underground car parks, where sheet piles constitute part of the structure. A good example of such a structure is a car park in Bristol constructed in 1999, where steel sheet piles were used as the walls. In Poland, steel sheet piles have been applied as permanent integral structural elements only recently.

3. General information on the TS-1 tunnel

The civil engineering structure discussed in this paper is the TS-1 tunnel with two overpasses. The project was completed in Kielce in 2011 [17]. It is part of the eastern exit from one of the Poland's main expressways, i.e. road S74. As was assumed at the design stage, the entire section of the exit from S74 was constructed to have parameters of an expressway, e.g. grade-separated junctions. The same rules were followed for the urban interchange of expressway S74 and national route DK73, where the TS-1 tunnel is situated. At this two-level interchange, road S74 was designed and constructed to run below road DK73. As a result, road DK73 includes two overpasses above the TS-1 tunnel. The solution to build an interchange in the form of a tunnel is not a novel concept; yet, the overpasses above the TS-1 tunnel in the form of single-span bridges comprising a system of T-beams made of prestressed concrete supported on steel sheet pile walls is definitely not a common solution in Poland. The entrance to the TS-1 tunnel is located in the middle of the Świętokrzyska section between Warszawska Street and Solidarności Alley. As mentioned above, the tunnel constitutes a section of road S74 above which there are two overpasses for road DK73 (figure 1a). The two-level interchange of roads S74 and DK73 was designed as two separate bridges: eastern overpass above the road in the direction of Warsaw and the western overpass above the road in the direction of Tarnów.

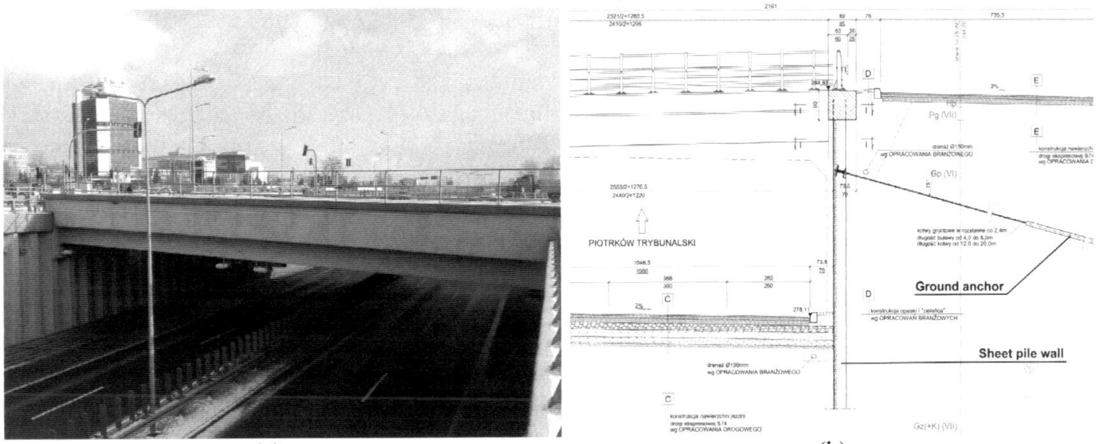

(a) (b)

Figure 1. (a) The TS-1 tunnel and the western overpass (view from the west); (b) Cross-section of the eastern overpass above the TS-1 tunnel with a detail of the supporting sheet pile wall [basing on 18].

4. Design assumptions – the use of steel sheet piles to construct the TS-1 tunnel walls

Different design and construction options were considered for the eastern exit from expressway S74 from Kielce. One of the basic problems encountered was complex soil conditions over the section including the TS-1 tunnel. The in situ tests showed that the soils used for the embankments included construction debris and subsoil (extending up to a depth of 1.8 m). Below that depth, the soil in the tunnel area is: non-cohesive (fine and medium sand, ranging from semi-compact to compact), cohesive (loamy sand, sandy clay, clay, and compact clay ranging from plastic to semi-solid) and rocky (at a depth of 13÷15 m) [19]. In most test holes, ground water was also present. Because of the complex soil and ground water conditions, it was necessary to install retaining walls with anchors placed in the soil to protect the tunnel excavation and to build a drainage system. The structure was classified as category II. The decision to use steel sheet piles to build the excavation wall resulted both from the soil and ground conditions and the economic aspects. The costs of construction of traditional retaining walls and bridge supports made of reinforced concrete were higher than the costs of construction of sheet pile walls. The function of the sheet pile walls was to support excavation during construction and to support the bridge above the tunnel. One of the main reasons for this solution was the high load-carrying capacity of steel sheet pile walls to withstand vertical loads. The ground water conditions required designing and building a tunnel drainage system including a drainage ditch along the entrance to the tunnel.

5. Details of the TS-1 tunnel structure

The two separate bridges, the eastern and western overpasses, constituting a section of national route DK73, were used as part of the roof of the TS-1 tunnel. The overpasses were single-span bridges with a horizontal clearance of 24.4 m, designed and constructed as a system of T27-type prefabricated prestressed concrete beams with a height of 110 cm spaced every 90 cm (figures 1a and b). The beams were supported on sheet pile walls and linked to them using caps (0.90 m in height and 0.85 m in width), fixed at the coping level, to form a frame system. The bridge deck with a thickness of 25 cm was made of reinforced concrete cast in-situ over the T27-type prefabricated pre-stressed concrete beams. C30/37 concrete and B500SP reinforcement steel were used for that purpose. The space between the eastern and western overpasses was shaped with the bridge deck; it was used to construct the inner corners of the spans and the retaining walls. The bridge entrance ramps were designed and constructed to have transition plates with a length of 5 m and a thickness of 25 cm. The entrance ramp is linked to the bridge using a bitumen expansion joint, whose role is to receive horizontal displacements ranging from –4.72 mm to +4.05 mm. The bridge deck was insulated using torch-applied modified bitumen membrane with a thickness of 5 mm. The pavement was designed to have two layers: the binder course, whose function was also to level irregularities at the roundabout, made of hard paving grade bitumen with a thickness of 5-8 cm and the surface course made of stone mastic asphalt (SMA) with a thickness of 4 cm.

According to the design and construction assumptions, the excavation was protected with steel sheet pile walls along the whole length of the TS-1 tunnel and the access roads. Cantilevered retaining walls were used at the initial sections of the tunnel entrance and exit, up to a depth of about 2.5 m. In the other sections, the walls were additionally reinforced by means of anchors placed in the soil.

Although steel sheet piles are used not only to protect excavation walls in different civil engineering structures but also as permanent structural elements, their application as abutment walls in bridges is a completely new solution in Poland, because of the size and function of the structure. Initially, it was assumed that the excavation walls and bridge abutments should be constructed using VL 604-type steel sheet piles. During installation, however, a decision was made to replace them with Z-type steel sheet piles (AZ 17 and AZ19 profiles shown in figures. 2 a, b and 3) produced by Arcelor Mittal.

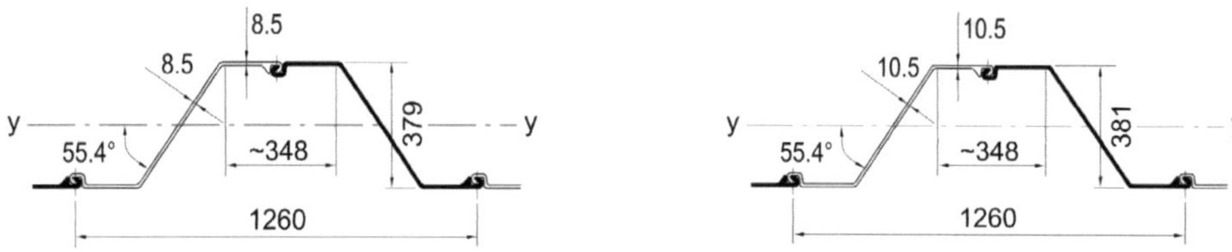

(a) (b)

Figure 2. Geometry of an Arcelor Mittal AZ sheet pile profiles: (a) AZ 17; (b) AZ 19 [19].

The reason for that was a number of benefits that AZ 17 and AZ19 profiles offer, and these include high strength parameters (because of the web continuity and symmetric distribution of joints), a high elastic section modulus to mass ratio, a high moment of inertia corresponding to lower deflections, faster assembly because of a large profile width and, finally, high durability. The values of deterioration in the form of thickness losses for steel sheet piles attributable to corrosion taking place in dry or watered ground was estimated according to the standard EN 1993-5 [20].

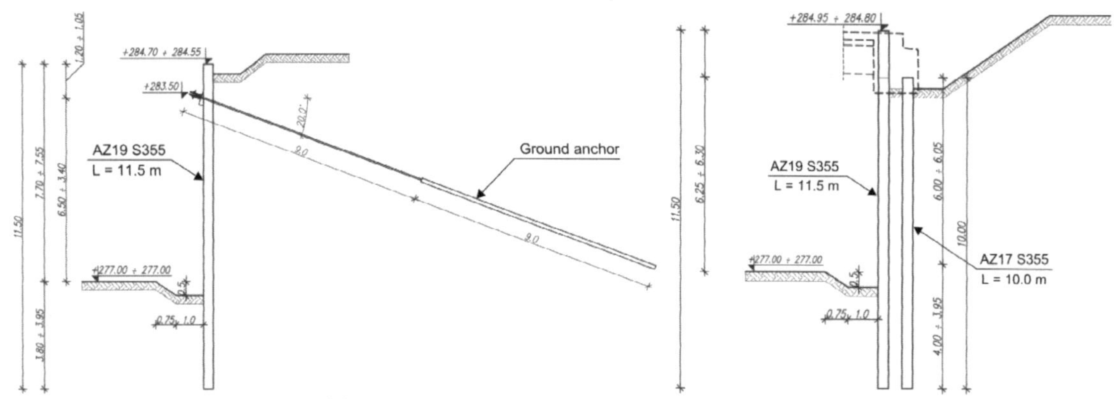

(a) (b)

Figure 3. Structural cross-sections of the sheet-pile walls of the TS-1 Tunnel: (a) sheet-pile excavation wall; (b) eastern overpass.

The driving technology was also very important. Steel sheet piles were driven using static jacking. An interesting problem from the design point of view is the load-carrying capacity of the steel sheet pile walls to withstand vertical forces; they are used as the support elements of the bridge to carry the considerable weight of the bridge itself (dead load) and the weight of the moving vehicles (live load). The selected sheet piles possess a high vertical load-carrying capacity, which enables the transfer of loads occurring along the width of the bridge. The load along the width of the abutment walls may amount to 881.9 kN per 1 m.

6. Conclusion

The TS-1 tunnel described in this paper is an example of one of the first Polish structures where steel sheet piles were used as durable structural elements for abutment walls. As bridges are subjected to higher and higher loads, resulting from an increase in road transport intensity, and they are more widely exposed to pollution, their condition may sometimes be unsatisfactory and they may be approaching the end of their service life. In the TS-1 tunnel project, one of the major structural problems to be solved was to reduce the number of places where destructive processes could occur. It was decided that continuous integral structural elements, i.e. sheet piles, should be used as supports. This solution can be regarded as a significant step forward in the development of integral bridges. It was essential, however, to predict the load-carrying capacity of the supports. The load-carrying capacity of steel sheet piles subjected to vertical loads has been studied since the last century. The engineers working on the TS-1 tunnel project not only approached the problem theoretically but also conducted experiments on laboratory- and real-scale models under test loads. The results were used to develop methods for determining the load-carrying capacity of sheet pile profiles. These findings can be employed to determine the load-carrying capacity of sheet pile walls used in the construction of bridges with any span length.

The TS-1 tunnel constructed as an integral bridge seems to be an optimal alternative for the future development of bridges. This structural solution should now be implemented on a larger scale; it should inspire both design and construction engineers to apply sheet piles in similar civil engineering projects.

References

[1] PN-EN 1990 EN 1990:2004 Eurocode - Basis of Structural Design
[2] Kossakowski P G 2013 *The Baltic J. Road and Bridge Eng.* **8(3)** 166–73.
[3] Burke M P Jr. 2009 *Wiley-Blackwell*
[4] Design Manual for Roads and Bridges. The Design of Integral Bridges. BA 42/96 2003 *London: Dep. of Transp.*
[5] Arsoy S, Barker R M and Duncan M J 2002 *Charlottesville: Virginia Transp. Res. Counc.*
[6] Faraji S, Ting J, Crovo D and Ernst H 2001 *J. Geotech. Geoenvironmental Eng.* **127(5)** 454-61.
[7] David T, Forth J and Ye J 2014 *J. Bridge Eng.* **19(6)** 04014012.
[8] Sherafati A and Azizinamini A 2014 *J. Bridge Eng.* **20(4)** 04014071.
[9] Feldmann M, Pak D, Hechler O and Martin P O 2011 *Struct. Eng. Int.* **21(3)** 311–9.
[10] Wasiu J and Popoola O O 2015 *J. Multidiscip. Eng. Sci. Tech.* **2(8)** 1994-9.
[11] England G L, Bush D I and Tsang N C M 2000 *London: Thomas Telford.*
[12] Furtak K and Warna B 2005 *Warszawa: Wydawnictwa Komunikacji i Łączności.*
[13] Masrom A 2014 *ESTEEM Acad. J.* **10(1)** 75-88.
[14] Kossakowski P G 2014 *J. Theor. Appl. Mech.* **52(4)** 995–1006.
[15] Kossakowski P G 2014 *Pol. Marit. Res.* **21(4)** 100-7.
[16] Kossakowski P G, Wciślik W 2014 *Key Eng. Mater. (Fract. Fatigue Mater. Struct.)* **598** 113–8.
[17] http://www.kielce-s74.pl
[18] Nowak W, Białas P, Dziurla M and Grosicki K 2010 *Kielce.*
[19] Arcelor Mittal Commercial RPS S.a r.l. Sheet piling 2012 *Steel Foundation Solutions for Projects Steel Sheet Piling.*
[20] PN-EN 1993-5:2009 Eurocode 3: Design of steel structures - Part 5: Piling

IOP Conf. Series: Materials Science and Engineering **507** (2019) 012004 doi:10.1088/1757-899X/507/1/012004

Tight Hammering for Plastic Zone of Materials

WANG Yingwei[1], MA Zhongnan[1], DONG Xin [1], YAN Qingguang[1], GAO Jinzhu[2], GAI Xiuying[2]

[1]Jilin University, Changchun 130022, China
[2]Institute of Metals, Chinese Academy of Sciences, Shenyang 110016, China

Abstract. Considering the failure due to cracks and peeling from curved surfaces of complex components, we used the tight hammering method to study the internal plastic deformation zones of curved-surface Q235 steel specimens after hammering shock and display relevant information. A Dino-capture microscope and a ZEISS LSM 700 laser scanning microscope were used to obtain and fit the morphologic data of rebounded zones and identify effective boundary points. Data of the boundary points were processed by least squares fitting. After curved-surface specimens of different curvature radii and planar specimens were hammered by the SR5.5 punch head and the plastic zones on the tight curved surfaces were all circularly-shaped at different impact strokes. The hammer traces were symmetrical spatially, indicating the internal plastic zones of the tight curved-surface specimens were approximately spherically-shaped. Deformations of plastic zones at external corners were large in magnitude and small in area, while those at internal corners were small in magnitude and large in area.

1. Introduction

The rapid industrial development has intensified the demand for high value industrial equipment and made it urgent to improve its fatigue life and reduce its manufacturing costs by appropriate surface treatment. Component surfaces can be strengthened by shot peening, rolling, hole expansion, explosion shock and friction strengthening [1, 2]. It is because any type of surface deformation processing will induce plastic deformation on metal surfaces and finally trigger high-density dislocation that can efficiently inhibit the initiation and propagation of surface microcracks [3-5]. So far, the main research methods of plastic zone include stress relaxation [6, 7], magnetic method [8, 9], X-ray diffraction [10], and ultrasound & shock indentation [11]. These methods all have cons and pros.

In recent years, there has been more studies in this field [12-19]. Underwood measured residual stress using surface displacement around indentations [15]. Chen et al. studied plastic deformation zones using impact indentation method and finite element numerical simulation [17-19]. Maximo et al. built a 3D single-hole expansion model and verified its practicability [20]. However, the analysis of plastic zone was dependent on finite element numerical simulation, in which the process model presetting and boundary condition setting largely affected the analytical results, as the evident indirectness of result estimation increased errors. Moreover, no novel research method or influential study has been reported in recent years.

Therefore, we innovatively proposed a tight hammering method for curved-surface plastic zones and succeeded in displaying the information of plastic deformation zone in curved surfaces. Specifically, plastic zones were observed under laser scanning confocal microscopy, and together with

IOP Conf. Series: Materials Science and Engineering **507** (2019) 012004 doi:10.1088/1757-899X/507/1/012004

mathematical theoretical computation. Their shapes and areas were studied. In particular, the effects of surface morphology and other factors were analyzed.

2. Experimental Section

2.1. Experiment materials

The tested material Q235 steel was made into strip specimens 170 mm × 20 mm × 30 mm (length × width × height). The hammered surface and tight surface of each strip were ground. After that, the surface roughness was 0.4 μm and hardness was 117HBW/10/3000.

2.2. Experimental methods

Metal-based pseudo-isotropic specimens were designed. Internal plastic deformations occurred after planar or curved surfaces were vertically hammered by a spherical punch head. These deformations would leave round-pit- shaped traces on material surfaces. These traces were the exposed surfaces of internal plastic zones, which were located beneath the traces and resulted from plastic rheological change under hammering compression.

Figure 1&2 show the schematic of hammer test and the specimens after hammering. The hammer test device consisted of a manual two-dimensional platform with a movement precision of 0.05 mm in both X- and Y-directions, a bracket, a straight-line guide rail, a slider and a 4.7 kg punch head. Two specimens were tightened together by a tightening device and simulated as one specimen. The tightening force should be large enough, which ensured instantaneous opening of tight fissures would not occur during the hammering process and deformation would occur immediately after the tight fissures were hammered by the punch head. After the tightening force was withdrawn, a rebounded convex zone (short for rebounded zone) would appear on the tight surface after the two tightened specimens were separated. The platform in the middle of this convex zone was the plastic deformation zone (short for plastic zone). The measuring points were set at appropriate distance, which avoided between-point interaction. Figure 3 shows the frontal hammer traces and rebounded zone on tight surfaces of separated specimens.

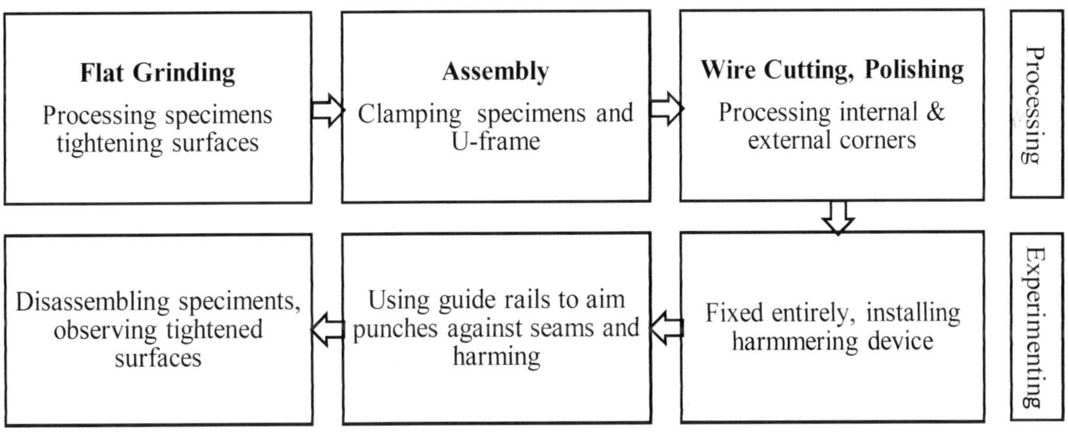

Figure 1. Schematic Illustration of Tight Hammering Method.

Considering the effects of hammering work, curved-surface shape and curvature radius on the results, we designed an SR5.5 hard alloy punch head with six impact strokes (25 mm, 50mm, 75 mm, 100 mm, 125 mm, 150 mm), internal corners and external corners of plane or curved-surfaces with five curvature radii (8 mm, 10 mm, 12 mm, 14 mm, 16 mm).

The specimens were observed and measured using a Dino-capture microscope 30×, which returned information about the surface shapes of rebounded convex zones. The rebounded convex zones were scanned using a ZEISS LSM 700 laser confocal microscope at z-axis resolution of 10 nm, and diagonal line of 18 mm in the scanning field, which returned precise 3D data of rebounded convex zones.

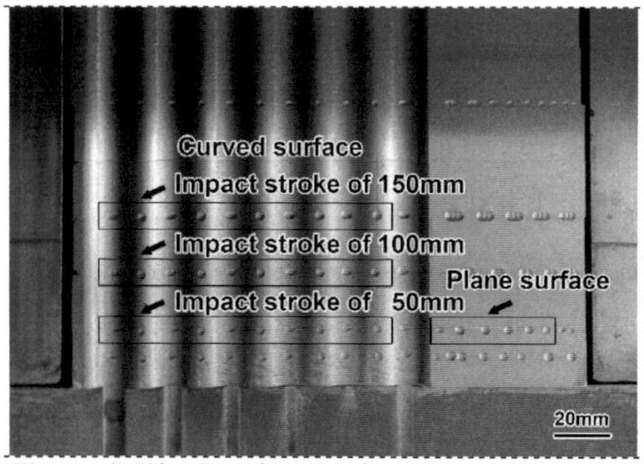

Figure 2. The Specimens after Hammering.

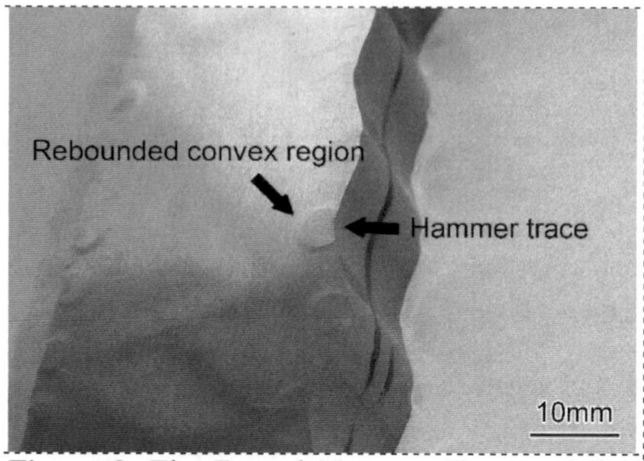

Figure 3. The Frontal Hammer Trace and the Rebounded Convex Region on the Tight Surface of Separated Sample.

2.3. Analytical methods

To accurately describe the shapes of plastic zones, we defined a plane perpendicular to the sample that divided the plastic zone into two equal parts. This plane, called the sagittal plane, had an intersecting line (called the sagittal line) with the surface of the rebounded zone. By adjusting the value of coordinate X, we could move the sagittal plane to right or left, forming a parallel sagittal plane, and the corresponding intersecting line was called the sagittal parallel line. Functions were detected using a laser confocal scanning microscope, which produced the 3D coordinates of 40 equidistant sagittal parallels passing through the rebounded convex zone. The 40 lines were fitted separately by Logistics [21]. The statistic R^2=0.957 (with the sagittal parallel line at X=3.677 as example) indicates the fitting effect is good and this model is ideal for determination of real morphology (1):

$$Z = h_2 + \frac{h_1 - h_2}{1+\left(\frac{Y}{Y_0}\right)^p} \quad (1)$$

Where Z is the rebounded height; Y is the y-axis coordinate; h_2 is height of undeformed zone; h_1 is the height of plastic zone; Y_0 is the position of largest deformation; p is the slope of turning point.

The fitted curves were processed by 3-order derivation against y, which resulted in zero-point Y_1 and Y_2 [22], where Y_2 is an effective boundary point of the rebounded zone. In this way, the effective boundary points of all sagittal parallel lines were determined.

Least squares fitting [23] of these points produced a fitted circular curve. The accuracy of the fitting results was estimated via χ^2 test to be $\chi^2=48.88<\chi^2_{0.05}(37)$ [24], indicating the boundaries of the plastic zone were circular- shaped.

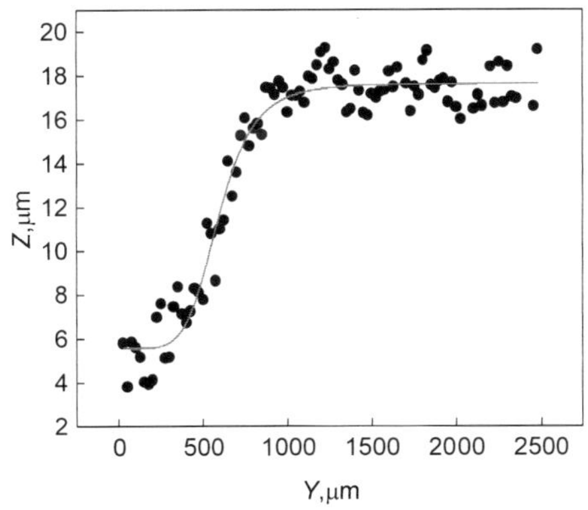

Figure 4. Sagittal Parallel Scatter Diagram Fitting of X=3.677.

Figure 4 shows the schematic diagram of fitting results in the xOy coordinate system. The projected area of the plastic zone (A_P) is the area enclosed by minor arc AB and major arc ACB. A_P in the plastic zone from the SR5.5 spherical punch head is 12.26 mm^2 at the impact stroke of 150 mm; the largest height H of hammered plastic zone is 3.92 mm, while the residual stress field depth formed by shot peening is less than 0.5 mm [25]. It is indicated the hammering method can produce a plastic zone deeper than that formed by shot peening surface strengthening, and ensures the precision-machined parts reserve a prepressing stress layer with appropriate depth, which contributes to prolonging the fatigue life of precision-machined parts.

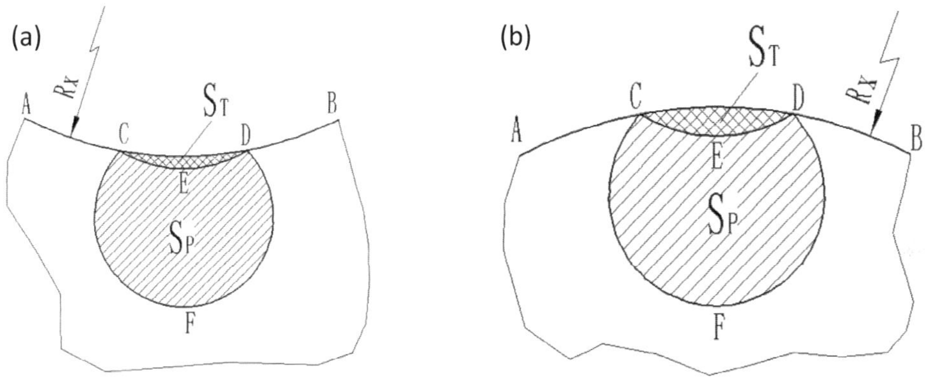

Figure 5. Projected Schematic Illustration of the Hammering Traces and the Plastic Zone at the Internal and External Corner.

3. Results and discussion

3.1. Results

Figures 6-8 show the rebounded convex zones generated with the SR5.5 punch head, internal corner with curvature radius R=10 mm, external corner with R=12 mm and impact strokes of 50 mm, 100 mm and 150 mm, respectively. The platform is clearly raised at the hampered place on the tight surface and surrounded by a ring-shaped transition zone, which is surrounded by the undeformed or original morphology. The rebounded zone appeared because the tightening force on the tight surface after hammering was large enough to inhibit the rebounding of the plastic rheological zone, and after the samples were separated, the withdrawal of the tightening force released the pressure stress, so the plastic zone rebounded and raised, forming a rebounded convex zone, which brought up the metals around the plastic zone, forming a transition zone.

IOP Conf. Series: Materials Science and Engineering **507** (2019) 012004 doi:10.1088/1757-899X/507/1/012004

Figure 12-14(a) shows the plastic zone laser confocal images and fitting results with the SR5.5 punch head at impact stroke of 150 mm, on plane; internal corner at impact stroke of 50 mm, curvature radius R =10 mm; external corner with impact stroke of 50 mm, R=12 mm. Clearly, plastic zones in these three situations are all circular-shaped, the position closer to the hammering place is redder, and the rebounded height after hammering is larger.

Figure 12-14(b) show the rebounded convex zones of three different shapes at different impact strokes and the fitting results of effective boundary points on sagittal parallel lines. As data of statistic χ^2 are all smaller than $\chi^2_{0.05}(37)$ in the table 1, which verifies the plastic zone boundaries are circular-shaped. Table 1 shows the fitting and calculated results of plastic zones with internal corner R=10 mm and external corner R=12 mm on different impact strokes.

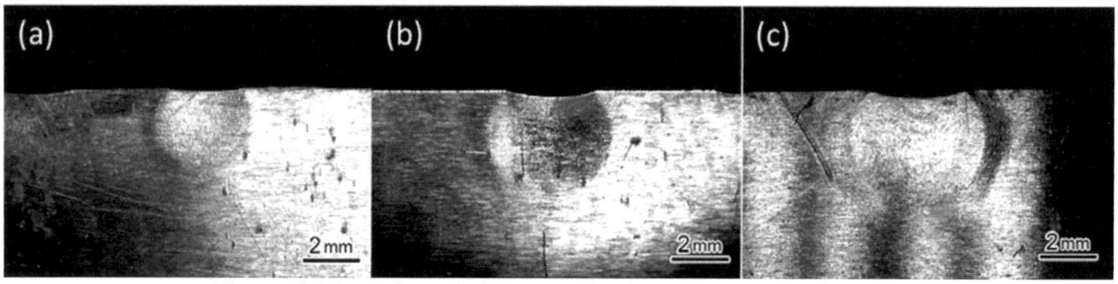

Figure 6. The Rebounded Convex Region of Impact Strokes of (a) 50 mm, (b) 100 mm and (c) 150 mm.

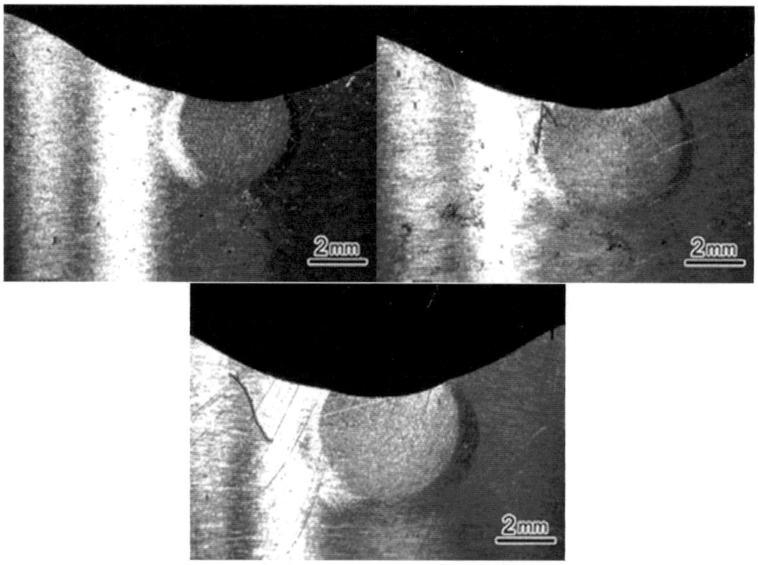

Figure 7. Rebounded Convex Region of Impact Strokes of (a) 50 mm, (b) 100 mm and (c) 150 mm of Curvature Radius R= 10 mm of Internal Corner.

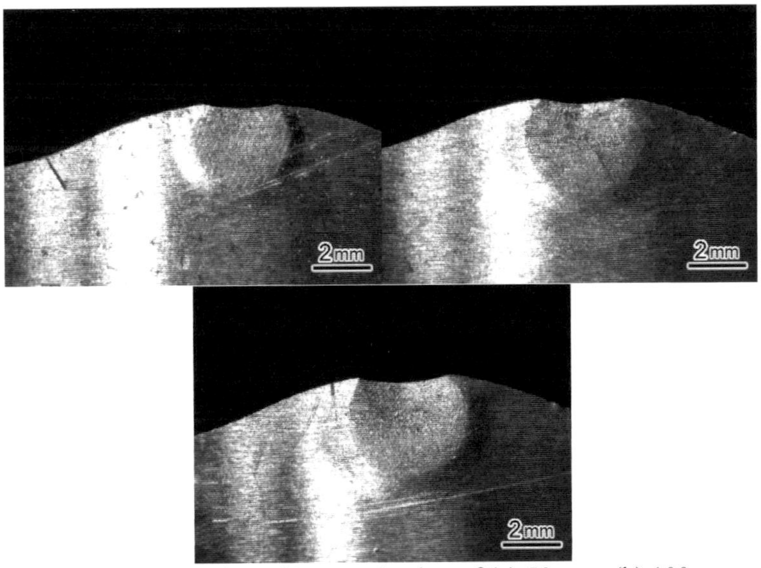

Figure 8. Rebounded Convex Region of Impact Strokes of (a) 50 mm, (b) 100 mm and (c) 150 mm of Curvature Radius R= 12mm of External Corner.

Table 1. Plastic Zone Fitting and Calculation Results of SR5.5 punch Head under Different Impact Strokes.

R_C / (mm)	L / mm	R / mm	H / mm	A_P/mm^2	χ^2
Pl	150	2.26	3.92	12.26	48.88
In10	50	1.87	3.41	10.52	47.64
Ex12	50	1.55	2.83	7.23	50.33

Note: R_C—curvature radius, *Pl*—plane, *In*—internal corners, *Ex*—external corners, *L*—impact stroke, *R*—radius, *H*—depth, A_P—area of plastic zone, χ^2—statistic

3.2. *Discussion*

The corresponding data were calculated using the method in Section 2.3. Figure 9 shows the changing trend of plastic zone area at impact strokes are 50 mm, 100 mm, 150 mm, internal corner (In) and external corner (Ex) with curvature radius of 8 mm, 10 mm, 12 mm, 14 mm, 16 mm. Clearly, as the curvature radius rises, the plastic zone area does not change significantly. For the same surface shape, the plastic zone area increases with the rise of impact stroke. For the same impact stroke, the plastic zone area formed at the internal corner is far larger than that at the external corner. Figure 10 shows the change trend of plastic zone area of plane (Pl), internal corner (In) and external corner (Ex) at impact strokes of 50 mm, 100 mm and 150 mm. It is indicated that the plastic zone areas are enlarged with the rise of impact stroke, and the areas at internal corners are bigger than those at internal corners and flat plane. Those plastic areas on the plane were changed rapidly with the increase of impact stroke, but those at the internal external corners were changed slowly. This is because the plastic strain energy of hammering affects the deformation degree and scope of the plastic zone, which are reflected by the rebounded height and plastic zone area, respectively. We divided the area value of the hammered pits from corresponded plastic zones, in order to describe the size relationship between them. The data thus calculated is called deformation degree (α). The plastic zone deformation at the external corner is large in degree and small in scope, but that at the internal corner is small in degree and large in scope.The deformation degree of the internal corner gradually enlarged with the increase of the radius of curvature, while the deformation degree of the external corner is on the contrary, as shown in figure 11.

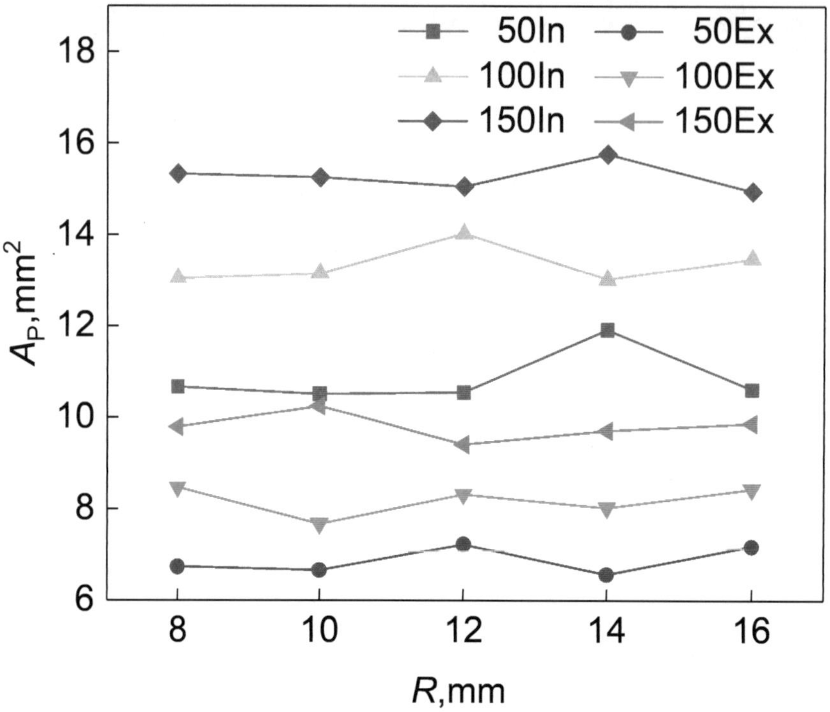

Figure 9. Change Trend of Plastic Zone Area of Different Curvature Radii of Internal Corner and External Corner.

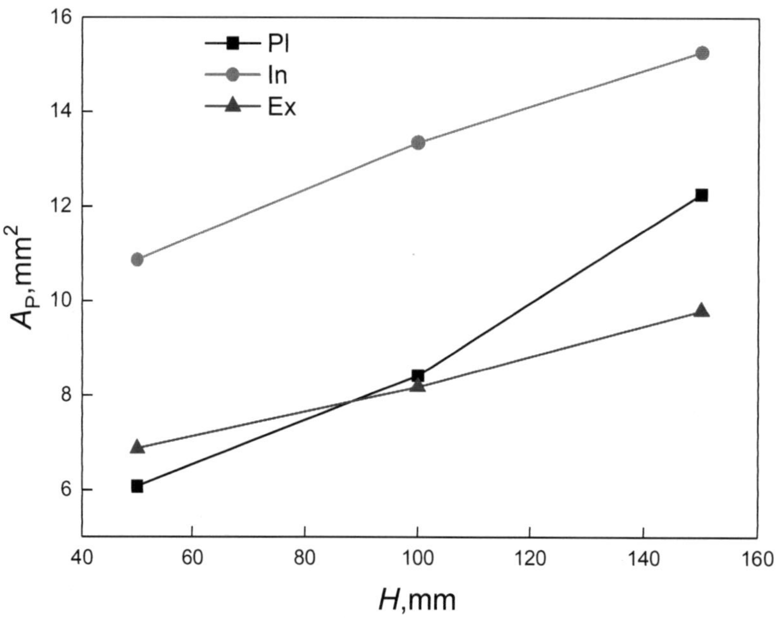

Figure 10. Change Trend of Plastic Zone Area of Plane, Internal Corner and External Corner under Different Impact Strokes.

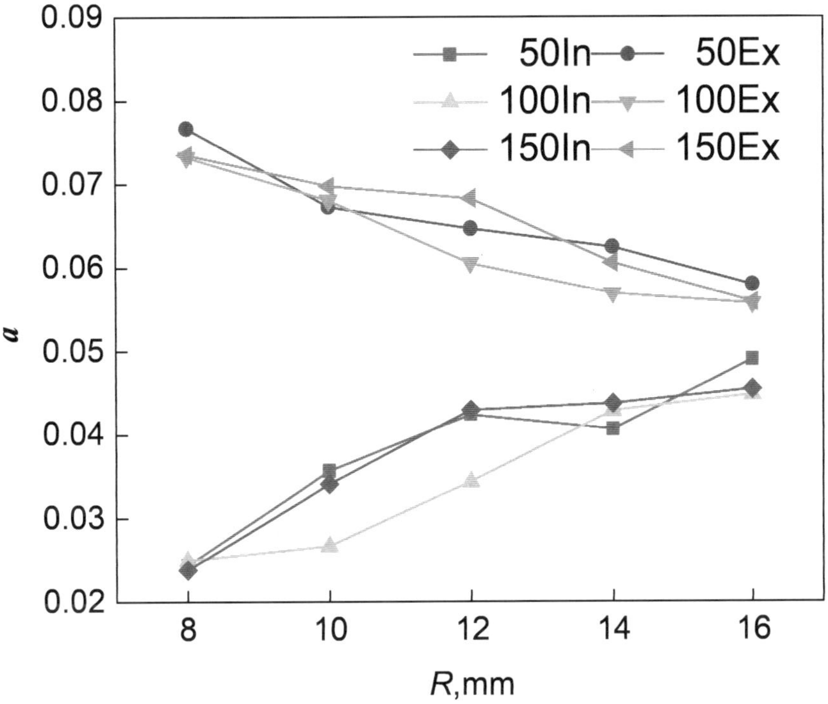

Figure 11. The Change Trend of Deformation Degrees of Internal Corner and External Corner under Different Impact Strokes.

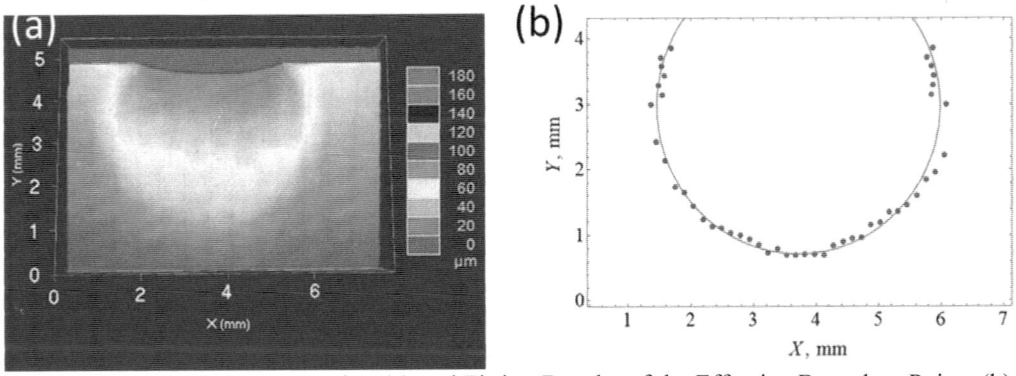

Figure 12. Rebounded Convex Region (a) and Fitting Results of the Effective Boundary Points (b) of Plane.

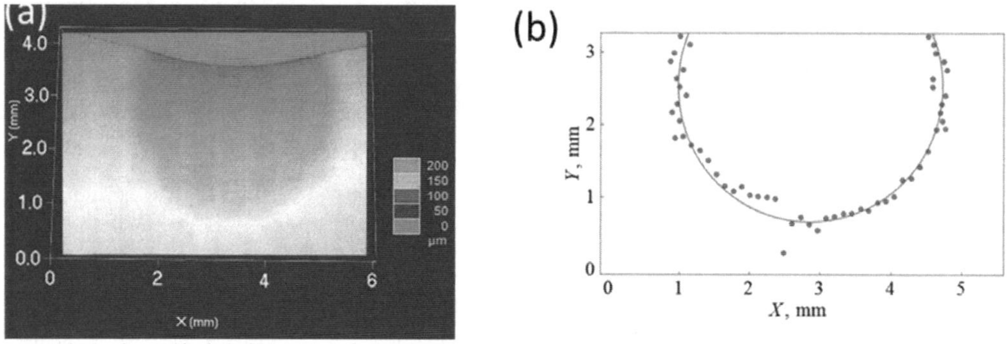

Figure 13. Rebounded Convex Region (a) and Fitting results of the Effective Boundary Points (b) of Curvature 10 mm of External Corner.

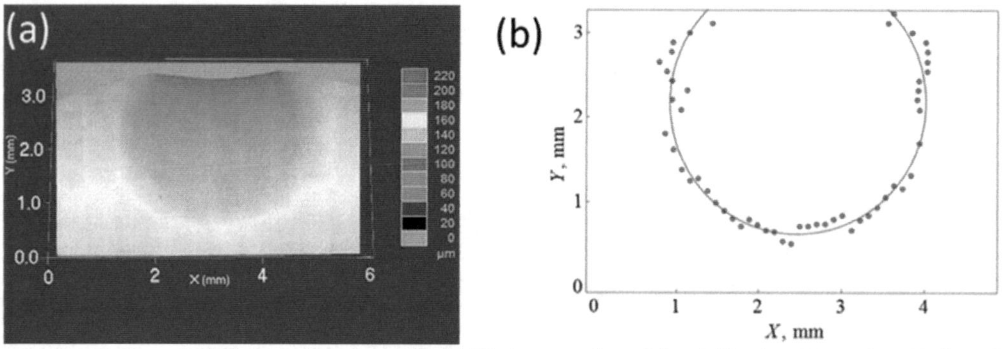

Figure 14. Rebounded Convex Region (a) and Fitting results of the Effective Boundary Points (b) of Curvature 12 mm of External Corner.

4. Conclusions

(1) The hammering method was used to study plastic deformation zone inside the Q235. Mathematical fitting method was used to calculate effective boundary points of each rebound plastic zone. The plastic zones which appeared at internal (In) and external (Ex) curved-surface corners and that appeared at plane-surfaces (Pl) were both circular-shaped, and the spaces around the hammered points were spherical-shaped.

(2) The maximum depth of the hammering plastic zone is 3.92 mm which means that there is still a certain depth of pre-stressed layer after the parts are final processed.

(3) The plastic areas were enlarged with the increase of impact stroke, and those at internal corner surpassed those at external corner with the same impact stroke. As the curvature radius rises, the plastic zone areas did not change significantly. Those plastic areas formed on the plane were changed more quickly than those at the internal external corners.

References

[1] Xiao X D 2015 *Modelling of shot peening stresses and simulation integral deformation of strip peening forming.* Xi'an: Northwest Polytechnical University.

[2] Ma X, Yuan Z T and Liu Y H 2017 *Light Ind. Sci. Technol.* **10** 55.

[3] Zhou Y Q, Zhao Y and Wang P 2010 *Mech. Manag. Dev.* **25(5)** 104.

[4] Hou X Z and Chen L 2017 *Technol. Appl.* **20** 102.

[5] Zou J L 2001 *Die Mould Ind.* **5** 44.

[6] Dong H B and Kang Y L 2011 *Trans. Mater. Heat Treat.* **32(3)** 103.

[7] Chen H N and Tang Y D 1993 *J. Mech. Strength.* **15(3)** 21.

[8] Meng X L 2007 *Numerical simulation of indentation-strain method under different residual stress fields.* Shenyang: Institute of Metal Research Chinese Academy of Sciences.

[9] Jin P and Li X D 2017 *J. Exp. Mech.* **32(5)** 645.
[10] Gai X Y 1993 *Study on yield behaviors of metal surface.* Shenyang: Institute of Metal Research Chinese Academy of Sciences.
[11] Chen H N, Lin Q H, Chen J et al 2001 *Study of plastic zone in residual stress measurement by impact indentation method.* Proc. 10th Nati. Welding Acad. Conf. Harbin: Heilongjiang People's Publishing House 230.
[12] Zhao Z, Chen J, Tan H, Lin X and Huang W D 2017 *Opt. Laser Technol.* **92** 36.
[13] Wang H M, Li R J and Zhou M 2017 *J. Eur. Ceram. Soc.* **37** 2705.
[14] Gu X, Qin J P and Zhang W C 2011 *Chi. Heavy Equip.* **3** 41.
[15] Underwood J H 1973 *Exp. Mech.* **9(1)** 373.
[16] Yu Z F, Zhao Y H, Chen H N et al 2001 *J. Shenyang Arch. Civ. Eng. Univ.* **17(3)** 200.
[17] Chen H N, Lin Q H, Li T R et al 2006 *Trans. Chin. Weld. Inst.* **27(8)** 27.
[18] Chen H N, Lin Q H and Qu P C 2005 *Several basic problems in measuring welding stress by indentation method.* Proc. 11th Nati. Welding Acad. Conf. Harbin: Heilongjiang People's Publishing House 116.
[19] Chen H N, Hu K X and Wu C Z 2010 *Chin. Meas. Teat.* **36(1)** 24.
[20] Maximov J T, Duncheva G V and Ganev N 2012 *J. Construct. Steel Res.* **74** 37.
[21] Hong N 1994 *J. Biomath.* **9(3)** 148.
[22] Cui D Q 2005 *J. Appl. Stat. Manag.* **24(1)** 112.
[23] Xu G W and Liao M C 2002 *J. Wuhan Polytech. Univ.* **4** 104.
[24] Fan L, Liu J H and Wang J L 2016 *J. N.W.T. For. Univ.* **31(3)** 215.
[25] Gao Y K 2004 *Chi. J. Nonferr. Met.* **14(1)** 60.

IOP Conf. Series: Materials Science and Engineering **507** (2019) 012005 doi:10.1088/1757-899X/507/1/012005

Experiment and Numerical Simulation of Non-Isothermal Bubble Growth in Polymer Foaming

Qingguang Yan[1], Huifang Wang[1], Rui Li[1], Dongdi Huang[1], Xiao Han[1]

[1]College of materials Science and Engineering, Jilin University, Changchun 130022, China

Abstract. A numerical simulation for non-isothermal bubble growth in polymer foaming were performed. The mathematical model based on the cell model was established to solve momentum equation, mass equation, diffusion equation and constitutive equation. We assumed the foaming polymer melt cooled at a set rate and took into account the effect of temperature on the rheological properties of the polymer-gas melt, Henry's law constant, diffusion coefficient and surface tension. A PS/CO_2 system was used herein as a case example to investigate the change of bubble growth behavior under non-isothermal circumstances and the effect of cooling speed on bubble growth. A batch foaming experiment with different cooling patterns was conducted to verify the simulation results. The results demonstrate that the main reasons to hinder the bubble growth are the increase of Henry's law constant and the decrease of diffusion coefficient during the cooling process. Increase the cooling rate is an effective way to produce foamed plastics with high bubble density and finer bubble size.

1. Introduction

A large number of bubbles can reduce the consumption of materials that improve the mechanical properties of products [1-3]. The foaming process of microcellular plastics mainly includes cell nucleation, growth and stability. The bubble growth processes is the key step to determine the final bubble structure and quality of foamed plastics. Thus, it is necessary to study bubble growth in the microcellular molding process, which can promote the development of microcellular foaming technology.

Various researchers have previously investigated the bubble growth phenomenon during the foaming process. The early researchers assumed that a single bubble grew in an infinite sea of polymer melt with an unlimited amount of dissolved gas [4, 5]. However, the amount of dissolved gas for bubble growth is finite and will eventually be depleted. Amon and Denson [6] introduced the well-known cell model to analyze a large amount of gas bubbles growth in close proximity to each other. The model can describe the actual foaming process more realistically because bubbles expand with finite dissolved gas in the model. Otsuki and Kanai [7] simulated the dynamic growth behavior of spherically symmetric bubbles with the diffusion of a foaming agent. They used a multimode Phan-Thien Tanner model to analyze the effect of viscoelastic characteristic on bubble growth. Xu [8] examined the growth of a spherical bubble with the cell model by using a finite-element-based numerical simulation method. A truncated power-law constitutive equation was used to describe the rheology of the melt. The effects of process parameters and the rheology of the melt on bubble growth were investigated. Ramesh [9] proposed a modified cell model. The model used the upper-Maxwell model to describe the viscoelastic of polymer/gas solutions. They studied the effects of dissolved gas content, temperature on rheology and other physical properties. Li [10] based on the cell model, a mathematical model for bubble growth was established under isothermal conditions. The established model was used to predict the influence of the physical properties and rheological properties on bubble growth dynamics.

All the researches above analyzed the isothermal bubble growth. However, in the process of polymer foaming, bubbles usually grow under non-isothermal condition. The properties of materials have change greatly during the cooling process. Thus, the non-isothermal bubble growth differs much from the isothermal one. In the paper, a mathematical model based on the cell model was established to simulate the non-isothermal bubble growth. We assumed that the polystyrene/carbon dioxide system cooled at a set speed. In the simulation process, the effect of temperature on the rheological properties of the polymer-gas melt, Henry's law constant, diffusion coefficient and surface tension were considered. The change of bubble growth behavior in non-isothermal was simulated with the model. In addition, CO_2 as blowing agent has been performed to study the effects of different cooling rates on final cell structures of foamed plastics. The batch foaming experiment was conducted to verify the simulation results.

2. Modeling of non-isothermal bubble growth

In this study, the cell model was used to establish the mathematic model. This model divided the polymer/gas melt into spherical unit cells composed of equal and finite amounts of dissolved gas and assumed that gas only diffuse in the unit cells. The following assumptions were made to model the non-isothermal bubble growth dynamics:

(1) The bubble is spherically symmetric throughout the growth process.
(2) The polymer/gas solution is incompressible.
(3) The effects of inertial forces and gravity are negligible.
(4) The initial dissolved gas distributes uniformly in the unit cell. The dissolved gas concentration at the polymer/gas solution interface obeys Henry's Law [11]:

$$c(R) = P_g K_h \tag{1}$$

where $c(R)$ is the dissolved gas concentration at the polymer/gas solution interface and K_h is the Henry's Law constant.

(5) The initial stress around the bubble is zero.
(6) Using the upper-convected Maxwell model to describe the viscoelastic properties of the polymer.

2.1. Bubble growth dynamics

Based on momentum equation, mass equation, diffusion equation and constitutive equation, a mathematical model can be established for the bubble growth. Considering the viscoelastic stress, surface tension, pressure in the bubble, and the ambient pressure, the momentum equation can be reduced to [11]

$$P_g - \frac{2\sigma}{R} - P_f + 2\int_R^S \frac{\tau_{rr} - \tau_{\theta\theta}}{r} dr = 0 \tag{2}$$

where P_f is the ambient pressure, σ is the surface tension at the bubble surface, r is radial position. τ_{rr} and $\tau_{\theta\theta}$ are the stress in the r and θ directions respectively, and S is the radius of the spherical unit cell, R is bubble radius.

Because the initial stress around the bubble is zero, the initial momentum equation can be expressed as

$$P_g(t=0) = P_f + \frac{2\sigma}{R} \tag{3}$$

where $P_g(t=0)$ is the initial pressure in the bubble.

To solve the momentum equation, the upper-convected Maxwell model was introduced to describe the viscoelastic properties of the polymer. The equations are expressed as follows [11]:

$$\tau_{rr} + \lambda\left(\frac{\partial \tau_{rr}}{\partial t} + \frac{R^2\dot{R}}{r^2}\frac{\partial \tau_{rr}}{\partial r} + \frac{4R^2\dot{R}}{r^3}\tau_{rr}\right) = -4\eta\frac{R^2\dot{R}}{r^3} \quad (4)$$

$$\tau_{\theta\theta} + \lambda\left(\frac{\partial \tau_{\theta\theta}}{\partial t} + \frac{R^2\dot{R}}{r^2}\frac{\partial \tau_{\theta\theta}}{\partial r} - \frac{2R^2\dot{R}}{r^3}\tau_{\theta\theta}\right) = 2\eta\frac{R^2\dot{R}}{r^3} \quad (5)$$

where λ is the relaxation time of the polymer/gas solution and η is the zero-shear viscosity.

The amount of gas in the bubble increases with the expanding of the bubble. The rate of the change of the mass in the bubble should be equal to the rate of the mass diffusing into the bubble. The mass conservation equation can be expressed as [12]

$$\frac{d}{dt}\left(\frac{4\pi}{3}\frac{P_g R^3}{ZR_g T}\right) = 4\pi R^2 D\left.\frac{\partial c}{\partial r}\right|_{r=R} \quad (6)$$

where D is the diffusion coefficient, Z is the compressibility factor, T is melt temperature, c is the dissolved gas concentration, and R_g is the universal gas constant.

Considering the non-ideal behavior of the gas under high pressure, the Peng–Robinson cubic equation of state is introduced to describe the properties of the gas. The equations are as follows [13]:

$$Z^3 - (1-B)Z^2 + (A - 2B - 3B^2)Z - (AB - B^2 - B^3) = 0 \quad (7)$$

where

$$A = 0.45724\alpha\frac{P_r}{T_r^2} \quad (7a)$$

$$B = 0.0778\frac{P_r}{T_r} \quad (7b)$$

$$\alpha = \left[1 + f(1 - T_r^{0.5})\right]^2 \quad (7c)$$

$$f = 0.37464 + 1.54226\omega - 0.2699\omega^2 \quad (7d)$$

To solver the equation (6), the distribution of the gas concentration in the unit cell should be obtained. Many studies described the dissolved gas concentration distribution in the form of a polynomial [14, 15]. In this study, the diffusion equation is used to calculate the distribution of the gas concentration. The equation can be expressed as [16]

$$\frac{\partial c}{\partial t} + V_r\frac{\partial c}{\partial r} = \frac{D}{r^2}\frac{\partial}{\partial r}\left(r^2\frac{\partial c}{\partial r}\right) \quad (8)$$

where V_r is the fluid velocity in the radial direction. According to the continuity equation for the flow around the bubble, V_r can be expressed as

$$V_r = \frac{\dot{R}R^2}{r^2} \quad (9)$$

The initial and boundary conditions are as follows:

$$c\left(t=0\right)=c_0 \tag{10}$$

$$c(r=R)=P_g K_h \tag{11}$$

$$\left.\frac{\partial c}{\partial r}\right|_{r=R}=0 \tag{12}$$

where c_0 is the initial dissolved gas concentration.

2.2. The properties of PS/CO_2 solution

To analyze the non-isothermal bubble growth, we assumed that the temperature decrease with a set rate. During the cooling process, the properties of the PS/CO_2 solution change greatly. Several models were utilized to describe the properties of the PS/CO_2 solution as a function of the temperature.

The temperature dependence of gas diffusivity of the system was assumed to obey the Arrhenius relationship [17].

$$D=D_0\exp\left(-\frac{E_a}{R_g T}\right) \tag{13}$$

where D_0 is the pre-exponential factor and E_a is the activation energy for the gas diffusion. The gas diffusivity at each temperature was calculated by the equation (13) with the experiment date by Sato [18].

The solubility of CO_2 in PS also changes with the decrease of temperature. The model for the solubility at each temperature is as follows [18]:

$$\ln K_h=6.4+2.537\left(\frac{T_c}{T}\right)^2 \tag{14}$$

where T_c is the critical temperature of CO_2.

Lee et al [19] proposed a generalized Arrhenius equation to estimate the zero-shear viscosity of PS/CO_2 system. The model took into account the effects of temperature, pressure, and CO_2 content. The equation is as follows:

$$\eta_0=0.8298\exp(\frac{1272}{T-341.1}+6.023P-66.51c) \tag{15}$$

where c is the dissolved gas concentration(wt/wt).

The model by Park [20] was introduced to calculate the temperature dependence of surface tension at the interface. The equation can be expressed as

$$\sigma=38.7032-0.0559P-0.01P+2.596\times10^{-5}TP \tag{16}$$

Because the unavailability of the relaxation time of PS/CO_2 system, the corresponding value for pure PS was used. The relaxation time of pure PS was estimated with the experiment date by Venerus [21].

2.3. Numerical simulation algorithm

A Lagrangian coordinate transformation [11] was used to reduce the governing equations for convenient of calculation. At early stage of bubble growth, the change in state was sharply near the bubble surface, so a non-uniform mesh, denser as the location got closer to the bubble interface, was used to discrete the governing equations. The finite volume method and the fourth-order Runge-Kutta method were employed to solve the governing equations and simulate the bubble growth dynamics. The diffusion equation was discrete with the finite volume method to form the implicit differential equation on the non-uniform mesh. In every time step, we calculated the rheological properties of the polymer-gas melt, Henry's law constant, diffusion coefficient and surface tension firstly. Then the viscoelastic stress was obtained with the equations (4) and (5). Lastly solve the equations (2), (6) and (8) to acquire the radius of the expanding bubble. The parameters for bubble growth simulation were obtained by Leung [12] and Lee [19]. The parameters were listed in table 1.

Table 1. Parameters for bubble growth simulation.

Ambient pressure	5MPa
Initial unit cell radius	50μm
Initial bubble radius	1μm
Initial temperature	200℃
Initial dissolved gas concentration	0.04wt/wt
The rate of cooling	50℃/s

3. Experiment

Materials: The PS used in the experiment was purchased from Petrochina Daqing Petrochemical Company in the form of pellets. The PS pellets were compression molded to 1 mm-thick sheets by pressing at 180℃; the sheets were cut into 2×2 mm pieces. The physical blowing agent was 99% pure carbon dioxide.

Foaming process: A depressurization batch foaming process was used to foam the PS. Foams were prepared in a 2L stainless steel hexagonal high-pressure autoclave purchased from Dalian Automatic Control Equipment Plant. The samples were placed in the autoclave. The vessel was preheated to a set temperature and then filled with CO_2 to the desired pressure. The sample was maintained for a prescribed period of time until saturation conditions were reached. Then the vessel was depressurized to ambient pressure to induce bubble nucleation and growth. At last the foamed sample was cooled and stabilization. In all the experiments, samples were saturated under 10 MPa of $ScCO_2$ for 6 h at 100℃ and the depressurization time is 10s. To investigate the effects of different cooling rates on foam process, three cooling methods were utilized to cooling the samples. In the first approach, once the pressure was completely released the 20℃ water was quickly injected into the vessel to cool the sample. In the second approach, the sample was maintained in the vessel for 3 minutes after depressurization and then was removed from the autoclave and cooled in the air. In the third approach, the sample was cooled to room temperature in the autoclave.

Characterization: The foams were characterized to determine the cell sizes and densities. All foamed samples were immersed in liquid nitrogen for 10 min and then fractured to assure that the microstructure remained intact. The cellular structures of the samples were characterized by scanning electron microscope.

4. Results and discussion

4.1. The effect of temperature on bubble growth

During the cooling process, the bubble growth behavior changes greatly. To analyze the non-isothermal bubble growth in detail, the influences of the change of the zero-shear viscosity, Henry's law constant, diffusion coefficient and surface tension on bubble growth dynamics were investigated respectively.

The zero-shear viscosity, varying with the temperature, is one of the retarding forces that affect the bubble growth behavior. Figure 1 illustrates the effect of the change of zero-shear viscosity on bubble growth. In the early process of bubble growth (0-0.7s), the effect of zero-shear viscosity on bubble growth is almost negligible. It probably resulted that the melt temperature is still high in the process, the resistance produced by zero-shear viscosity is small. In the later growing process (0.7-1s), with the decrease of the temperature, zero-shear viscosity increases greatly, and viscoelastic stress around the bubble gradually accumulates. Therefore, the bubble growth rate under non-isothermal conditions will be reduced. In general, the effect of zero-shear viscosity is not pronounced.

IOP Conf. Series: Materials Science and Engineering **507** (2019) 012005 doi:10.1088/1757-899X/507/1/012005

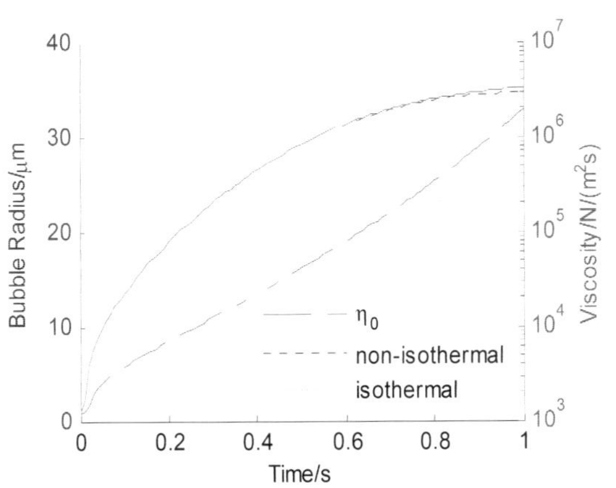

Figure 1. Effect of zero-shear viscosity on bubble growth.

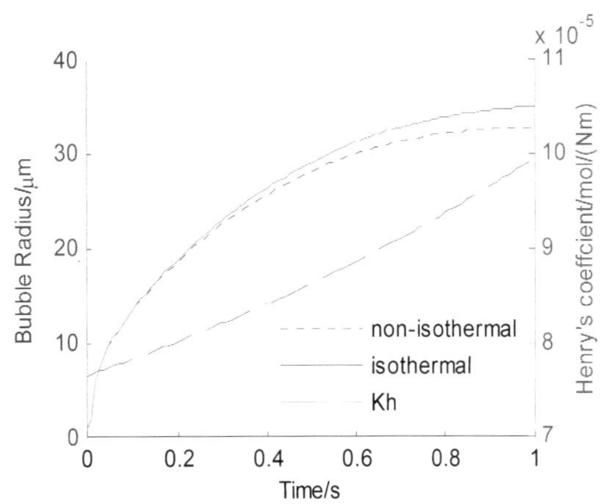

Figure 2. Effect of Henry's law constant on bubble growth.

Figure 2 indicates that the effect of Henry's constant on non-isothermal bubble growth. From 0 to 0.2s, the effect of Henry's constant on bubble growth is almost negligible. From 0.2 to 1s, the bubble radius increases slowly under non-isothermal conditions, and the difference between the two conditions increases gradually. It probably resulted that the increase of Henry's constant with the decrease of the temperature means the polymer has a higher capability of dissolving gas. The gas concentration of the bubble interface decrease faster and the diffusion of CO_2 gets slower. Therefore, the bubble expands slowly in non-isothermal condition.

Another factor that will affect the bubble growth rate is diffusion coefficient. A higher diffusion coefficient means a faster diffusion of gas molecule into the bubble. So the bubble growth rate increases. As shown in figure 3, at the later stage of bubble growth (0.3-1s), the bubble growth rate will reduce under non-isothermal condition because of the decreasing of diffusion coefficient. As the diffusion coefficient decreases, the difference between the two conditions becomes larger.

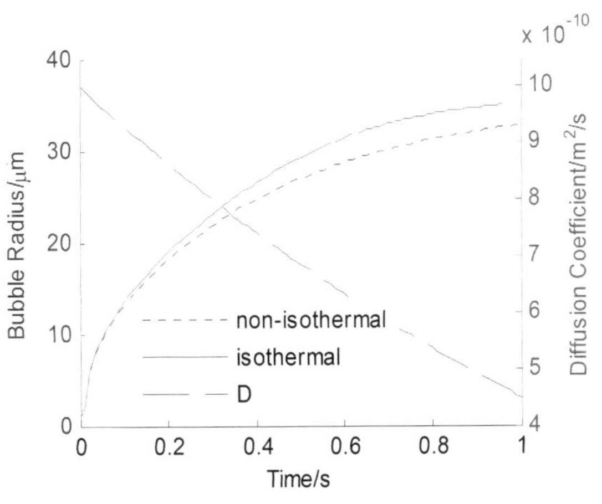

Figure 3. Effect of diffusion coefficient on bubble growth.

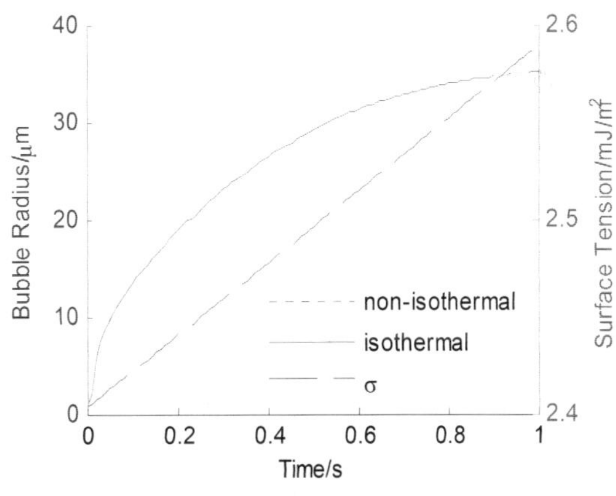

Figure 4. Effect of surface tension on bubble growth.

Figure 4 illustrates the effect of surface tension on non-isothermal bubble growth. The surface tension increases with the decreasing of the temperature. However, the bubble growth states in the two different conditions are almost the same. This indicated that the surface tension has little effect on bubble growth.

From the discussion above, it's clear that the bubble growth rate in non-isothermal condition gets slower than the isothermal situation. The increasing of Henry's constant and the reduction of diffusion coefficient are the main reasons for the decreasing of bubble growth rate during the cooling process. The effects of viscosity and surface tension are negligible.

4.2. The effect of cooling rate on bubble growth

The reduction of temperature has significant effect on the bubble growth dynamics. The effect of cooling rate on bubble growth is analyzed in detail. As shown in figure 5, the bubble growth rate decreases with the increasing of cooling rate. A higher cooling rate leads to Henry's constant increasing faster and diffusion coefficient decreasing more rapidly. So the bubble growth rate get slower and the final cell size will be finer. In order to investigate the effect of cooling rate on the cellular structures of the foams, a series of experiments was carried out in which the cooling rate was varied while holding the other conditions constant. Figure 6a, 6b and 6c show the microstructures of PS foams prepared with three different cooling approaches respectively. It is observed that the final cell size of the foam decreases with the increasing of cooling rate while the final cell density increases. Stabilized at a fast cooling speed can suppress bubble growth, so the final cell size of the foam cooled by water is smallest compared the other two foams. On the other hand, immediately cooling the foam after bubble nucleation can prevent cell coalescence which is an important cause for reduction of cell density. Cell coalescence can be observed in the foam by furnace cooling, the bubble surface is rough, and the phenomenon of bubble coalescence is serious. This resulted that the interface strength of the bubble-polymer melt is low, and the stability of the bubble nuclei is weakened, which is disadvantageous for microcellular foaming. This shows that increasing the cooling rate can effectively suppress the expansion of the bubble, and obtain high density and small size cell structure.

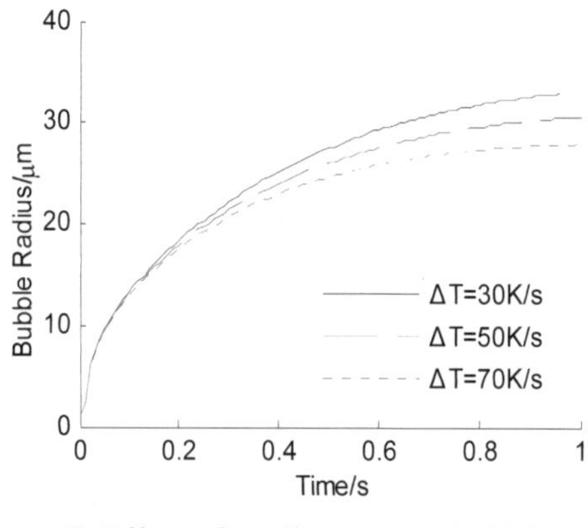

Figure 5. Effect of cooling rate on bubble growth.

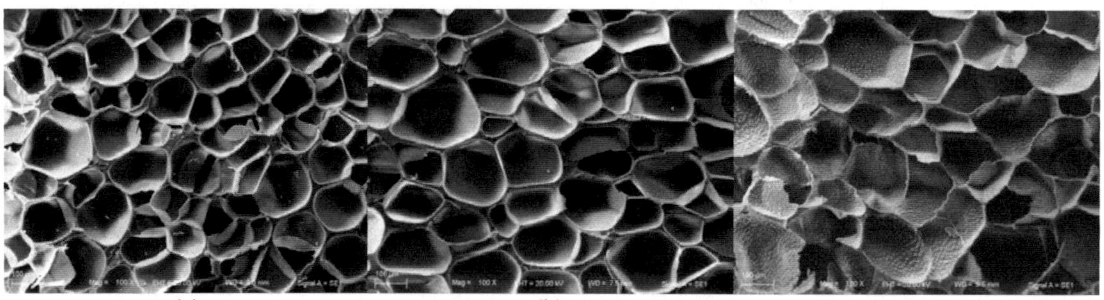

(a) (b) (c)

Figure 6. SEM micrographs showing cellular structures in polystyrene foams prepared with different cooling methods: (a) water cooling (fast speed); (b) air cooling (intermediate speed); (c) furnace cooling (slow speed).

5. Conclusion

A mathematics model was established to simulate the non-isothermal bubble growth. The model took into account the changes of the rheological properties of the polymer/gas solution, Henry's law constant, diffusion coefficient and surface tension during the cooling process. A PS foam experiment with different cooling methods was conducted to verify the simulation results. The results show that Henry's constant and diffusion coefficient have a significant effect on bubble growth. With the decrease of the temperature, the increasing of Henry's constant and the reduction of diffusion coefficient can reduce the bubble growth rate compared with the isothermal condition. Zero-shear viscosity changes greatly during the cooling process. At the later bubble-growth process, the increase of zero-shear viscosity will suppress the bubble growth. However, surface tension has no significant effects on bubble growth. Increasing the cooling rate can increase the cell density and reduce the cell size. So adopting a more rapid cooling approach is an effective way to produce foamed plastics with high bubble density and finer bubble size.

References

[1] Sauceau and Fages J. 2011 *Prog. Polym. Sci.* **36(6)** 749-66.
[2] Jacobs L J M, Kemmere M F and Keurentjes J T F 2008 *Green Chem.* **10(7)** 731-8.
[3] Suh N P 2003 *Movcromol. Symp.* **201(1)** 187-202.
[4] Street J R., Fricke A L and Reiss L P 1971 *Ind. Eng. Chem. Fundamen.* **10(1)** 54-64.
[5] Venerus D C, Yala N and Bernstein B 1998 *J. Non-Newton. Fluid Mech.* **75(1)** 55-75.
[6] Amon M and Denson C D 1984 *Polym. Eng. Sci.* **24(13)** 1026-34.
[7] Otsuki Y and Kanai T 2005 *Polym. Eng. Sci.* **45(9)** 1277-87.
[8] Xu X M, Zhao G Q and Li H P 2010 *J. Appl. Polym. Sci.* **116(3)** 1264-71.
[9] Ramesh N S, Rasmussen D H, Campbell G A 1991 *J. Polym. Eng. Sci.* 31(23): 1657-1664.
[10] Li Y, Yao Z, Chen Z H 2011 *J. Chem. Eng. Sci.* 66(16):3656-3665.
[11] Arefmanesh A and Advani S G 1991 *Rheol. Acta* **30(3)** 274-83.
[12] Leung S N, Park C B and Xu D 2006 *J. Ind. Eng. Chem. Res.* 45(23): 7823-7831.
[13] Poling B E, Prausnitz J M and O'Connell J P 2001 *McGraw-Hill*
[14] Favelukis M 2004 *Polym. Eng. Sci.* **44(10)** 1900-6.
[15] Taki K 2008 *Chem. Eng. Sci.* **63(14)** 3643-53.
[16] Welty J R, Wicks C E and Wilson R E 2001 *Wiley*
[17] Osorio A and Turng L S 2004 *J. Polym. Eng. Sci.* 44(12): 2274-2287.
[18] Sato Y, Takikawa T, Takishima S and Masuoka H 2001 *J. Supercrit. Fluids* **19(2)** 187-98.
[19] Lee M, Park C B and Tzoganakis C 1999 Tzoganakis C. *Polym. Eng. Sci.* **39(1)** 99-109.
[20] Park H, Thompson R B, Lanson N et al 2007 *J. Phys. Chem. B* **111(15)** 3859-68.
[21] Venerus D C, Zhu S H et al 1999 *J. Rheol.* **43** 795-813.

IOP Conf. Series: Materials Science and Engineering **507** (2019) 012006 doi:10.1088/1757-899X/507/1/012006

Design Optimization of A Rotating Flywheel Under High Centrifugal Forces using Response Surface Method

Noor Fawazi[1, 2], C M Hasrizam[1], A Hafiz[1] and Kazuma Sekiguchi[2]

[1]Intelligent Dynamic System Ikohza, Department of Mechanical Precision Engineering, Malaysian Japan International Institute of Technology (MJIIT), Universiti Teknologi Malaysia Kuala Lumpur, Jalan Sultan Yahya Petra, 54100 Kuala Lumpur, Malaysia
[2]Tokyo City University, Tokyo 158-8557 Japan.

Abstract. Flywheel is a mechanical device specifically designed to efficiently store rotational energy. Flywheels resist changes in rotational speed by their moment of inertia. In automotive clutch system, flywheel rotates at high speed under the influence of rotational force which may result in severe crack. In this study, finite element analysis was used to predict equivalent stress of the flywheel at different rotational speed. Taking 8000 rpm as the maximum rotational speed, the layout of the flywheel was then modified. This is done not only to reduce the maximum equivalent stress, but also reduce the mass of the flywheel. The comparison between the original layout of the flywheel and the modified one is shown in this work.

1. Introduction

An automotive component in clutch system like flywheel may experience high rotational motion [1-3]. The induced severe radial and circumferential stress may result in crack and damage. It is the intention of this work to investigate the maximum equivalent stress using Finite Element Analysis. The original design was then modified by introducing a series of slot segment within the flywheel with the aim to reduce the original equivalent stress beside reducing the weight of the flywheel.

Previous work [3] investigated the stresses and deformation of the frictional clutch components using finite element analysis. This work clearly suggested the importance of the optimal design of the flywheel in order to reduce stress. Otherwise, improper design may lead to initial crack and breakage due to high rotational speed. Other than that, the weight of the flywheel was also considered [4] to suggest optimal design of a flywheel.

In this work, response surface method was used to propose new modified design of the flywheel by considering the equivalent stress. By introducing the slot segments within the flywheel, the weight of the flyweight can be reduced besides reducing the equivalent stress at certain rotational speed. In this work, the optimal dimensions of the flywheel are based on the maximum 8000 rpm speed rotational motion.

2. Finite element model and response surface method

In this study, finite element analysis is used to analyze the flywheel under different rotational speeds. The objective is to investigate the changing of the stress with respect to the rotational speeds. The pattern is expected to follow the trend in [5]. The flywheel design is to withstand at 8000 rpm to simulate the high rotational speed condition. The boundary condition in the finite element model is similar to Ref. [2]. The material properties of the flywheel are shown in table 1.

IOP Conf. Series: Materials Science and Engineering **507** (2019) 012006 doi:10.1088/1757-899X/507/1/012006

Table 1. Material Properties.

AISI 316 Stainless Steel	
Elastic Modulus	196 GPa
Poisson Ratio	0.27
Yield Stress	205 MPa
Density	7999 kg/m

In order to propose the optimal design of the flywheel, the response surface method is used [6-8]. Central Composite Designs is adopted in the design process using quadratic model. The mathematical model is created that best fits to the obtained data using FEA. The objective function is constructed as a second order degree polynomial.

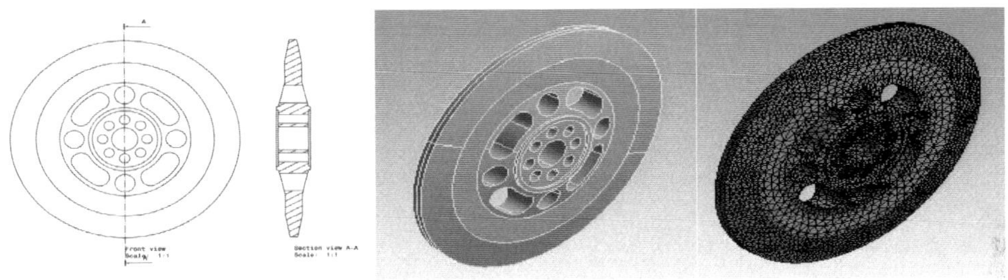

(a) (b) (c)

Figure 1. Original Flywheel Design before modification
(a) cross section, (b) CAD model (c) FEA model.

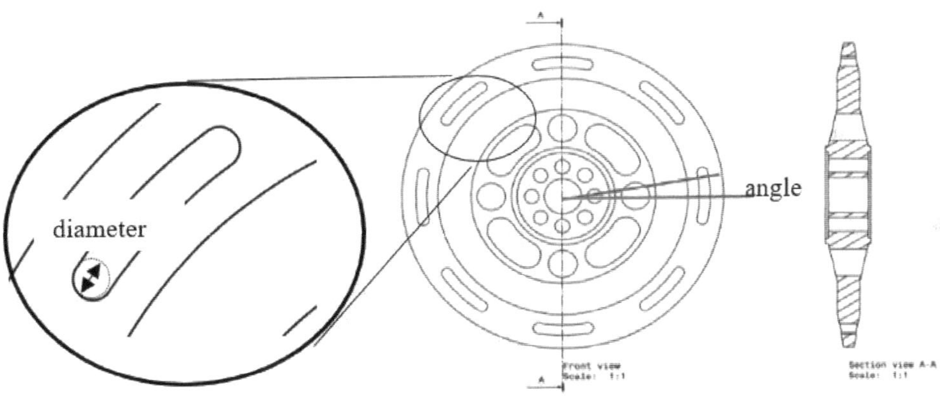

Figure 2. Modified flywheel.

The simulation results provide data of equivalent stresses for each design. In this study, the original flywheel is re-designed with a series of slot profile. The diameter of the slots is purposely varied from the range of 6mm to 12mm. The profile of the slots is also defined with respect to the given angle from the central point between the range of 10 degrees to 20 degrees. The target design can be formulated as follows;

Objective	Minimize Maximum Equivalent Stress	(1)
Subject to	6mm < diameter of slot < 12mm	(2)
	10 degree <Slot size (angle from center point) < 20 degree	(3)
Constraint	High Rotational Speed 8000 rpm	(4)

The response surface method is used in this study by adopting central composite design method. The objective function is defined as a function of two second-order quadratic terms as follows;

$$y = \beta_0 + \beta_1 x_0 + \beta_2 x_2 + \beta_{12} x_1 x_2 + \beta_{11} x_1^2 + \beta_{22} x_2^2 + \varepsilon \quad (5)$$

This second order model is the basis for the response surface designs. This model provides a good approximation to the surface near the maximum or a minimum. For $k = 2$ there will be a 22 design with center points, which is required for the first order model. The star or axial points are, in general, at some value α and -α on each axis. The 22 design create a box, and adding the axial points outside of the box gives a spherical design where $\alpha = \sqrt{k}$ as shown in figure 3. It is an advantage to use central composite design. This computation is often used when the design plan calls for sequential experimentation because these designs can include information from a correctly planned factorial experiment.

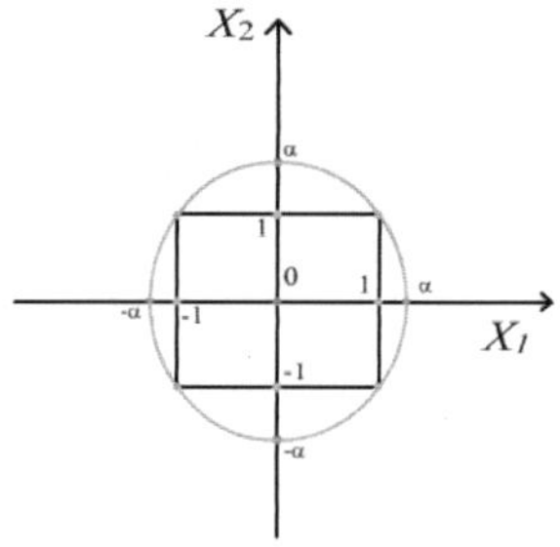

Figure 3. Central Composite Design.

3. Analysis and discussions

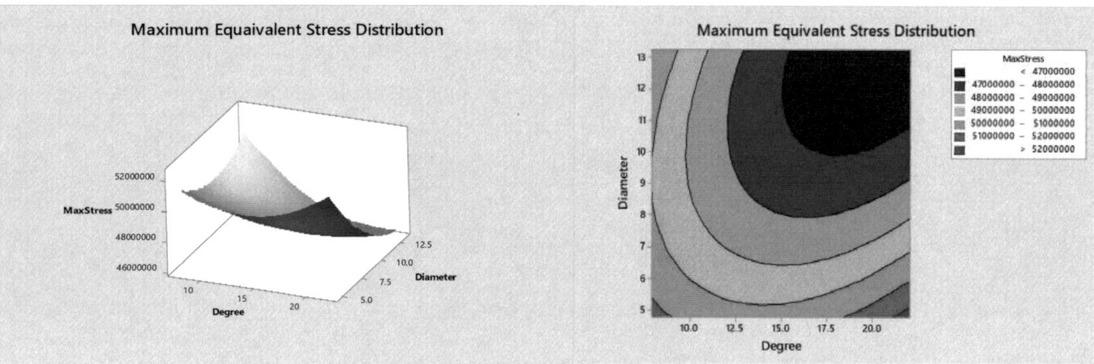

Figure 4. 3D and 2D surface plots for variations in diameter ($X1$) and angle ($X1$) at constant maximum 8000 rpm.

The result obtained from the equation (5), were analyzed and shown in figure 4 and figure 5. The 3d surface plots for variation of diameters and the angle profile (in degree) of the slot as shown in earlier figure 2, show the changing trend of the maximum equivalent stress or Von Misses Stress. It can be observed that the maximum equivalent stress tends to decrease with the increasing of the diameter and angle profile. It can be easily concluded that by increasing the diameter and the angle profile will decrease the maximum equivalent stress. However, in this optimization process, the design is subjected to the limitation as defined in equation 2. Not only that, it is also can be seen from figure 5 that by simply increasing the diameter and the angle profile, the maximum equivalent is not continuously decreasing. The changing trend start to show an increment after certain point of minimum value of maximum equivalent stress. The interaction of the changing parameters is also shown in the same figure.

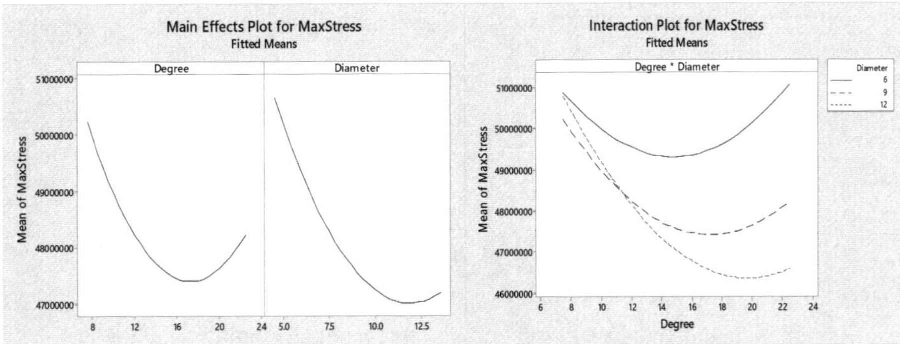

Figure 5. The effect of Maximum Equivalent Stress with the changing of the size of slot (degree) and diameter of the flywheel.

By searching the optimum value of the diameter and the angle profile of the slot, the optimal dimensions of the slot is decided which satisfy the target design. The comparison of the maximum equivalent stresses at different rotational speeds are also show on figure 6. During the optimization process, even though the high rotational speed is constrained, the maximum equivalent stress will also decrease at different rotational speeds.

Table 2. Comparison between the original and Modified Design of flywheel.

Comparison	Volume m^3	Mass kg	Slot Profile	Maximum Stress at 8000 rpm
Original Design	$2.6411(10^{-4})$	2.1126	No Slot	50908000 Pa
Modified Design	$2.3061(10^{-4})$	1.8447	Diameter = 12mm, Degree = 19.7°	45422000 Pa

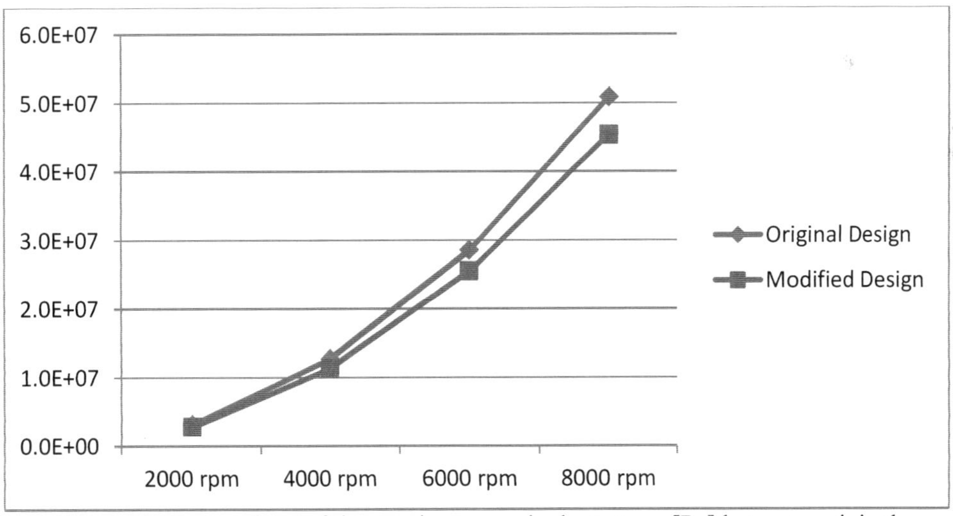

Figure 6. Comparison of the maximum equivalent stress [Pa] between original and modified design at different rotational speed [rpm].

4. Conclusion

The Response Surface Method is useful for the design of experiments investigating the effect of the two evaluated factors (diameter of the slots and slot size) on the response parameter (equivalent stress). This computational experimental design not only investigates the variation of the changing response, but also able to identify the optimal dimensions of the evaluated factors with respect to the equivalent stress. This structural modification not only reduces maximum equivalent stress but also the weight of the flywheel itself.

Acknowledgments
Authors wishing to acknowledge assistance or encouragement from colleagues, special work by technical staff or financial support from Tokyo City University and Universiti Teknologi Malaysia.

References

[1] Danfelt E L, Hewes S A and Chou T W 1977 *Int. J. Mech. Sci.* **19** 69-78.
[2] Imer C, Kartel S and Genc M O 2017 *Proc. of the 3rd World Congress on Mech. Chem. and Mater. Eng. ICMIE* Stress Behaviour Improvement of Automobile Flywheel Under High Centrifugal Forces **120** pp. 1-6.
[3] Abdullah O I and Schlattmann J 2013 *Tribol. Ind.* **35(2)** 155-62.
[4] Maninder and Bhardwaj S 2015 *Int. J. Innov. Sci. Res.* **4(8)** 361-5.
[5] Trudell J 2001 *6th Annual International Symposium on NDT for Health Monitoring and Diagnostic (California)* Structural Analysis of Composite Flywheels: an Integrated NDE and FEM Approach pp 4-8.
[6] Asadollahzadeh M, Tavakoli H, Torab-Mostaedi M, Hosseini G and Hemmati A 2014 *Talanta.* Response surface methodology based on central composite design as a chemometric tool for optimization of dispersive-solidification liquid–liquid microextraction for speciation of inorganic arsenic in environmental water samples Design **123** 25-31.
[7] Edward L 2016 *Materials.* **9** 427.
[8] Ahmet Dag, 2016 *Adv. Mater. Sci. Eng.* **2349476** 1-13

IOP Conf. Series: Materials Science and Engineering **507** (2019) 012007 doi:10.1088/1757-899X/507/1/012007

Design of a torque standard component of an integrated calibration equipment

Abstract. This paper describes the establishment of a torque standard component (TSC) of an integrated calibration equipment (ICE). The TSC was developed for on-site automated calibration of several hand torque wrenches (like as setting type torque wrenches, indicating type torque wrenches and hand torque screwdrivers). Based on integrated idea and lightweight design, The TSC was optimized to weight only 5 kg, which saved much room, reduced gross weight of the ICE itself, and brought great benefit to on-site calibration. The relative expanded uncertainty of torque realized by the TSC was estimated to be 0.9% from 0.5 Nm to 60 Nm, with the coverage factor k=2. Performance of the TSC was examined by carrying comparison experiments with a reference torque standard (RTS) at our lab. As a result, torque realized by the TSC was shown to be equivalent to that realized by the RTS within the calmed uncertainties.

1. Introduction

Hand torque wrenches are widely used to accurately control torque in assembly and fastening operations in mechanical manufacturing, chemical engineering, automotive industry, ship-building, aerospace, power, railway and other assembling lines. In order to maintain consistent accuracy, calibration or verification of torque wrenches or screwdrivers on a periodic basis would be advisable. However, this is only part of the requirement. Data obtained from the calibration of hundreds of wrenches over several year period indicates that there is still a big probability that wrench failure occurs prior to periodic recalibration, especially those overused. Those disqualified wrenches may cause bolting related failure, even catastrophic accidents. Consequently, it is reasonable that wrenches themselves need to be calibrated on the site of customer just before applying whenever necessary. It is more and more being realized that for the quality control of assembling, those torque tools need to be on-site calibrated. In consequence, Engineers have begun developing portable torque wrench testers with which laboratories or factories would be able to calibrate wrenches whenever needed. In addition, there are many commercial portable torque wrench testers available for on-site calibration of hand torque tools. However, most of the commercial portable torque wrench testers are still manually handling, which was normally a much time consuming procedure and a laboured task. Furthermore, those so called 'portable testers' usually weight dozens of kilograms to cover a large calibration range, most of which are seldom used.

Besides, in recent years the demand for the calibration of multiple instruments with an integrated calibration equipment (ICE) has increased sharply. To meet the customized requirements, we developed an ICE with the aim of on-site calibration of several geometric instruments, thermal instruments, and mechanical instruments, in which a torque standard component (TSC) was designed for on-site calibration of several commonly used hand torque tools. The relative expanded uncertainty of torque realized by the TSC was evaluated in the 0.5~60 Nm torque range and was estimated to be 0.9%, with a coverage factor, k, being equal to 2. The TSC was compared with a reference torque standard (RTS) at our lab. The calibration results obtained by these two devices were coincided within the claimed uncertainties. Details about the TSC will be described in the following section, followed by uncertainty evaluation and finishing with comparison results of the TSC and RTS.

2. Design

2.1. Establishment of equipment

The ICE was designed to perform calibrating various geometric, thermal and mechanical instruments, by taking into account several standard components working as transfer standards, all of which are traceable to national standards. The TSC is one of the standard components to calibrate several hand torque wrenches, namely setting type torque wrenches (STWs), indicating type torque wrenches (ITWs) and hand torque screwdrivers (STWs). A schematic diagram of the ICE is shown in figure 1, on top of which is part of the TSC as illustrated in figure 2.

The TSC was designed on the basis of a portable torque wrench tester [1] by means of integrating to the ICE and was optimized to weight only 5 kg by integrated design and finite element analysis. Mechanical parts of the TSC mainly consists of a motor, a gearbox, two standard torque transducers, a driver and clamping structures as depicted in figure 3. Two alternative torque transducers covering the calibration range from 0.5 Nm to 60 Nm were employed to measure torque value of the calibrated wrench. Various easily detachable square drives were prepared in different sizes to enable torque wrenches of various sizes and shapes to be installed on either of the two torque transducers. A calibrated torque wrench was mounted on a torque transducer through a square drive and supported by a rod for loading point at its handle. Unlike our previous independent version, guiders of the portable torque wrench tester in figure 1 at literature [1] were omitted, and instead, part of framework of the ICE as shown in figure 2 was designed as the sliding guide. Thus the sliding block has been improved to a much simplified one. A slightly loose fit between an adapter fixed to the output shaft of gearbox and the input of transducer so as to measure and transfer torque, simultaneously. A gearbox with a magnification coefficient of 50 was chosen such as to amplify the torque outputted from the motor and lower the loading speed. A motor with a rated torque of 1.4 Nm is driven by a driver controlled by a control unit in the ICE, which receives the commands from a laptop via USB interface. The laptop evaluates signals from the torque transducer and controls the motor in a closed loop.

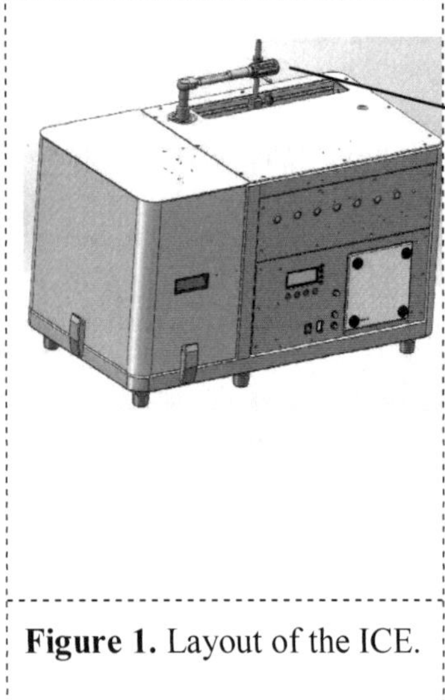

Figure 1. Layout of the ICE.

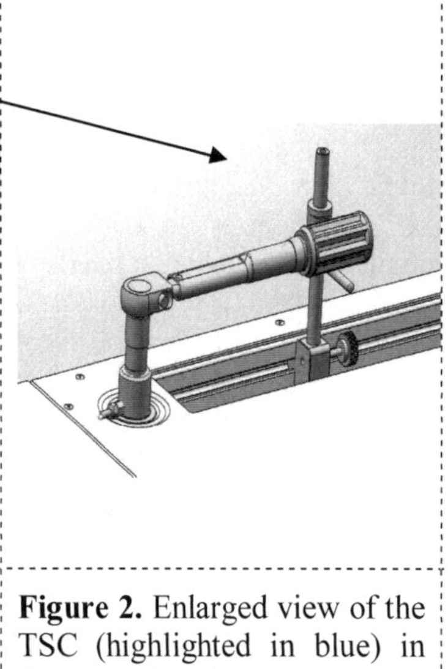

Figure 2. Enlarged view of the TSC (highlighted in blue) in figure 1. (For interpretation of the references to color in this figure legend, the reader is referred to the web version of this article.)

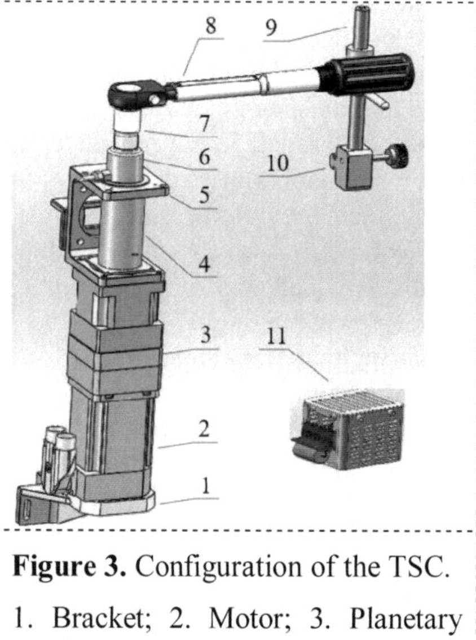

Figure 3. Configuration of the TSC.

1. Bracket; 2. Motor; 3. Planetary gearbox; 4. Adapter; 5. Frame; 6. Torque transducer; 7. Square drive; 8. Calibrated Torque wrench; 9. Rod for loading point; 10. Sliding block; 11. Driver.

2.2. Calibration procedures
A laptop with a graphical user interface installed was employed to run automated calibration procedures by means of controlling rotary movements of the motor via giving command to the control unit in the ICE. As a final product, the program elaborates the calibration certification, automatically. Different calibration procedures were designed for each type of wrenches due to diversity of operating principles. For more details about the calibration procedures one can refer to our prior work [1].

3. Uncertainty evaluation

3.1. Uncertainty of the calibration results
The performance of the new developed TSC were evaluated according to GJB 2749A [2] whereas uncertainty of calibration was evaluated according to ISO/IEC guide 98-3 [3], and evaluation methods are described as following. The relative expanded uncertainty U_r of calibration results for every measuring point is expressed by the following equation:

$$U_r = k \cdot u_c = k \cdot (u_{rep}^2 + u_{pro}^2 + u_{ind}^2 + u_{sta}^2 + u_{res}^2)^{1/2} \quad (1)$$

Where u_c is the relative combined standard uncertainty of calibration. A coverage factor of $k = 2$ was used in calculation of all expanded uncertainties in this paper, which correspond to the level of confidence of approximate 95 %. u_{rep} is uncertainty contribution of repeatability without a change in the mounting position, u_{pro} is that for the reproducibility with a change in the mounting position, u_{ind} is uncertainty due to indication difference from the reference value, u_{sta} is uncertainty due to long stability of transducer sensitivity, u_{res} is that due to the resolution.

3.2. Repeatability in the same mounting position
The repeatability with the unchanged mounting position u_{rep} was evaluated by certain number of repeated tests with the wrench, rod for loading point, square drive and torque transducer in the same mounting position. The relative repeatability with the unchanged mounting position is calculated as:

$$u_{rep} = \frac{1}{\overline{S_i}} \left(\frac{\sum_{i=1}^{n} \left(S_i - \overline{S_i} \right)^2}{n-1} \right)^{1/2} \quad (2)$$

Where, i and n are the indexes of cycles and the total number of cycles (here n=10), respectively. S_i is the measurement value obtained from each cycle and $\overline{S_i}$ is the mean of S_i.

3.3. Reproducibility with changing mounting position
In consideration of the influence of mounting position on the calibration results, Reproducibility with changing mounting position u_{pro} was evaluated by a certain number of repeated tests in a way that after each test, the wrench and rod for loading point were reinstalled and square drive and torque transducer were rotated to a different position. The relative reproducibility with different mounting positions is calculated according to equation (3) as the experimental standard deviation for the measurement values of each mounting position:

$$u_{pro} = \frac{1}{\overline{S_j}} \left(\frac{\sum_{j=1}^{k} \left(S_j - \overline{S_j} \right)^2}{k-1} \right)^{1/2} \quad (3)$$

Where, j and k are the indexes of different mounting position and the total number of mounting positions (here k=10), respectively. S_j is the measurement value obtained from each mounting position and $\overline{S_j}$ is the mean of S_j.

3.4. Deviation due to indication

The relative deviation due to the indication u_{ind} is calculated as the deviation between reference values indicated by a reference torque wrench or a weight-bar-system of higher grade and the indicating results of TSC. Components related to this term might be calculated from the certificate.

3.5. Stability

The TSC was calibrated four times by using a torque standard of higher grade over approximately one year in order to investigate its long-term stability. Relative standard uncertainty due to stability u_{sta} is calculated as:

$$u_{sta} = \frac{S_{max} - S_{min}}{d_m \cdot S_{mean}} \tag{4}$$

Where, S_{max}, S_{min} and S_{mean} are the maximum, minimum and mean of the 4 calibration results, respectively. Here, d_m equals to 2.06 ascribed to 4 calibration results [2].

3.6. Resolution

The resolution r of the indication is defined as the smallest fraction of a scale division that is readable in the case of analogue scale. In the case of digital scale, r is considered to be one increment of the last active number of the numerical indicator, provided that the indication does not fluctuate when the TSC is under the non-loading condition. Relative standard uncertainty due to resolution u_{res} is calculated according to equation (5), by determining r as a half-width of the rectangular distribution:

$$u_{res} = \left(\frac{2}{3}\right)^{1/2} \cdot \frac{r}{2T_i} \tag{5}$$

Where, T_i means applied toque. u_{res} is multiplied by $(2)^{1/2}$ considering that the measurement value is obtained as the difference between the indicated values of the loaded torque step and non-loading zero step before starting the cycle (double readings) [4].

Table 1. Uncertainty budget.

Torque	0.5 Nm	2 Nm	5 Nm	30 Nm	60 Nm
u_{rep}	0.12%	0.11%	0.09%	0.08%	0.05%
u_{pro}	0.27%	0.25%	0.24%	0.23%	0.15%
u_{ind}	0.29%	0.29%	0.29%	0.29%	0.29%
u_{sta}	0.16%	0.15%	0.11%	0.12%	0.08%
u_{res}	0.08%	0.02%	0.01%	0.00%	0.00%
u_c	0.45%	0.43%	0.40%	0.40%	0.34%
U_r	0.90%	0.85%	0.80%	0.79%	0.68%

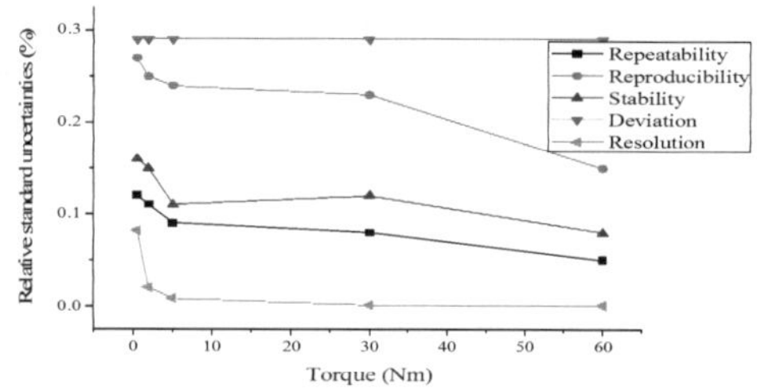

Figure 4. Standard uncertainty contributions.

3.7. Results of uncertainty evaluation

Uncertainty of the calibration results was evaluated at 0.5 Nm, 2 Nm, 5 Nm, 30 Nm and 60 Nm. The relative expanded uncertainty U_r of every point is calculated by equation (1) and summarized in table 1. In all the evaluated cases, the expanded relative uncertainty of calibration results within 0.9% could be obtained. Thus, the relative expanded uncertainty of torque realized by the TSC was estimated to be 0.9% in the range from 0.5 Nm to 60 Nm, with the coverage factor k=2. To make it more intuitive, standard uncertainties were plotted as a function of measuring points in figure 4 to visualise contributions each factor makes to the overall combined standard uncertainties. In general, relative standard uncertainties are larger in lower torque cases since it's harder to yield accurate calibration results in the lower range of the standard transducer as compared to the results of higher range. It was obvious that u_{ind} was dominant uncertainty contribution in all cases. In addition, Reproducibility due to changing mounting position have profound effect on the combined uncertainty since remounting a wrench or those moving clamping parts at the same positions is scarcely possible. While uncertainties due to resolution are sufficiently small compared with other terms which can be ignored, especially for larger rated capacity.

4. Verification of the uncertainty evaluation results

4.1. Verification method

In order to investigate the validity of uncertainty estimation of torque realized by the TSC, intra-laboratory comparison was conducted between the TSC and a RTS at our lab using several torque wrenches as transfer devices. The E_n ratio between TSC and the RTS for each measuring point was evaluated using the following equation:

$$E_n = \frac{S_1 - S_2}{\left(U_1^2 + U_2^2\right)^{1/2}} \tag{6}$$

Where, S_1, S_2 denote calibration results of the same point of a transfer device given by TSC and the RTS, and U_1, U_2 are the associated expanded uncertainties.

4.2. Results of comparison

Table 2 summarizes the results of the E_n number evaluation between TSC and the reference standard. The E_n numbers were all far less than one over the calibration range from 0.5 Nm to 60 Nm. The comparison results showed good agreement within the uncertainties in all cases. Thus, the torque realized by the TSC was shown to be equivalent to that achieved by the reference standard.

Table 2. Results of E_n evaluation.

Torque	0.5 Nm	2 Nm	5 Nm	30 Nm	60 Nm
S_1	0.498	1.995	4.776	30.135	63.061
S_2	0.499	1.998	4.782	30.102	63.126
E_n	0.18	0.13	0.11	0.10	0.09

5. Summary

In this study, a torque standard component of an integrated calibration equipment is proposed that allows the on-site calibration of several commonly used hand torque tools. The relative expanded uncertainty was estimated to be 0.9% for the calibration range from 0.5 Nm to 60 Nm, with the coverage factor k=2. A comparison experiment has been performed between the torque standard component and a reference torque standard at our lab to confirm the validity of uncertainty evaluation. As a result, sufficiently small deviations were obtained and the consistency between those two devices has been verified under the claimed uncertainties.

References

[1] Wang Y, Zhang Q, Gou C and Su D 2018 *IOP Conf. Ser.: Mater. Sci. Eng.* **317** 012031.
[2] GJB 2749A 2009 *General requirement of establishment and conservation of measurement standard for military metrology* (Chinese)
[3] ISO/IEC guide 98-3 2008 *Uncertainty of measurement-Part 3*
[4] Ogushi K, Nishino A, Maeda K and Ueda K 2011 *Sice Conference* (Tokyo, Japan) pp 411-416.

IOP Conf. Series: Materials Science and Engineering **507** (2019) 012008 doi:10.1088/1757-899X/507/1/012008

Cassava starch/carboxymethylcellulose biocomposite film for food paper packaging incorporated with turmeric oil

FA Mustapha[1], J Jai[1]*, ZIM Sharif[1] and NM Yusof[1]

[1]Faculty of Chemical Engineering, University Teknologi MARA, 40450 Shah Alam, Selangor

E-mail: junejai@salam.uitm.edu.my

Abstract. Issues related to the performance and processing remain a challenge when managing materials from renewable and natural resources. This study aims to show the potential of biopolymer film composite from cassava starch and carboxymethylcellulose to replace the synthetic coating and the area of studies are its barrier, mechanical and chemical properties. Fourier Transform Infrared analysis identified that only film containing CMC could act as antimicrobial agent carrier. The biopolymer film coated on the kraft paper with the presence of 30% v/v glycerol and 10% w/w CMC had the highest tensile strength, at 15 MPa. The film-coated paper also exhibited a hydrophobic surface with reduction of the contact angle from 104.95° to 96.33° for 120 seconds. Meanwhile, addition of turmeric oil into developed BF 3 showed insignificant changes in tensile strength but reduced the contact angle to less than 90°. Thus, the formulated coating materials can be used as an alternative green technology for paper coating used in food industry and can be classified as a form of active packaging.

1. Introduction

Most of the researches have been conducted with the goal of infusing food additives into the food packaging to maintain the quality of food and elongate the usability period of these foods by inhibiting the microbial growth [1, 2]. This system of packaging is recognized as active packaging, where it inhibits the microbial development in non-sterilized food and pasteurize food, while avoiding food contamination during the post-processing [3]. Recent studies discovered that additives extracted from plants with medicinal properties possess an inhibitory effect on various types of microorganism [4]. The examples include Curcuma longa L., Labisia pumila, Morinda citrifolia L. etc. This antimicrobial property can be attributed to the large number of secondary metabolites found in spices and herbs which have the ability to impede the growth of the infectious organisms [5, 6].

Cassava starch has a very good potential in term of domestic demand. A study on cassava starch by Parra [7] discovered that films based on cassava starch have good flexibility and low water permeability. This conforms with the study done by Maran et al. [8] where the authors indicated that cassava starch has great potential as the starch material for the production of biopolymer film and coating because it exhibits film-forming properties with great biodegradability. However, film with a high concentration of starch may recrystallize and become brittle during long-term storage, hence reducing its value [9]. This is in agreement with study by Kester [10], where starch-based films possess poor mechanical strength after a period of time due to the retrogradation process of the molecular chain. The drawback of starch as a biopolymer base can be overcome by adding a stabilizer or by mixing it with other biopolymer bases as mentioned by Kester [10].

Fortunately, this problem can be solved by adding plasticisers to improve the biopolymers workability and realise the full potential of starch [11–13]. The council of International Union of Pure and Applied Chemistry (IUPAC) has defined plasticiser as a material or substance that improves workability and flexibility of material such as plastic or elastomer. Hydrophilic compounds with low molecular weight such as glycerol and sorbitol are commonly used as biodegradable plasticisers in starch films. From the literature review, glycerol is the most incorporated component in hydrocolloid films including films based on cassava starch [11–14]. Through their experiments, Ahmad et al. [15] shown that tensile strength of thermoplastic sago starch increases with increasing amount of glycerol. Vieira et al. [9] have also stated that the crystallisation of potato starch containing glycerol can significantly affect the mechanical properties of the plastic sheet. This can be attributed to the formation of the entangled starch matrix and starch chain-to-chain associations as plasticiser content is increased.

Ghanbarzadeh et al. [16] stated that single based biopolymer films from a single source of biopolymer may possess good properties in some aspects but poor in others. This issue will be a great obstacle in a full-scale industrial application. Hence, researchers suggested the biopolymer to be mixed with other biopolymers in order to enhance the performance of bio-based coatings [17]. The films produced by the blending of biopolymers from polysaccharide with protein, polysaccharide with lipid, or even polysaccharide with polysaccharide sources are known as biocomposite films. A study by Alves et al. [18] on corn starch film had proven that the films by themselves are unsuitable for packaging purposes as they possess poor mechanical properties and are highly sensitive to water [18]. Meanwhile, a study by Guo et al. [19] on composite films by blending corn starch with carboxymethylcellulose (CMC) shown that the film mechanical properties have improved twice compared to the corn starch film itself [20]. CMC is suitable for film blending with other biopolymers due to its non-toxic, non-allergenic and high viscosity properties as well as its ability to act as a filler to improve barrier and mechanical properties [16, 21].

Cassava starch as film based has been tested in multiple studies on essential oil of cinnamon, lemongrass, ginger and Curcuma xanthorrhiza Roxburgh as antimicrobial agents [11-14]. A study by Dani et al. [25] on cassava starch biopolymer film incorporated with lemongrass oil stated that Trichoderma and Penicillium will not appear until the third day of incubation and the films are capable of reducing the microbial count on meat during storage. However, adding more lemongrass oil will lead to reduced tensile strength and roughness of the biopolymer film. In addition, Rojas et al. [25] also highlighted that the presence of essential oil from plants can significantly modify the tensile properties but would not affect the permeability of oxygen and water vapour of the films. Therefore, in this study, the performance of the composite biopolymer film incorporated with turmeric oil was characterised on its tensile strength, surface wettability and stability in releasing turmeric oil by using Fourier Transform Infrared (FTIR) analysis.

2. Materials and methodology

2.1. Materials

Commercial cassava starch (Cap Kapal ABC.Co, Thailand) was applied as primary biopolymer, glycerol (Merck, Germany) as plasticizer, carboxymethylcellulose (CMC) as composite biopolymer which was kindly provided by Waris NOVE Sdn. Bhd., Pahang, Malaysia and turmeric oil as antimicrobial agent sources. Unbleached kraft paper (Muda Packaging, Malaysia) was used as the paper food packaging material.

2.2. Biopolymer film preparations

Film-forming solutions was prepared by cassava starch gelatinization (5%) at a temperature of 75°C ± 5°C for 20 mins under continuous stirring at 500 rpm before being plasticized with glycerol (30% w/w of starch). Meanwhile, 10% w/w CMC was dissolved in 100 mL distilled water at room temperature. Then, both of the solutions were mixed and slowly stirred at 250 rpm for 20 minutes at 40°C to ensure the mixture was completely mixed. Turmeric oil (0.15% v/v) was added and stirred for another 20 minutes. The film-forming solutions were then casted at volumes of 30 mL on kraft paper (30 cm length × 15 cm width) using an automatic film casting machine at a speed of 1 cm/s for blade distance of 1.25 mm, respectively. After that, the wet-coated paper was dried at 50°C for two and half hours and conditioned in a desiccator at 25 + 5°C for 24 hours prior to sample analysis.

2.3. Fourier transform infrared analysis

The disruption on the functional group of starch-based film affected by the different content of CMC on the starch based film were analyzed by using the Fourier Transform Infrared (FTIR) spectroscopy (Perkin Elmer, USA) at wave number ranging from 400 to 4000 cm-1. The sample preparation of the biopolymer films was done by poured 30 mL of film-forming solutions in disposable Petri dish (90 mm × 15 mm). Next, the solutions were dried at 50°C until the weight of the samples was constant. The sample port was cleaned by using acetone to remove any variation caused by contamination on the scanning. The spectrum yields from the sample scanning were collected and studied.

2.4. Tensile strength measurements

Tensile strength measurement studied if the strength of the film-coated paper was sufficient to apply as a food packaging. The mechanical test was conducted using the Tinius Olsen Universal Testing Machine, Model H50KT (USA). Samples were cut into the dimensions of 75 × 25 × 2.5 mm (length × width × thickness) before placing them onto the sample holder. The bottom and upper gaps were tightened. The crosshead speed was set at 50 mm/min with a gauge length of 50 mm. The test was repeated three times. The data were tabulated and plotted using Microsoft Excel. The sequence of procedure is according to ASTM D3500.

2.5. Wettability surface measurements

Wettability analysis was done to identify the hydrophobicity surface of the CS/CMC films by measuring the contact angle of water in contact with the film-coated paper using Video Contact Angle Goniometer (AST Products, INC.). Wettability of film was measured using the contact angle method as suggested by Liu et al. [26]. The samples were cut in rectangular strips with a dimension of 1 cm × 1 cm (width × length). The rectangular strip samples were then directly placed on the horizontal moveable stage and 10 µL of water was dropped onto the film surface using a 500 µL precision micro syringe. For each sample, three measurements were taken and the average values were calculated.

3. Results and discussions

3.1. FTIR analysis of biopolymer films

From the spectre of cassava starch in figure 1, it can be seen that C-OH and C-O-C groups of cassava starch wavenumbers were seen at wavelength 3290 cm-1 and 927 cm-1, respectively. It was expected that the crosslinking reaction of the polymeric chain would occur at these two regions as suggested by Ahmad et al. [15] in a study of sago starch as a biodegradable polymer. The authors suggested that interaction of hydrogen bond formation of the starch with glycerol occurred in the C-OH group and the C-O-C group where the wavenumbers for the C-OH group and the C-O-C group were reduced; indicating that the OH group of starch took part in hydrogen bond formation. The existence of a hydroxyl group at wavenumber of 3281 cm-1 also can be seen in the spectre of CMC, which is expected to contribute in formation of various kinds of intermolecular and intramolecular hydrogen bonds with the hydroxyl group in cassava starch, glycerol and turmeric oil. The formation of hydrogen bond could be attributed to the strength of the film as suggested by Fan et al. [27].

Intermolecular interaction of biopolymer chain with turmeric oil was analysed by using FTIR analysis and shown in figure 1. Referring to figure 1 (a) and (b), there were no presence of phenol group indicated that the intermolecular interaction of turmeric oil with cassava starch film and cassava starch with glycerol film occurred at O-H stretch region which then forming the C-O-C stretch. In contradiction, it can be seen in figure 1 (c) that the intermolecular interaction occurred at regions of C=O stretch and intermolecular bonding of C-OH as peaks at 1737 cm-1 and 1231 cm-1 were disappeared with addition of turmeric oil into film-forming formulation. At the same time, a new peak was formed at region of phenol group. This finding indicated that presence of CMC preserve the functional group of phenol which also the major groups of antimicrobial compounds from turmeric oils.

Besides that, the structural property which was displayed as FTIR spectre of cassava starch film and cassava starch with glycerol film significantly deviated with increasing of storage time. This may attributed by formation of hydrogen bond between the polymeric chain with moisture as both glycerol and starch was hydrophilic compounds with high intensity of hydroxyl group. Meanwhile, the structural properties of film containing CMC approximately same as can been observed from the FTIR spectres with increasing of storage time. This was attributed by the formation of intermolecular and intramolecular hydrogen bonds of CMC with the polymer chains was stronger than the hydrogen bond formed with the moisture content. Thus, it can be stated that only cassava starch film with presence of glycerol and CMC could act as antimicrobial agent carrier as it can behold the turmeric oil and allowed it to release at right stipulated period of time such as when there was presence of microbes. This finding is in agreement with study done by Fajardo et al. [28] where they stated that controlled released of additives from polymer matrix was based on the structural characteristic of the film. The functional group of the respective regions was justified accordingly to Coates [29], Gokel [30], Kizil et al. [31] and Fan et al. [27].

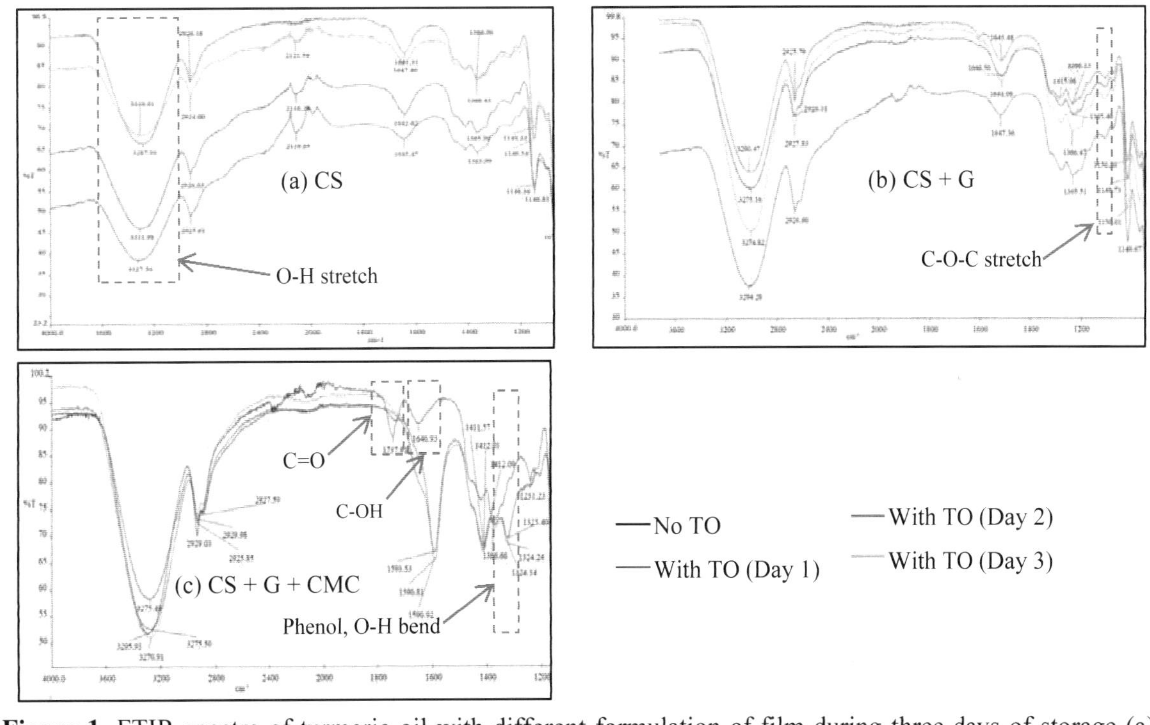

Figure 1. FTIR spectre of turmeric oil with different formulation of film during three days of storage (a) Cassava starch, (b) Cassava starch with glycerol and (c) Cassava starch with glycerol and carboxymethylcellulose film.

3.2. Tensile strength measurements

Referring to figure 2, non-plasticized cassava starch film-coated paper (BF 1) had the lowest value of tensile stress, which was 12 MPa, compared with plasticized film-coated paper (BF 2), whose tensile stress value was 12.9 MPa. A similar behaviour regarding the effect of glycerol on film strength was found in a study by Bonilla et al. [32]. The authors stated that generally, as the plasticizer concentration increased, the tensile strength was also increased. It can also be seen in figure 2 that addition of CMC significantly enhanced the tensile strength. Presence of CMC in starch complex chain contributed in stable dense of biopolymer film as filler. This finding was also supported by Hong [20] in a study on composite films by blending corn starch with CMC. He identified that the film mechanical properties had been improved twice compared with the corn starch film itself. As depicted in figure 2, there is a significant discrepancy between the tensile strength of biopolymer film coatings on the kraft paper without the presence of turmeric oil and those with turmeric oil present. This is because turmeric oil reduced intermolecular interaction of polymeric chain. Similar findings have been found by Maria et al. [25], wherein authors found that the incorporation of essential oil into biopolymer films caused a noteworthy reduction in the tensile strength of biopolymer films.

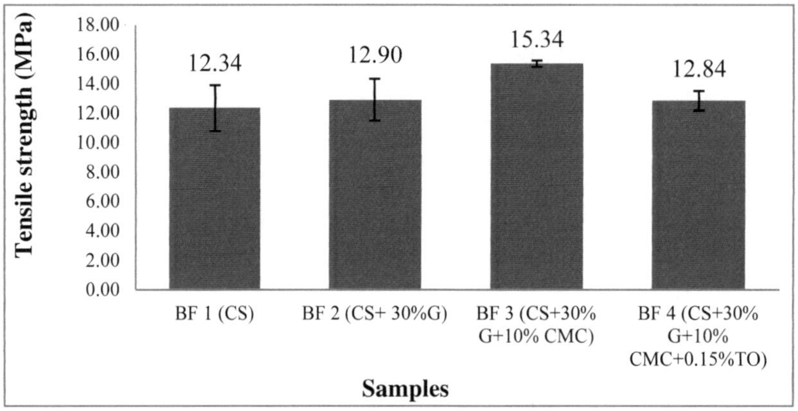

Figure 2. Effects of glycerol, carboxymethylcellulose and turmeric oil on tensile strength of cassava starch film-coated on the kraft paper

3.3. Wettability measurements

Figure 3 show that BF 1 had the lowest contact angle distributions throughout 120 seconds, which is lower than 90°. This finding indicates that cassava starch-based film exhibit hydrophilic properties which are not desirable in this study as it increased the affinity of moisture on the film surface. Presence of glycerol reduced the hydrophilicity of the cassava starch-based film with the increase in contact angle throughout 120 seconds of observation. This indicates that the surface affinity of the biopolymer films for water decreased when glycerol was added, which then reduced the water uptake. In light of the results obtained, this is believed to have occurred because the biopolymer film surface of BF 2 remained dense as glycerol filled the intermolecular chains of the biopolymer. A similar behaviour was found in a study by Basiak et al. [33]. The authors stated that the higher the contact angle, the lesser the affinity of water for the film surface as the contact angle represents surface equilibrium among water, air and a biopolymer film surface.

The behaviour of water droplets on the surface of BF 3 with the presence of CMC clearly changed from BF 2 (0% CMC). In addition, only BF 3 exhibited hydrophobic properties, as the contact angle remained higher than 90° over 120 seconds, whereas BF 1 exhibited hydrophilic properties as the contact angle remained lower than 90°. As for BF 2 and BF 4, hydrophobic properties were apparent only when the water was initially dropped, and it started to spread or adsorb into the film with time, exhibiting hydrophilic properties at the end of 120 seconds. Based on the contact angle against water on different sources, the materials were classified as hydrophilic for contact angle values lower than 90° or hydrophobic for contact angle values higher than 90°as reported by a number of researchers [24-26]. This is in agreement with a similar finding reported by Hong et al. [20]. With a lower content of CMC, CMC acted as filler, which allowed it to disperse well in a starch matrix and thus improved the water barrier of the film. Decreased in the contact angle of a water droplet on biopolymer film was observed in figure 3 with increased in volume of turmeric oil. This is mainly due to an increase in hydrophilicity of the biopolymer film. This decrease could be due to reduction in molecular interaction of the biopolymer film with the presence of turmeric oil. This is in agreement with the results reported by Souza et al. [36], in which the authors identified that incorporation of cinnamon essential oil decreased the water barrier properties of biodegradable film. This was because introduction of essential oil reduced the molecular interaction between polymeric chains and thus allowed water uptake on the film.

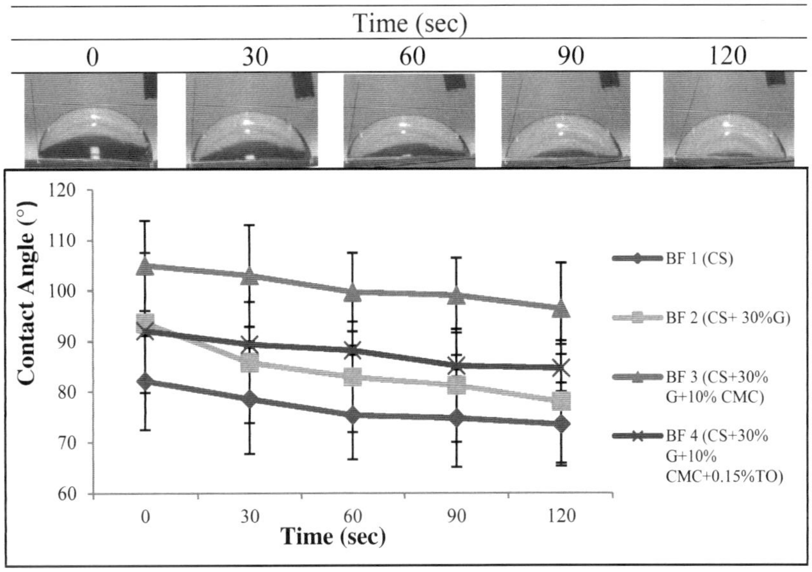

Figure 3. Effects of glycerol, carboxymethylcellulose and turmeric oil on wettability of cassava starch film-coated on the kraft paper

4. Conclusions

The performance of the composite biopolymer film was evaluated on its tensile strength, wettability and the ability of the biopolymer film to act as additive carrier. Tensile strength and wettability measurements indicated that the biopolymer film containing 30% (v/v starch) glycerol and 10% (w/w) CMC labelled as BF 3 had the highest tensile strength (15 MPa) and exhibit hydrophobic properties with contact angle higher than 90° throughout 120 seconds. Meanwhile, addition of turmeric oil into developed BF 3 showed insignificant changes in tensile strength but reduced the contact angle to less than 90°. FTIR analysis on incorporation of turmeric oil in each of starch film, starch with glycerol film, and starch with glycerol and CMC identify that presence of CMC stabilize the intermolecular interaction of the turmeric oil with the complex polymeric chain. Therefore, the results suggested that applying the biopolymer film containing turmeric oil as kraft paper coating may improve its functionality in prolongs the shelf life of foods. Thus, it can be stated that the develop biopolymer film could replace the currently used petroleum based derivative plastic in the food industry.

Acknowledgements

The authors would like to acknowledge University Technology MARA for the financial support to carry out this research through Institute of Postgraduate Studies (IPSis).

References

[1] Appendini P and Hotchkiss J H 2002 *Innov. Food Sci. Emerg. Technol.* **3** 113.
[2] Doores S 1993 Organic acids ed Davidson P M and Branen A L (New York: Marcel Dekker Inc.) 117–119.
[3] Shit S C and Shah P M 2014 *J. Polym.* **1** 1.
[4] Beth A C, Carter M and Zuoxing Z 2005 *Food Safety Magazine, Manhattan* 1.
[5] Tana H, Matthews K K and Di R 2015 *Biomed Res. Int.* **1** 1.
[6] Tiwari B K, Valdramidis V P, Donnell C P O, Muthukumarappan K, Borke P and Cullen P J 2009 *J. Agric. Food Chem. Rev.* **57** 5987.
[7] Parra D, Tadini C, Ponce P and Lugao A 2004 *Carbohydr. Polym.* **58** 475.

[8] Maran J P, Sivakumar V, Sridhar R and Thirugnanasambandham K 2013 *Carbohydr. Polym.* **92** 1335.
[9] Gurgel M, Altenhofen M, Oliveira L and Beppu M M, 2011 *Eur. Polym. J.* **47** 254.
[10] Fennema O R, Kamper S L and Kester J J 1990 Patent No:**4**,915,971.
[11] Navia D P and Villada H S 2012 Thermoplastic Elastomers ed Prof. Adel El-Sonbati (Croatia: InTech,) 24–36.
[12] Paes S S, Yakimets I and Mitchell J R 2008 *Food Hydrocoll.* **22** 788.
[13] Shanks R and Kong I 2012 Thermoplastic Elastomers ed Prof. Adel El-Sonbati (Croatia: InTech,) 24–36.
[14] Oriani V B, Molina G, Chiumarelli M, Pastore M and Hubinger M D 2014 *J. Food Sci.***79** 189.
[15] Ahmad Z, Anuar H, and Yusof Y 2011 *Key Eng. Mater.* 397.
[16] Ghanbarzadeh B, Almasi H and Entezami A A 2010 *Innov. Food Sci. Emerg. Technol.* **11** 697.
[17] Rastogi V K and Samyn P 2015 *Coatings* **5** 887.
[18] Alves V, Costa N, Hilliou L, Larotonda F, Goncalves M and Sereno A 2006 *Desalination* 331.
[19] Guo J, Liu M H and Gong H X 2013 *BioResources* **8** 6510.
[20] Hong K M 2013 "Preparation and characterization of carboxymethylcellulose from sugarcane bagasse".
[21] Suderman N, Isa M I N and Sarbon N M 2016 *Int. Food Res. J.* **23** 75.
[22] Utami R, Kawiji, Nurhartadi E and Putra A Y T 2014 *Int. Food Res. J.* **21** 413.
[23] Praseptiangga D, Nurhartadi E, Manuhara G J, Khasanah L U, Azzahra F A and Indrayati F 2016 *J. Teknol.* **78** 193.
[24] Dani Supardan M, Annisa Y, Arpi N, Satriana and Wan Mustapha W A 2016 *Int. J. Adv. Sci. Eng. Inf. Technol.* **6** 216.
[25] Rojas-Grau M A, Avena-Bustillos R J, Olsen C, Friedman M, Henika P R, Martin-Belloso O, Pan Z and McHugh T C 2007 *J. Food Eng.* **81** 634.
[26] Liu Y, Cai Y, Jiang X, Wu J and Le X 2016 *Food Hydrocoll.* **52** 564.
[27] Fan M, Dai D and Huang B 2012 *Fourier Transform - Mater. Anal.* **1** 45.
[28] Fajardo P, Martins J T, Fuciños C, Pastrana L, Teixeira J A and Vicente A A 2010 *J. Food Eng.* **4** 349.
[29] Coates J 2000 *Encycl. Anal. Chem.*10815.
[30] Gokel G W 2004 Dean's Handbook of Organic Chemistry vol 2 (Washington: McGRAW-HILL) 6.21-6.51.
[31] Kizil R, Irudayaraj J and Seetharaman K 2013 "Characterization of irradiated starches by using FT-Raman and FTIR spectroscopy".
[32] Bonilla J, Vicentini M, Dos R M C, Bittante Q B and Sobral P J A 2015 Mechanical properties of cassava starch films as affected by different plasticizers and different relative humidity conditions **4** 116.
[33] Basiak E, Debeaufort F and Lenart A 2016 *Food Chem.* **195** 56.
[34] Balan T, Guezennec C, Nicu R, Ciolacu F and Bobu E 2015 *Cellul. Chem. Technol.* **49** 607.
[35] Skurtys O, Acevedo C, Pedreschi F, Enrione J, Osorio F and Aguilera J 2010 *Food Science and Technology (Santiago: NOVA Publisher)* 1–34.
[36] Souza A C, Goto G E O, Mainardi J A, Coelho A C V and Tadini C C 2013 *LWT - Food Sci. Technol.* **54** 346.

Optimization of Starch Composite Edible Coating Formulation on Fresh-Cut "Fuji" Apple through Surface Tension, Wettability and FTIR Spectroscopy

Z I M Sharif[1], J Jai[1], I Subuki[1], N A M Zaki[1], F A Mustapha[1], N Mohd Yusof[1] and S A Idris[1]

[1]Faculty of Chemical Engineering, University Technology Mara (UiTM), 40450 Shah Alam, Selangor, Malaysia

Abstract. Fresh-cut "Fuji" apples are easily oxidised and deteriorated. Therefore, starch composite edible coating formulation were formulated to extend its life span. In this study, the effect of different amount of carboxymethyl cellulose (CMC) (1, 2, 3 and 4% (w/v)), glycerol (1, 2, 3 and 4% (v/v)) and turmeric oil (TO) (50, 175 and 300ppm) were evaluated through surface tension and wettability analysis. Furthermore, the molecular interaction between raw materials of the edible coating film were analysed through FTIR analysis. The results showed that formulation using 2% (w/v) CMC and 2% (v/v) glycerol were the best amount to formulate composite edible coating. Meanwhile, the range of turmeric oil were 50, 175 and 300ppm suitable to be used for further analysis on fresh-cut "Fuji" apples as it does not exceed the general rule of surface tension. The findings for intermolecular reaction of raw materials revealed that there were interactions occurred when citric acid (CA), glycerol and CMC were added; further enhanced functionality of the starch. Besides that, FTIR showed that the functional group in turmeric oil that acts as an active compound to inhibit the oxidation process and microbial activities were present in the edible coating film.

1. Introduction

Maintaining fresh-cut "Fuji" apples quality and organoleptic properties is difficult due to elimination of the apple skin and the cutting process [1]. This could lead to oxidation process and growth of microbial, thus, decrease the life spans of the fresh-cut "Fuji" apples. Edible coating is one of the preservation method that able to preserve the quality of the fresh-cut "Fuji" apples by controlling the water permeability, exchange of gases, oxidation process, microbial growth etc. Edible coating is made from hydrocolloid (polysaccharide, protein and lipid). Composite edible coating are formed when two or more types of hydrocolloid are used as the base material that provide benefits to each other along with reducing their drawbacks [2]. According to Cothran *et. al.* [3], the surface tension value should be in the range of 28 to 38 dyne/cm to formulate coating for fruits with uniform surface. Meanwhile, the great wettability should be high in work of adhesion (W_a), low in work of cohesion (W_c) and negative value that near to zero for spreading coefficient (W_s) [4]. The edible coating properties could be enhanced with addition of preservatives such as turmeric oil due to the presence of active compounds which are phenolic compound, curcumin, sesquiterpenes etc. [5]. Therefore, the aims of this research are to i) optimize the best amount of raw materials in forming the starch composite edible coating formulation through surface tension and wettability, and ii) determine the molecular interaction that occurred during formation of starch composite edible coating film.

IOP Conf. Series: Materials Science and Engineering **507** (2019) 012009 doi:10.1088/1757-899X/507/1/012009

2. Material and methodology

2.1. Materials

Fresh 'Fuji' apples were acquired from local market in Shah Alam, Selangor. Cassava flour (Cap Kapal ABC Co., Malaysia) and CMC (Novelecell, Malaysia) were used as base materials for composite coating formulation. CA (Merck, United State) served as the cross-linking agent. Glycerol (Merck, United State) was used as plasticizer and turmeric oil (Soul, Malaysia) was used as antioxidation agents. Below figure 1 shows the flowchart of the research process.

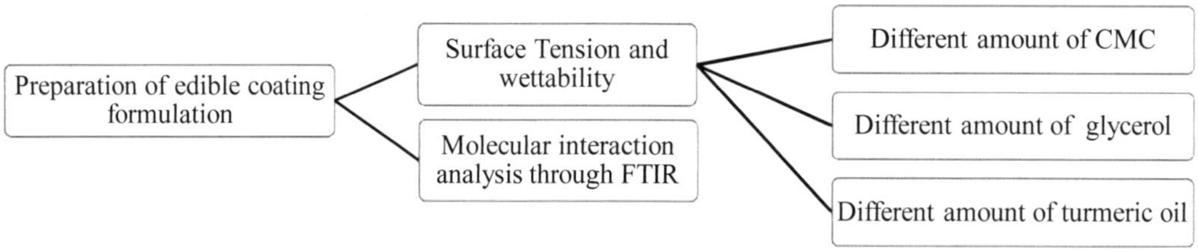

Figure 1. The flowchart of research process

2.2. Preparation of edible coating formulation

The 6% (w/v) cassava starch was diluted in 100 mL of distilled water (DW) for 40 minutes at temperature 80°C. Then, 0.5% (w/v) CA was added in the formulation as the cross-linking agent. Glycerol (1, 2, 3 and 4% (v/v) glycerol) was added in the formulation as plasticizer. The CMC (1, 2, 3 and 4% (w/v) CMC) was solubilized in 100 mL of distilled water for 30 minutes at 75°C. pH of the formulation was titrated with standard alkaline solution to pH 5.6 [6]. The formulation was cooled down at 40°C and stirred slowly for 20 minutes to release the air bubbles. Lastly, TO (50, 175 and 300ppm) was added into the formulation.

2.3. Surface tension

Surface tension analysis was used in order to determine the number of energy needed to raise the surface per unit area [7]. Sessile drop method was used to analyse the contact angle by using Contact Angle Goniometer (AST Products, INC.). The formulation was dropped on the surface of fresh-cut "Fuji" apples and the process were recorded by Contact Angle Goniometer video. Surface tension was calculated using the surface energy software (SE2500). This process was replicated for four times for each of the formulation.

2.4. Wettability

Wettability is to ascertain the emulsion ability in spreading, its adhesion on solid surface and also their cohesion work. Wettability analysis was calculated in order to determine the value of spreading coefficient (Ws). The closest W_s value towards zero indicates the best in spreading activity and its adhesion between emulsion and surface fresh-cut 'Fuji' apple. This analysis was done based on J. Viera method. W_s was calculated according to below equation. Where W_a is work of adhesion, W_c is work of cohesion, W_s is spreading coefficient and γ_L is surface tension.

$$W_s = W_a - W_c \quad (1)$$

$$W_a = \gamma_L (1+\cos(\Theta)) \quad (2)$$

$$W_c = 2\,\gamma_L \quad (3)$$

2.5. Fourier transform infrared (FTIR) spectroscopy

FT-IR spectroscopy was used to determine the molecular interaction between raw materials during formation of edible coating [9]. The changes of the wavelength and transmittance were resulted from the molecular interaction between the raw materials in forming the edible coating film.

3. Results and discussion

3.1. Surface tension and wettability analysis

3.1.1. Different amount of carboxymethyl cellulose (CMC). CMC was used as filler in starch composite edible coating formulation and was represented as C. Analysis from table 1 shows that the surface tension value decreased with increased amount of CMC added which were 37.49 dyne/cm, 37.92 dyne/cm, 30.18 dyne/cm and 28.34 dyne/cm, respectively. Therefore, it can be considered that CMC has surface active properties due to increase in formulation viscosity and tends to have high water vapour permeability. It can be analysed that C2 formulation had W_s value near to zero (-25.04 ± 25.72 dyne/cm) and had the highest W_a (40.66 ± 30.68 dyne/cm) compared to others. This was due to its high energy loss when spreading on the surface of fresh cut 'Fuji' apples flesh. This finding was in agreement with Choi *et. al.* [10] studies that reported that high value work of adhesion attributed to huge loss of energy during spreading process. Table 1 also shows the W_s value for C3 formulation was low due to high CMC content which was less effective as it would flock and disturb the spreading process. Tongdeesoontorn *et. al.* [11] also reported similar findings which the use of higher CMC content could lead to congregate. Meanwhile, the W_c value decreased as the amount of CMC increased. Even though W_c for C3 and C4 were better than C2, they were not compatible to be used in the formulation of edible coating emulsion due to the high amount of CMC and low W_a and W_s. Therefore, it can be concluded that C2 was the best amount of CMC to be used in formulating the starch composite edible coating.

3.1.2. Different amount of glycerol. Table 1 also shows the surface tension and wettability for formulation that consists different amount of glycerol (G1 – G4). The result shows that the surface tension of G4 has exceeded the general range for surface tension. Thus, only G1, G2 and G3 were considered as the suitable amount of glycerol to be added in formulation. Furthermore, G2 had the highest value of W_s and W_a compared to others. This finding was observed due to the high ability in adhesion and spreading. The similar findings were found from Casariego *et. al.* [12] which reported that concentration of glycerol influence the surface properties of coating. Increased in glycerol concentration will lead to decrease of W_a and W_s. W_c was decreasing with increasing amount of glycerol. According to Garcia *et. al.* [13], the maximum amount of glycerol that is suitable to be used with starch edible coating formulation is 20 g/L. Therefore, volume of glycerol in formulation of G3 was not compatible to be used. Higher amount of plasticizer used will lead to the formation of soggy coating film due to the higher amount of glycerol in the hydroxyl group compared to the one in the starch and it tends to bond with the hydroxyl group that present in moisture of surrounding [14]. Thus, the optimum amount of glycerol in G2 was used to formulate the starch composite edible coating formulation.

3.1.3. Different amount of turmeric oil (TO). Table 1 also tabulates the analysis of surface tension and wettability for different concentrations of turmeric oil (T1 – T4). There was a slightly decrease in surface tension, W_a and W_c values as the concentration of turmeric oil increased by not more than 1 dyne/cm. This finding was in agreement with Yusof *et. al.*[15] which reported that that increase in turmeric oil concentration will decrease the surface tension value. T2 has the highest W_a value which is 49.53 dyne/cm compared to T3 which is 47.28 dyne/cm. This shows that formulation T2 (49.53 dyne/cm) has the best W_a on the surface of fresh-cut 'Fuji' apples flesh followed by T3, T4 and T1. The W_s value appeared to decrease with increasing concentration of turmeric oil content, which is likely due to addition of oil (lipid) in the emulsion that leads to expansion of the emulsion droplet. Therefore, it can be concluded that addition of turmeric oil was not significant to the surface tension, W_a, W_c, and W_s. Hence, the concentration range of turmeric in this study can be used in the formulation.

Table 1. The surface tension and wettability analysis on different amount of carboxymethyl cellulose (CMC), glycerol and turmeric oil.

	CMC (%w/v)	Glycerol (%v/v)	Concentration of TO (ppm)	Surface Tension (dyne/cm)	Wettability (dyne/cm)		
					W_a	W_c	W_s
C1	1	2	0	37.49	26.01±21.57	74.99±1.56	-48.99±21.76
C2	2	2	0	37.92	40.66±30.68	75.84±1.83	-25.04±25.72
C3	3	2	0	30.18	15.70±10.04	60.35±1.59	-44.65±11.01
C4	4	2	0	28.34	33.19±11.53	56.68±3.50	-27.99±10.12
G1	2	1	0	34.69	35.43±25.49	69.39±2.68	-33.97±26.11
G2	2	2	0	37.92	40.66±30.68	75.84±1.83	-25.04±25.72
G3	2	3	0	28.63	22.37±26.77	57.26±3.7	-34.89±25.27
G4	2	4	0	27.64	8.94±12.98	55.29±1.34	-46.33±13.4
T1	2	2	0	37.92	40.66±30.68	75.84±1.83	-25.04±25.72
T2	2	2	50	37.46	49.53±1.16	74.91±1.26	-25.37±2.42
T3	2	2	175	37.38	47.28±20.11	74.75±0.61	-27.47±19.50
T4	2	2	300	37.11	44.49±39.16	74.22±2.71	-29.73±38.49

3.2. Molecular interaction analysis through FTIR

Figure 2 shows, with the additional of citric acids (CA) spectra (B), the cross-linking between CA and starch occurred. It can be seen that a new peak produced at 1716.28 cm^{-1} and the absorbance reading was reduced at 3306.20 cm^{-1} to 3291.17 cm^{-1}. They were reduced due to esterification reaction occurred. Besides that, the hydrolysis reaction that occurred at the glycosidic linkage was one of the factors that caused the peak to decrease. Similar findings were found by N. Reddy [16].

When glycerol was added into the formulation in spectra (C), further decreased was occurred at the peak of –OH groups which was from 3291.17 cm^{-1} to 3271.78 cm^{-1}. This is because of the esterification reaction occurred further between the -COOH group in CA with –OH groups of glycerol. The peak of 1716.28 cm^{-1} and 1591.53 cm^{-1} were disappeared due to the additional CA and glycerol together will lead to the high moisture content in the coating film as water absorption wavelength appeared at 1642.72 cm^{-1}. Increase in peak occurred at 2931.93 cm^{-1} and 1412.31 cm^{-1}. This indicates high content in hydrogen bond concentrations and hydrogen interaction between glycerol, starch and CA. Furthermore, it occurred due to the plasticization effect and this was also reported by Liu *et. al.* [17].

CMC was added to the mixture to stabilize the mixture due to higher moisture permeability after the addition of citric acid and glycerol as shown in spectra in figure 2 (D). The addition of CMC caused the peak of water absorption at 1642.72 cm^{-1} to disappear. This similar finding was also found out by Tongdeesoontorn [11]. When the turmeric oil was added as shown in spectra (F), it can be observed that there was presence of phenol compounds at the peak 3271.35 cm^{-1} (O-H stretch) and 1325.45 cm^{-1} (C=C). The turmeric oil active compounds were present in the edible coating film when it was incorporated with TO. The active compounds present could postpone the oxidation process and growth of microbial.

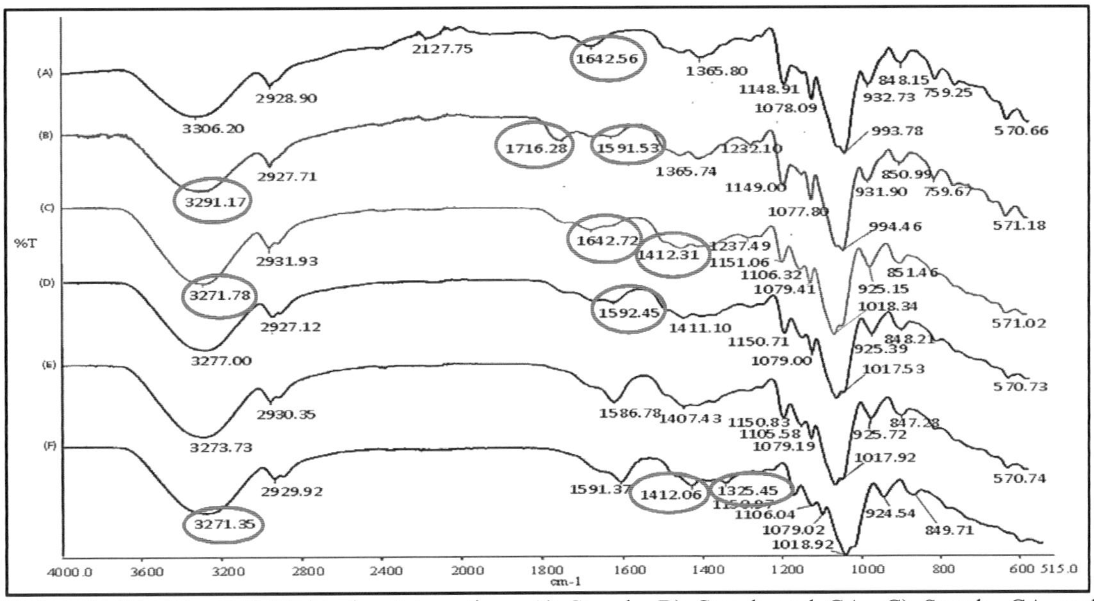

Figure 2. Spectrum of Molecular Interaction: A) Starch, B) Starch and CA, C) Starch, CA and Glycerol, D) Starch, CA, Glycerol and CMC, E) Starch, CA, Glycerol, CMC and Sodium Hydroxide F) Starch, CA, Glycerol, CMC, Sodium Hydroxide and TO.

4. Conclusion

Results obtained from the research show that the formulation of starch composite edible coating could be formulated by using 6 % (w/v) starch, 2% (w/v) CMC, 2% (v/v) glycerol and 0.5% (w/v) CA. This formulation was possibly could contributing to prolong the life span of fresh-cut "Fuji" apples. From the FTIR analysis, the molecular interaction occurred between raw materials. This can be observed due to the presence of ester bonds which represent the cross-linking process. It also can be observed that plasticization process occurred when concentration of hydrogen bond for alkene were increased and the moisture of the starch composite edible coating film was decreased when CMC was added as filler. This can be recommended that to used CMC, glycerol and CA as the raw materials to be used for edible coating for fresh-cut fruits due to their compatibility in forming the coating film.

Acknowledgement

The authors are grateful to Universiti Teknologi MARA for providing the grant research (600-IRMI/DANA KCM 5/3/LESTARI (131/2017)) throughout this study.

References

[1] Ahmad N A, Sharif Z I M, Jai J, Yusof N M and Mustapha F A 2018 *IOP Conf. Ser. Mater. Sci. Eng.* **334** 1–9.
[2] Sharif Z I M, Mustapha F A, Jai J, Yusof N M and Zaki N A M 2017 *Chem. Eng. Res. Bull.* **19(2017)** 145–53.
[3] Cothran C D W 1945 Treatment of fresh fruits and vegetables in preparation for market *US patent* 2,383,451
[4] Cerqueira M A, Lima Á M, Teixeira J A, Moreira R A and Vicente A A 2009 *J. Food Eng.* **94(3-4)** 372–8.
[5] Mustapha F A, Jai J, Hamidon F, Sharif Z I M and Yusof N M 2017 *Chem. Eng. Res. Bull.* **19** 57–66.
[6] Mu C M O and Laurindo B 2008 *Carbohydr. Polym.* **72(1)** 82–7.
[7] Moncayo D, Buitrago G and Algecira N 2013 *Ing. Inv.* **33(3)** 11–6.
[8] Vieira J M M 2014 *Thesis Universidade de Coimbra*
[9] Singh R and Mendhulkar V D 2015 *J. Chem. Pharm. Res.* **7(6)** 205–11.

IOP Conf. Series: Materials Science and Engineering **507** (2019) 012009 doi:10.1088/1757-899X/507/1/012009

[10] Choi W Y, Park H J, Ahn D J, Lee J and Lee C Y 2002 Wettability of chitosan coating solution on ' Fuji ' apple skin *J. Food Sci.* **67** 3–7.
[11] Tongdeesoontorn W, Mauer L J, Wongruong S, Sriburi P and Rachtanapun P 2011 *Chem. Cent. J.* **5(1)** 6.
[12] Casariego A, Souza B W S, Vicente A A, Teixeira J A, Cruz L and Dıaz R 2008 *Food Hydrocoll.* **22(8)** 1452–9.
[13] García M A, Pinotti A, Martino M N and Zaritzky N E 2009 *Edible films and coating for food applications* **6** 169–209.
[14] Gurgel M, Vieira A, Altenhofen M, Oliveira L and Beppu M M 2011 *Eur. Polym. J.* **47(3)** 254-63.
[15] Yusof N M, Jai J, Hamzah F, Yahya A and Pinijsuwan S 2017 *J. Phys. Conf. Ser.* **885** 1–6.
[16] Reddy N and Yang Y 2010 *Food Chem.* **118(3)** 702–11.
[17] Liu K, Yuan C, Chen Y, Li H and Liu J 2014 *Sci. Hortic.* **176** 45–53.

IOP Conf. Series: Materials Science and Engineering **507** (2019) 012010 doi:10.1088/1757-899X/507/1/012010

Study on Properties of PP Resin Composites Modified by Aluminum Powder

Weiguo Yao[1], Dongyang Zhang[1], Yanli Dou[1], Dongbo Guan[1]

[1]College of Materials Science and Engineering, Key Laboratory of Automobile Materials of Ministry of Education, Jilin University, Changchun 130025, PR China

Abstract. The aluminum powder was used to blend with PP to impart a metallic luster to the PP matrix for achieving the effect of coating, and the ionic liquid (1-octyl-3-methylimidazolium tetrafluoroborate) was added into PP to improve the compatibility between aluminum powder and PP matrix. The results were analyzed by observing the different changes in thermal properties, crystallinity, rheological properties, mechanical properties, and sectional structures of the composite after filling different amount of aluminum powder. The results show that when adding 2phr aluminum powder, the crystallinity is high, the initial decomposition temperature of the composite is 295°C, the shear viscosity is 61.52Pa·s, the tensile strength is 23.15MPa, and the hardness is 55.6N/mm2. In this ratio, the composite material has an outstanding performance.

1. Introduction

Polypropylene (PP) is a thermoplastic resin, usually seem like a colorless, translucent solid. The industrial polypropylene is widely used in automobile parts, pipes, and commodity due to its well-defined structure, high melting point, outstanding thermal resistance and mechanical properties. In recent years, there have been many researches on the properties of various aspects of PP, such as mechanical properties, flame retardancy, and so on [1-4]. There are also many methods for modifying PP, which are divided into physical modification and chemical modification. He et al. modified the PP/wood fiber composites with surface-treated nano-SiO2 to improve the mechanical properties and hydrophobicity of the materials [5]. Liu Ting et al. summarized the chemical modification of different methods [6]. In order to improve the appearance of plastic products, the surface is often endowed an appointed colour and gloss by spraying. However, the method of spray is cumbersome, the paint will distribute unevenly that lead a low scrap rate and lead the environment pollution. Therefore, the spray-free process has become a hot topic in recent years [7, 8].

Ionic liquid (IL) has been applied in many domains due to its excellent properties, such as good stability, compatibilization and settability", especially for the modification of polymer materials [9-11]. The mechanism of ionic liquid used as a solubilizer between resin matrix and filled particles is as follows [12]: On the one hand, inorganic particles distributed more evenly in the matrix under the action of ionic liquids. On the other, ILs' special molecular structure contains both polar and non-polar groups, which act as the linker between organic molecules and inorganic particles; Moreover, the adding of ILs can increase the binding force through chemical reation among groups. The aluminum powder-filled PP resin blended and modified using the ionic liquid (1-octyl-3-methylimidazolium tetrafluoroborate). The molecular structure depicted in the following Scheme 1.

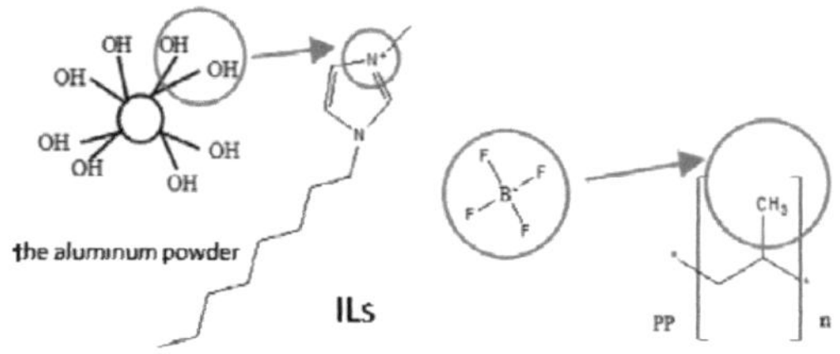

Scheme 1. The molecular structure of the ionic liquid.

The ionic liquid contains both polar and non-polar groups, which can respectively attract the methyl groups in the PP matrix and the hydroxyl groups on the surface of the aluminum powder, thereby combining the two phases and improving the interface environment between the PP and the aluminum particles, that boost the even-distribution and a better compatibility with the matrix. Then, the properties of the composite were studied.

2. Experiment

2.1 Materials
PP; aluminium powder: particle size of 7-8um, purity 99.0%, silver white powder, Dongguan Ou Peide Chemical Pigment Co., Ltd.; ionic liquid (1-octyl-3-methylimidazolium tetrafluoroborate): relative molecular weight 282.13, purity 99%, Shanghai Chengjie Chemical Co.Ltd.

2.2 Preparation of composites
(1). The PP pellets and aluminum powder were baked in a constant temperature drying oven at 100 °C for 12 hours to remove surface moisture and low molecular substances.
(2). The PP particles and the aluminum powder were mixed at a mass ratio of 100:0, 100:1, 100:2, 100:3, and 100:4. Then adding 3 phr of ionic liquid and mixed well.
(3). Five batches of the mixture in step 2 were separately mixed into an internal mixer for ten minutes, the temperature was set to 220 °C, and the rotation speed was set to 40-50 r/min. After 10 minutes of mixing, the material was discharged, cooled and crushed with a crusher, and then dry them at 100 °C for 4 hours.
(4). The mixture was preformed by a vulcanizing press, the experimental temperature was set to 220 ° C, and the load was applied at 3 MPa. After the pressure was applied for two minutes, the sheet was taken out to make samples.

2.3 Characterization

2.3.1. Scanning electron microscopy (SEM). The samples were frozen under liquid nitrogen for 24 hours and then brittle. After sections were sprayed with gold, all the samples were examined under a ZEISS EVO18 (Germany) scanning electron microscope (SEM) with an accelerating voltage of 20 kV.

2.3.2. Differential scanningcalorimetry (DSC) measurement. Crystallinity of the material were measured by Diamond DSC differential scanning calorimetry tester from Perkin Elmer, USA. Scanning temperature is 30-190 °C at heating rate of 10°C/min.

2.3.3. Thermogravimetric Analyses (TGA). TGA were carried out at 10 °C /min under nitrogen flow (flow rate: 40ml/min) with a NETZSCH STA 449F3 (Germany) thermogravimetric analyzer. In each case, the samples (2-5mg) were heated from 50 °C to 800 °C at the rate of 20 °C /min.

2.3.4. Rheological characteristics analysis. The rheological properties of the samples were characterized by MLW-400B capillary rheometer from Changchun Intelligent Instrument Equipment Co.Ltd.

2.3.5. Mechanical Properties Test

Impact property. The impact strength test was carried out in accordance with GB/T1843-1996 using the JJ-2 type memory impact tester of Changchun Intelligent Instrument Co.Ltd.

Tensile property. The tensile strength test was carried out in accordance with GB1040-2006 using the WSM-5KN universal testing machine of Changchun Intelligent Instrument Equipment Co.Ltd.

Surface hardness. The hardness test was carried out according to the standard GB/T3398.1-2008 using a QYS-96B plastic ball hardness tester.

3. Result and Discussion

3.1 Microstructure of composite materials

Shown in figure 1 are the scanning electron microscope images of fracture surface with different contents aluminum powder. As shown in figure 1(a), the sample without aluminum powder is relatively smooth and indicates a completely brittle fracture. In contrast, it can be seen that the cross section of the samples after adding a certain amount of aluminum powder are irregularly rough. In general, an incomplete interface will directly affect the macroscopic performance of material, especially the reduction of mechanical properties. However, there are many factors affecting the interface structure. We can consider them mainly from the interface combination and the agglomeration of filled particles. From the picture, many holes left in the interface after the aluminum particles coarse fell off the section, the phenomenon suggesting a weak interfacial binding force between the particle and the pp matrix. When 1 phr of aluminum powder added, as the aluminum powder randomly dispersed in the resin matrix, the continuous structures of the resin matrix destroyed, resulting in irregular distribution of the cross-sectional structure. When 2 phr of aluminum powder added, the distribution of aluminum particles in the matrix was in a relatively uniform saturated state, and the stress concentration inside the matrix is small, which makes the section protrusions relatively smooth. When the aluminum powder fraction was increased to 3 phr, the aluminum powder particles were agglomerated due to supersaturation in the resin matrix to cause stress concentration, resulting in a very irregular scaly shape of the cross section. It will appear more pronounced stress concentration, resulting in a rougher profile after adding 4phr aluminum powder.

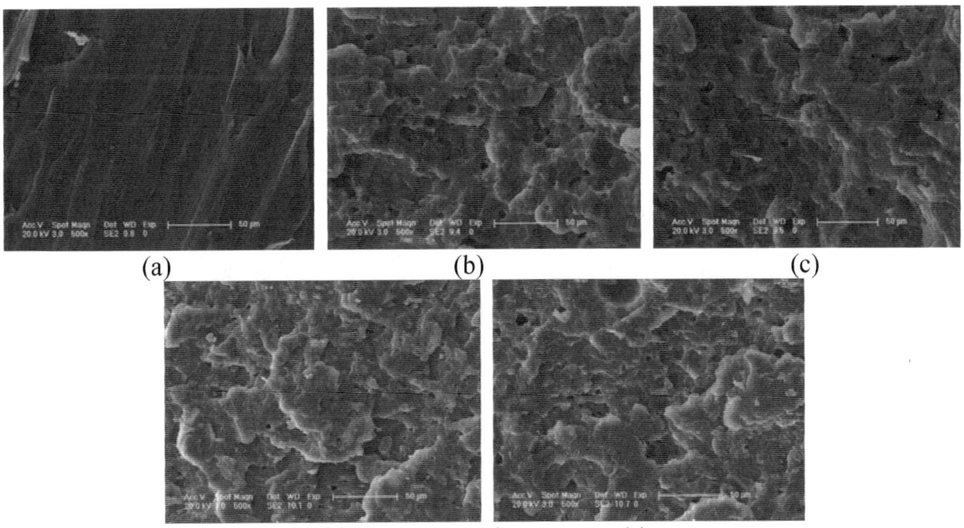

(a) (b) (c)

(d) (e)

Figure 1. SEM images of composites fracture surfaces
(a) Al=0phr; (b) Al=1phr; (c) Al=2phr; (d) Al=3phr; (e) Al=4phr.

IOP Conf. Series: Materials Science and Engineering **507** (2019) 012010 doi:10.1088/1757-899X/507/1/012010

3.2 Thermal properties and crystallization properties

Figure 2(a) shows the DSC test results of composites with different aluminum powder contents. The peak area in the image can reflect the melting enthalpy (ΔHf) of the polymer, and the crystallinity can be calculated by formula (1). ΔH_0 found to be 207J/g.

$$X_c = \frac{\Delta H_f}{\Delta H_0} \times 100\% \tag{1}$$

In the formula: Xc - the crystallinity (%) of the polymer, ΔH_f - the heat of fusion (J/g) of the polymer, ΔH_0 - the heat of fusion of the polymer at 100% crystallinity (J/g).

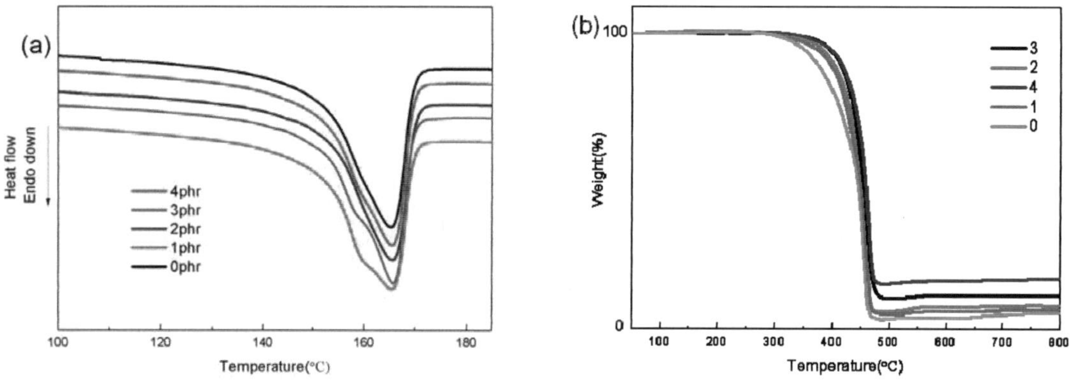

Figure 2. (a) Melting behavior of PP/Al/IL blends with different Al phr. (b) The TG curves of PP/Al/IL system.

The crystallinity of the filler shows in table 1. It can be seen from the table that the crystallinity of the composite first increases and then decreases. The addition of ionic liquid can destroy the entanglement effect between the macromolecules of the polypropylene, which causes a low crystallization of the composite. In addition, tiny aluminum particles act as heterogeneous nucleation centers, appropriate addition can promote crystallization. When 2 phr of aluminum powder added, the effect of heterogeneous nucleation was obvious, and the arrangement of the polypropylene macromolecules ordered as the uniform distribution of filling particles, two factors make the crystallinity of the composite improve. Once the aluminum powder exceeds 2 phr, the excess particles cause the entanglement effect of the polypropylene macromolecule to be inconspicuous. Moreover, if the matrix is broken, the crystallization will consume more energy, making it difficult to crystallize, so as the fraction of the aluminum powder increases, the crystallinity of the filler material begins to decrease.

Table 1. The crystallinity of samples.

phr of Al powder	0	1	2	3	4
ΔH_f(J/g)	93.09	85.46	88.78	85.54	78.32
X_c	44.97	41.28	42.89	41.32	37.84

Figure 2 (b) is the TG curve of the samples which reflects the thermal stability of the composite. As shown in the figure 2(b), when the aluminium powder content of the composite is gradually increased, the carbon residue rate is also increasing. The decomposition temperature of the samples (as shown in table 2) showed a tendency to increase after the addition of aluminum powder, which is related to the thermal conductivity of melt. On the one hand, the hindering action of filling particles can delay the depolymerization and decomposition of macromolecular chain. On the other, aluminum has good thermal conductivity, which can make the melt heated evenly. Besides, when the aluminum powder content reaches 3-4 phr, the interface of aluminum particles contact each other, the heat conduction effect will be more obvious.

Table 2. Decomposition-temperature of samples.

phr of Al powder	0	1	2	3	4
Decomposition-temperature(°C)	312.7	275	295	281.4	279

3.3 Rheological analysis

Figure 3 shows the rheological properties of composites at a constant shear rate. It can be seen from the figure that the shear strength and shear viscosity of the composites decreased with the increase of aluminum powder content. The addition of ionic liquid improves the compatibility of the aluminum powder, and increasing the amount of aluminum powder added will lead to an increase in the thermal conductivity of the material [13], thereby lowering the viscosity of the melt. On the other hand, an increase in the aluminum powder content increases the collision probability between the aluminum particles, and the particle surface has a lubricating property, so that the shear strength of the composite material is lowered.

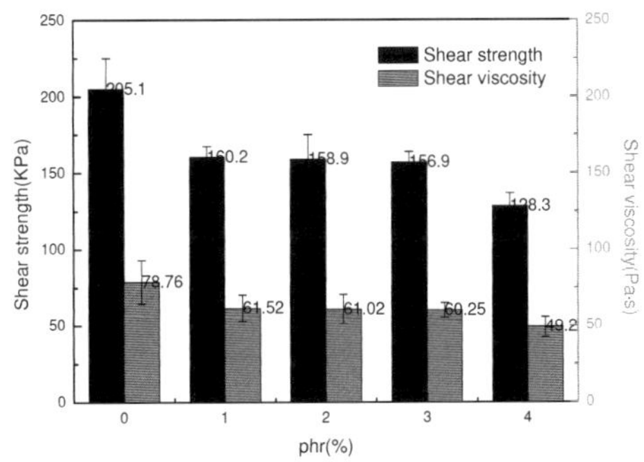

Figure 3. The result of shear strength and viscosity.

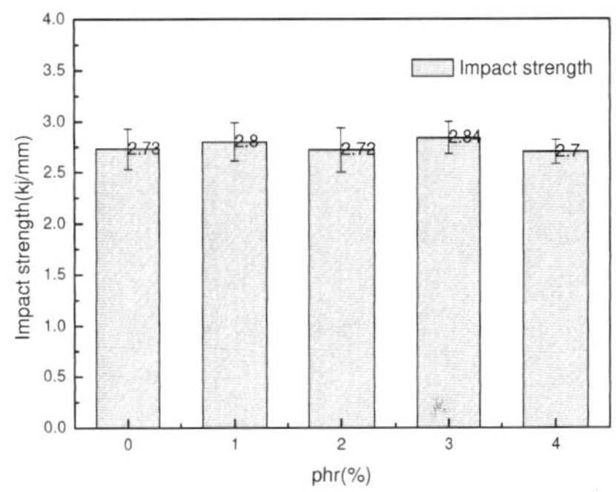

Figure 4. The impact strength of sample.

3.4 Mechanical properties analysis

Figure 4 shows the impact strength of different samples. Normally, the filling of inorganic particles will toughen the matrix [14] but the addition of aluminum powder destroys the continuous structure of the PP matrix and causes poor interface conditions, which reduces the mechanical properties of the composite. Since the aluminum particles are rigid particles, which will prevent cracks from propagating when the material was impacted [15], this is good for improving the impact strength of material. Therefore, the impact performance of the composite material does not change significantly base on correlation factor.

Figure 5 (a) shows the tensile strength of different patterns. The tensile properties of the samples after adding aluminum powder reduced to different extents compared with the pure component samples, but with the increase of the addition amount, the trend of increasing first and then decreasing was observed. This is the fact that the resin matrix blocked after the addition of the filler particles, forming a relatively continuous network structure. The mesh structure makes the tensile stress of the composite material relatively uniform and is not easy to cause stress concentration. Once the aluminum powder is excessive, the larger agglomeration formed will produce a large interface with poor affinity between the resins. When the tensile stress is applied, debonding occurs, so the tensile strength is decline.

Figure 5(b) shows the surface hardness of different patterns. As can be seen from the figure, the change in hardness is also a tendency to increase first and then decrease. There are two main factors

IOP Conf. Series: Materials Science and Engineering **507** (2019) 012010 doi:10.1088/1757-899X/507/1/012010

affecting the hardness of the filler material. Firstly, the addition of rigid particles increases the hardness of the material. Secondly, the filler particles destroy the crystallinity of the resin matrix and reduce the hardness. The amount of filler added in this experiment is relatively small, so the hardness is mainly dominated by the crystallinity of the material. The data in table 1 shows that the crystallinity of the material increases first and then decreases with the addition of aluminum powder, which is consistent with the matrix of hardness change.

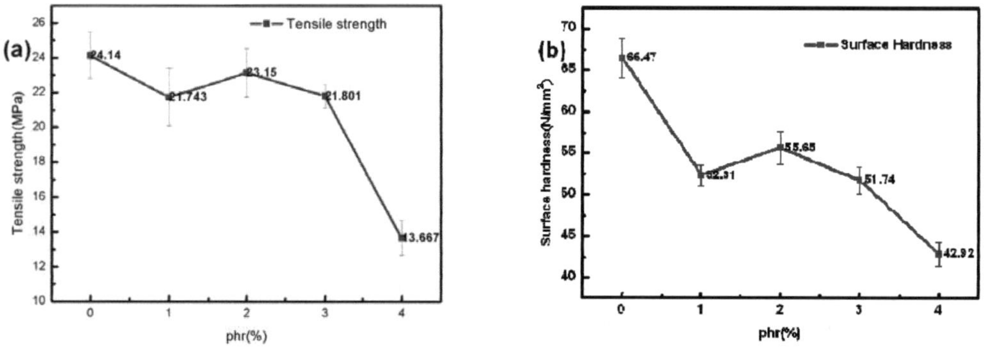

Figure 5. (a) The tensile strength of samples. (b) The surface hardness of samples.

4. Conclusions

This research investigated the variations in crystallization properties, thermal stability, rheological energy, and mechanical properties combined microstructure of the material after adding aluminum powder. The study shows that the addition of aluminum powder can improved the thermal properties of the material and change the crystallinity. Besides, the lubrication and thermal conductivity between aluminum particles reduce the viscosity of PP to a certain extent. The results of mechanical properties analysis show that the addition of aluminum powder in the PP matrix would reduce the tensile properties and hardness. When excess aluminum powder is added, there is a sharp decrease. That attributed to its damage to the matrix structure and poor interface environment. The unobvious change of impact property is the result of toughening effect of toughening effect of rigid particles and coarse interface environment

References

[1] Salmah H, Faisal A, Kamarudin H et al. 2011 *Polym. Mater.* **60(7)** 429-40.
[2] Peng P P, Yang Z H, et al. 2013 *Appl. Polym. Sci.* **130(6)** 3952-60.
[3] Ju Y Q, Wang J H, Wang X L 2015 *Plastics* **44(2)** 18-25.
[4] Zheng Z, Zhang L, et al. 2016 *Appl. Polym. Ses.* 23(4) 74.
[5] Park S J, Jin F L, Lee J R. 2004 *Mater. Sci. Eng. A*, **374(1)** 109–14.
[6] Liu T, Sheng X Y, Zhang X J 2010 *Plast. Add.* **6(84)** 20-3.
[7] Yu D X, Liu Q Q 2017 *J. Green Sci. Tech.* **1(2)** 109-10.
[8] Hu G 2015 *China Plast. Ind.* **43(9)** 117-20.
[9] Li J X, Liu J L, Guo L Y, et al. 2014 *Polym. Mater. Sci. Eng.* **30(11)** 10-4.
[10] Liu L H, Xue F, Ding E Y 2014 *Polym. Mater. Sci. Eng.* **30(10)** 95-9.
[11] Tang H O, Ding Y S, Li B X. 2009 *Acta Polym. Sin.* **2(2)** 191-4.
[12] Yousfi M, Livi S, Dumas A J et al. 2015 *Royal Soc. Chem.* **5** 46196-205.
[13] Ye P, Mamunya V V, Davydenko P, Pissis et al. 2002 *European Polym. J.* **38(9)** 1887-97.
[14] Arefazar A. 2013 *Polym. Plast. Technal. Eng.* **52** 629-650.
[15] Yang Y, Wu S H 2007 *Plastics* **36(6)** 5-7.

IOP Conf. Series: Materials Science and Engineering **507** (2019) 012011 doi:10.1088/1757-899X/507/1/012011

Effect of Coating Materials on the Properties of Chitosan-Starch-Based Edible Coatings

Noorsuhana Mohd Yusof[1], Junaidah Jai[1], Fazlena Hamzah[1]

[1]Faculty of Chemical Engineering, Universiti Teknologi MARA, Malaysia, 40450 Shah Alam, Selangor, Malaysia

Abstract. Edible coating is a thin layer on the surface that is used to extend shelf life of the fruit. The characteristics of the materials used and the properties of each substance in the coating solution will affect the quality of the edible coating. In this study, the use of chitosan at different concentrations has been assessed to produce an effective edible coating. Chitosan-starch edible coating with 1.5% (w/v) chitosan contents were chosen to give the best surface tension and wettability to coat fruit. Their structural and functional properties were also characterized through FTIR. Outcome from this study found that the shifting of peaks and the changes in the spectra were resulted due to the hydrogen bonding between OH and NH3+ of chitosan and OH group of starch. These shifts shown that the addition of glycerol promoted the hydrogen bonding interactions and plasticization effect of glycerol due to the hydrophilic nature. Additional of turmeric essential oil in the edible coating could enhance more its function to protect fruit and serves as antimicrobial agent due to the presence of phenolic group in its structure.

1. Introduction

Foodborne outbreaks, precocious decaying and economic losses are common problems related to fresh fruit retention. These problems have led to the advance of novel technologies and systems for food protection such as edible coating based on natural source materials. There are various methods of preservation that can be applied to extend shelf life of fruit. Currently, the used of edible coating in maintaining the fruits organoleptic properties are getting wide attention. These are due to its properties that able to control the moisture transfer, respiration process and oxidation process [1]. Furthermore, the used of edible coating are safe to be consume by consumer as it is made up from natural based [2].

Commonly, edible coating were made up of three types of hydrocolloids which are polysaccharides, proteins and lipids as its main bases [3]. In spite of using a type of hydrocolloid, the mixture of those three types hydrocolloids also can be used in producing the edible coating and this edible coating is known as composite edible coating [4]. Edible coating had the ability to be used with the presence of food preservatives such as antimicrobial agent, antioxidation agent, and chelating agents [5]. In this research study, turmeric oil was used as the antimicrobial agent. Turmeric oil consists of secondary metabolites that can act as antimicrobial agent such as turmerone, Ar-turmerone, curlone, cumene and many others [6]. The addition of turmeric essential oil in the edible coating film could enhance more its function as to protect fruit from deteriorate that causes by microbial and fungus.

Composite edible coating was used as the edible coating based which are the mixture of starch and chitosan. Starch is made up of amylose and amylopectin that are connected by hydrogen bond and formed the semi-crystalline granules naturally. Presence of water and heat could swelling and hydrating the granules which leads in increasing the viscosity of the solution [7]. Starch has poor mechanical properties which it has brittle and fragile characteristic. Thus, the presence of chitosan could improve the edible coating of cassava starch based film due to its structural compatibility [8]. Chitosan is one of the biomaterials developed particularly for food and packaging applications. It is a linear polysaccharide consisting of β-(1→4)-linked 2-amino-2- deoxy-D-glucose residues, originating from deacetylate derivative of chitin, which is the second most abundant polysaccharide in nature after cellulose. It could effectively control food decay as it has strong antimicrobial and antifungal activities. Chitosan coating can reduce the respiration rate of food by controlling its gas permeability [9]. Chitosan could modify the starch granule size six times larger compared to the original size. Besides that, the hydroxyl group that present in starch will interact with the amino groups that present in chitosan. Furthermore, this interaction could reduce the starch degradation [10]. The mechanical properties of this composite edible coating are improved more by adding plasticizer such as glycerol and sorbitol.

2. Materials and methods

2.1. Materials

Local commercial brand starch (Brand Kapal ABC. Co) was used as the biopolymer for edible coating formulation. Food grade chitosan with 97% deacetlylation (Nacalai Tesque, Japan) was used as composite coating. Acetic acid (Friendemann Schmidt, Australia) was added to dissolve the chitosan while glycerol (Merck, Germany) was added as plasticizer for coating formulation. Tween 80 (Systerm, Lab Sciences Engineering Sdn. Bhd., Malaysia) were used as emulsifier and turmeric was bought from the local market in Selangor.

2.2. Preparation of turmeric essential oil

Hydrodistillation extraction was used to extract turmeric essential oil. Fresh turmeric rhizomes of 500 g were washed to remove any dirts and dried using a paper towel. They were then being grated and placed in a 5 L round bottom flask with 1250 mL of distilled water for the extraction process [11]. After 4 hours of boiling, the oil was collected, protected from light and stored at 4 C.

2.3. Edible coating preparation

The edible coatings were prepared as described by [13]. Chitosan solution (100 ml) was prepared by dissolving chitosan 1.5% (w/v) in a 0.5% (w/v) acetic acid. To ensure that the chitosan is completely dissolved, the solution was stirred for 24 hours at 350 rpm until all the chitosan flakes disappear. A starch solution (100 ml) was then prepared by dispersed starch in a 0.5% (w/v) glycerol water solution to obtain (w/v) suspensions, which were heated on a hotplate at 75 ℃ to a complete starch gelatinization [14]. The solution was stirred for 30 minutes at 350 rpm to overcome the starch from being coagulated. The blend coatings were obtained by the addition of chitosan-acetic acid solution to gelatinized starch-glycerol solutions. Tween 80 with concentration of 0.1% (w/v) was added as a surfactant to increase wettability [15]. After that, turmeric essential oil was added to the mixture.

2.4. Surface tension and wettability

In order to determine the adhesion properties of the coating to the surface, it can be done by measuring the contact angle of a standard liquid on the surface [16]. In this study, analysis of surface tension for all the samples has been done using Contact Angle Goniometer (AST Products, INC.) using sessile drop method to measure the contact angle. It was done by dropping the coating solution on the surface of the samples and recorded by Contact Angle Goniometer. Surface tension was calculated using the surface energy software (SE2500) while wettability was determined using the equation as reported by [17]. The spreading coefficient (W_s) may be defined by equation:

$$W_s = W_a - W_c \quad (1)$$

where, W_a and W_c are the work of adhesion and cohesion, as defined by equations (2) and (3), respectively, while (Θ) is the contact angle and (γ) is the surface tension.

$$W_a = \gamma L\ (1+\cos(\Theta)) \quad (2)$$

$$W_c = 2\ \gamma L \quad (3)$$

2.5. Fourier Transform Infrared (FTIR)
The spectra obtained were used to determine possible interactions of functional groups of chitosan on the starch-based coating by using Fourier Transform Infrared (FTIR) spectroscopy (Perkin Elmer, USA) at wave number ranging from 400 to 4000 cm^{-1}. The sample port was cleaned by using acetone to remove any variation caused by contamination on the scanning. The spectrum yields from the sample scanning were collected and studied.

3. Results and discussion

3.1. Surface tension and wettability analysis
Surface tension analysis involves the interaction between a solid-liquid surface. Chitosan-starch coating solution with different amount chitosan at 0.5, 1.0, 1.5 and 2.0% (w/v) were used. Figure 1 shows surface tension, W_a, W_c and W_s for the designed coating formulations determined above. The increased amount of chitosan will affect the surface tension to reduced gradually. As reported by Cothran [18], suitable surface tension for fresh fruit coating solution was in the range of 28 to 38 dynes/cm. It can be seen that 1.5 and 2.0% (w/v) were in the acceptable range. The higher the amount of chitosan in the formulation, the lower the surface tension, reduced the hydrophilicity, water spreading and absorption [19]. The wettability (W_a, W_c) of the solution on the surface were decreased while for W_s the trend was rising up. As mention by Cerquira et. al., better spreading of solution on the surface and the W_s value should be near to zero and in negative formed [20]. Since the surface tension was in the range with W_s value nearest to zero among all, 1.5% (w/v) of chitosan was selected as a coating material in coating solution.

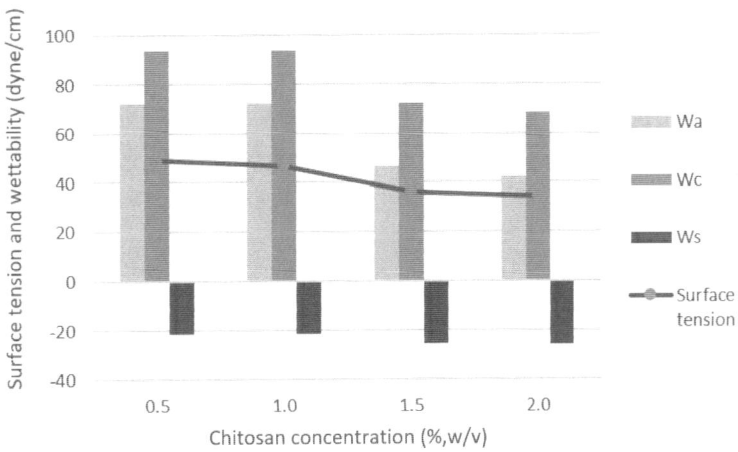

Figure 1. Effect of chitosan concentration (%, w/v) on surface tension and wettability W_a, W_c and W_s (dyne/cm).

3.2. Fourier Transform Infrared (FTIR) analysis
The infrared spectra of raw materials are presented in figure 2. Infrared spectra of chitosan figure 2(a) at 3286.1 cm^{-1} was the OH stretching, which overlaps the NH stretching in the same region. The peak at 2872.68 cm^{-1} is typical C-H stretch. A small peak at 1641.53 cm^{-1} was due to the C=O stretching (amide I), and the peak at 1554.91 cm^{-1} and 1377.37 cm^{-1} have been reported as amide II and III peaks respectively. The sharp peaks at 1411.23 cm^{-1} correspond to the CH_3 symmetrical deformation mode. The peak at 1151.63 cm^{-1} indicates the saccharide structure and the band at 1062.74 cm^{-1} was due to the C-O stretching vibration in chitosan vibrations [7].

In the spectrum for starch in figure 2(b), the broad band of starch at 3287.69 cm^{-1} is due to the hydrogen-bonded hydroxyl groups that contribute to the complex vibrational stretches. The sharp band

IOP Conf. Series: Materials Science and Engineering **507** (2019) 012011 doi:10.1088/1757-899X/507/1/012011

at 2924.28 cm^{-1} is characteristic of C-H stretches associated with the ring methane hydrogen atoms. The bands at 1636.46 and 1359.43 cm^{-1} are assigned to the δ (O-H) bending of water and CH_2, respectively [15]. The bands from 759.44 to 1148.83 attributed to the C-O bond stretching.

The spectra of glycerol in figure 2(d) has a wide band caused by stretching of O–H bonds with either inter or intra-molecular hydrogen bonds, that could also be formed by water molecules, at 1704.74 cm^{-1}. An additional O–H vibrational mode was also seen at 1211.93 cm^{-1}. The peak located at 2879.52 cm^{-1} was due to the C-H stretching. It is used as an indicator of the interaction of glycerol with chitosan, starch or more glycerol molecules [22].

Tween 80 shows many sharp absorption peaks that are due to the different functional groups present as shown in figure 2(e). Methyl group (–CH_3) shows absorption band at 2922.31 cm^{-1}, while the band at 2856.54 cm^{-1} is due to –CH_2 -stretching. The band at 1735.34 cm^{-1} can be attributed to carbonyl C=O. Hydrophilic groups in this compound are polyethers which are polymers of ethylene oxide. These polymers are amphiphilic and soluble in water to help in binding the molecules of water in the emulsion. Carneiro-da-Cunha discussed the effect of adding Tween 80 to a coating solution due to its surface-active properties, aiming at improving the wettability of the coatings/films [23].

The major bioactives in turmeric are polyphenols, including curcumin, which is well known for its strong antioxidant activity [24]. As can be seen in figure 2(f), the infrared spectra at 3456.26 cm'1 is due to O-H stretching vibrations, while at 2961.29 cm^{-1} and its doublet at 2928.88 cm^{-1} are due to asymmetric and symmetric stretching vibrations of the CH_2 group. Infrared spectra at 1376.77 cm^{-1} was the phenol or tertiary alcohol, O–H bend in which the properties of turmeric essential oil.

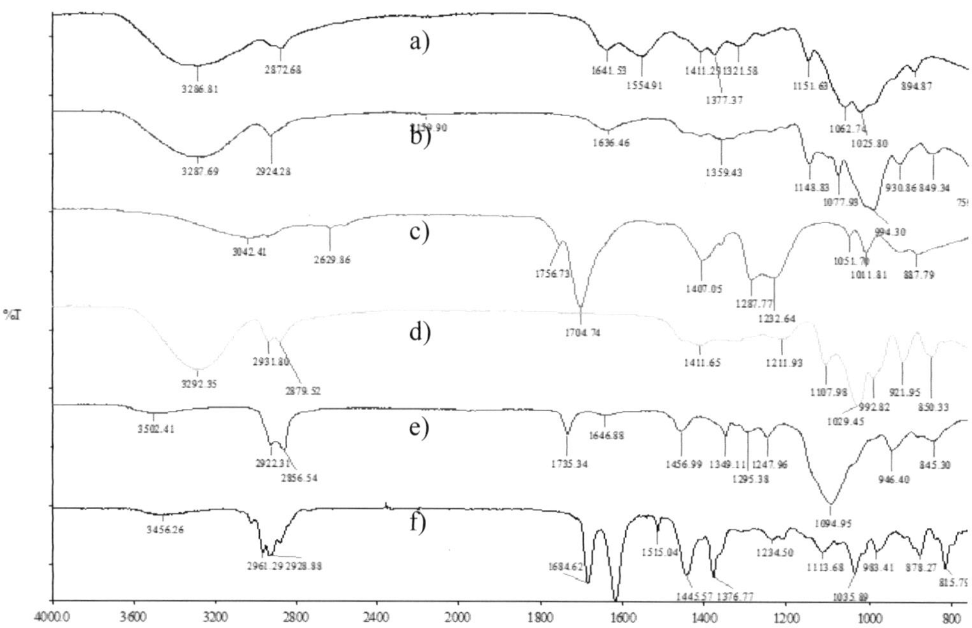

Figure 2. FTIR of materials in chitosan-starch edible coating incorporated with turmeric essential oil; a) chitosan b) starch c) acetic acid d) glycerol e) Tween 80 f) turmeric essential oil.

The blends film can be seen in figure 3. Changes in the spectra as well as the shifting of peaks due to the hydrogen bonding between OH and NH^{3+} of chitosan and OH group of starch was observed in the starch-based incorporated with chitosan. Yin et al. [25] reported that when two or more substances are mixed, physical blends versus chemical interactions are reflected by changes in characteristic spectral peaks [12]. The carbonyl, C = O-NHR, amine, NH_2 and ammonium, NH_3 + bands were situated in the region between 1400 cm^{-1} and 1700 cm^{-1}. The carbonyl, C = O-NHR band was observed at 1645 cm^{-1}, and the amine NH_2 band at around 1560 cm^{-1} appeared as a small shoulder in pure chitosan. The wavenumber for C-OH group and C-O-C were reduced showing that OH group of

IOP Conf. Series: Materials Science and Engineering **507** (2019) 012011 doi:10.1088/1757-899X/507/1/012011

starch took part in hydrogen bond formation [14]. Due to stability and compatibility with hydrophilic biopolymeric packaging chain, glycerol being chose as the most popular plasticizers used in edible coating due to the presence of hydroxyl group [23]. The addition of turmeric essential oil in the edible coating could act an effective antimicrobial activity due to the presence of phenolic group in its molecular structure.

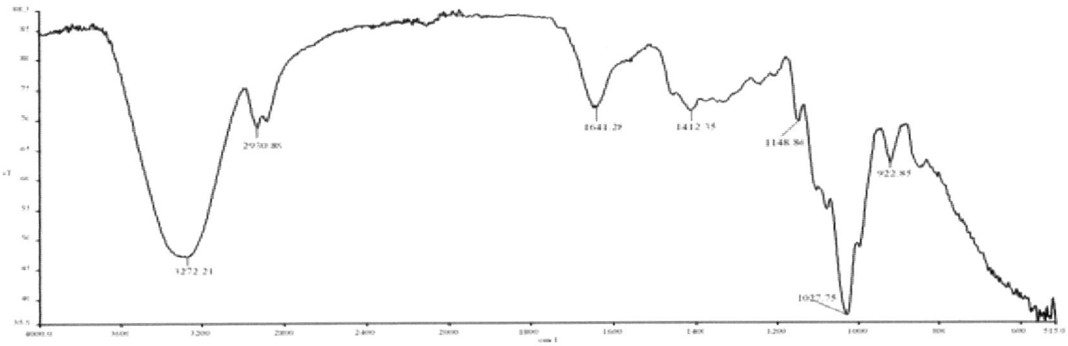

Figure 3. FTIR of blend film consists of chitosan, starch, acetic acid, glycerol, Tween 80 and turmeric essential oil.

4. Conclusion

Coating material greatly affects the surface tension and together with the wettability of the coating solution on fruits. Chitosan shows a decreasing pattern of surface tension when different masses of it being added in 100 mL of the coating solution. Since this edible coating formulation for hydrophobic surfaces such as surfaces of fruits with natural protection layer, it is good to observe the reduction of the surface tension of the coating solution. Edible coating formed by blending of two polymers result in modified physical properties compared to edible coating made of the individual components. Chitosan-starch edible coating showed certain shift of peak patterns in FTIR. It resulted to exhibit good coating property due to the existence of their own functional groups which contribute to the characteristics in coating properties.

Acknowledgement

The authors would like to acknowledge Universiti Teknologi MARA for the financial support to carry out this research through 600-IRMI/MYRA 5/3/LESTARI (021/2017). Authors also thankful to the staff of Faculty of Chemical Engineering, Universiti Teknologi MARA, Shah Alam for their help in performing laboratory work.

References

[1] Sharif Z I M, Mustapha F A, Jai J, Yusof N M and Zaki N A M 2017 *Chem. Eng. Res. Bull.* **19(Special Issue on Conference)** 145–53.
[2] Phani Tej R N 2012 *Effect of composite edible coatings and abiotic stress on post harvest quality of fruits*, McGill University.
[3] Bourtoom T 2008 *Int. Food Res. J.* **15(3)** 237–48.
[4] Baraiya N S, Rao T V R, and Thakkar V R 2016 *J. Agric. Sci. Technol.*, **18** 93–107.
[5] Sabharwal P K, Garg M, Sadhu S D, Khas H, and Delhi N, *World J. Pharm. Life Sci. WJPLS*, **2(3)** 160–70.
[6] Awasthi P K and Dixit S C 2005 *Pharmacognosy* **1(4)** 312–6.
[7] Azeredo H 2012 *"Edible Coatings," in Advances in Fruit Processing Technologies,* S. Rodrigues and F. Fabiano Andre Narciso, Eds. CRC Press, 345–62.
[8] García M A, Pinotti A, Martino M N, and Zaritzky N E 2009 *Characterization of Starch and Composite Edible Films and Coatings,* 1900.

[9] Davis S P 2011 *Chitosan: Manufacture, Properties, and Usage (Biotechnology in Agriculture, Industry and Medicine)*, UK. United Kingdom.
[10] Yongjia D, Xu S, Wenting S, Zhongkai Z, Zhiwei W, and Paiyun Z 2017 *J. Food*, **15(3)** 327–35.
[11] Avanço G B *et al.* 2017 *Food Control* **73** 806–13.
[12] Samadi M, Abidin Z Z, Yunus R, Awang Biak D R, Yoshida H and Lok E H 2017 *Chinese J. Chem. Eng.* **25(2)** 216–22.
[13] Azevedo A N *et al.*, 2014 *Food Control*, **43** 1–9.
[14] Vásconez M B, Flores S K, Campos C A, Alvarado J, and Gerschenson L N 2009 *Food Res. Int.* **42(7)** 762–9.
[15] Santos N S *et al.* 2012 *Food Microb.* **(32)** 345–53.
[16] Moncayo D and Buitrago G 2015 *The surface properties of biopolymer-coated fruit : A review The surface properties of biopolymer-coated fruit : A review Propiedades de superficie en frutas recubiertas con biopolímeros* no. April.
[17] Moncayo D, Buitrago G and Algecira N 2013 *The surface properties of biopolymer-coated fruit,* A review Propiedades de superficie en frutas recubiertas con biopolímeros **33(3)** 11–6.
[18] Aug P and Cothran C D W 1945 *KiET.*
[19] Glycerol H and Films E 2018 *How Glycerol and Water Contents Affect the Structural and Functional Properties of Starch-Based.*
[20] Cerqueira M A, Lima Á M, Teixeira J A, Moreira R A and Vicente A A 2009 *J. Food Eng.* **94(3–4)** 372–8.
[21] Liu P *et al.* 2011 *Carbohydr. Polym.*, **85(1)** 180–7.
[22] Liu H, Adhikari R, Guo Q and Adhikari B 2013 *J. Food Eng.* **116(2)** 588–97.
[23] Cervera M F *et al.*, 2004 *Eur. J. Pharm. Biopharm.* **58(1)** 69–76.
[24] Miquel J, Bernd A, Sempere J M, Díaz-Alperi J and Ramírez A 2002 *Arch. Gerontol. Geriatr.* **34(1)** 37–46.
[25] Vinet L and Zhedanov A 2010 *Polym. Int.* **433(November 1998)** 429–32.

A Technical Solution to Improve the Existing River Dikes for both Flood Prevention and Transportation Purposes

N H Hue[1, 3], D C Huong[2] and N H Thanh[3]

[1]Faculty of Civil Engineering, Thuyloi University, Hanoi, Vietnam
[2]Bac Ninh Department of Agriculture and Rural Development, Bac Ninh, Vietnam
[3]Institute of Civil Engineering, Thuyloi University, Hanoi, Vietnam

Abstract. In the demand for socio-economic development, today, many river dikes in Vietnam are planned to be used as roads, and this is an indispensable trend. However, these river dikes have general characteristics of being filled with various types of soil which are heterogeneous, and not being compacted following the criteria before filling, so they need to be reinforced for both ensure the flood prevention safety and meet the technical requirements of traffic roads. In Vietnam and in the world, there are no specific studies on the re-use of the existing weak soil of embankments with adhesives, namely fly-ash and cement for treatment of foundation that meet the technical requirements of the upper base layer when transportation is involved. In this study, the authors focus on the solution of using the existing weak soil of embankments with fly-ash and cement to strengthen the stability of the river embankment served as the upper base layer in the traffic load affected zone, and the river embankment sub-structure combined with transportation purpose. By using methods of laboratory experiment, the authors propose the appropriate river embankment sub-structure to guarantee flood prevention and transportation combination.

1. Introduction

The Vietnam river dike system has been through a long history of formation and development. Currently, the total length of the river dike system in the Red River Delta is about 3,000 km, including 2,417 km length of dike in Northern Delta, and 420 km length of dike in the Thanh - Nghe area. The Red River system has 1,667 km of dikes and 750 km of dikes are in the Thai Binh river system. These river dikes were formed through different periods, the fill soil layers were heterogeneous and were filled in many stages with flood prevention requirements. Therefore, when there are traffic vehicles on the dike surface, it is hardened resulting in cracking and settlement. Dike-road combination is an indispensable demand in the current stage in order to create a completed connection to the transport network and promote the socio-economic development and serve people. Several river dikes in Vietnam have been planned (partly or entirely) as roads and severed transportation for many years such as Huu Duong river dike, left embankment of Duong river dike, Thai Binh river dike, right embankment of Cau river dike, etc. Specifically, the inter-provincial routes such as ĐT276; ĐT283; ĐT291; ĐT280 of left and right bank of Duong river dike have been planned for traffic road corresponding to class IV road. In the process of improving these river dikes for both flood prevention and transportation purposes, they were mainly implemented according to experiences or on the basis of current standards and regulations on traffic road without a specific study. Many questions are raised around the issue of dike-transportation combination as what will affect the stability of dikes when transportation is involved; are there horizontal cracking, vertical cracking, sliding? There have been a number of studies on technical solutions to enhance the strength and stability of river dikes, but

IOP Conf. Series: Materials Science and Engineering **507** (2019) 012012 doi:10.1088/1757-899X/507/1/012012

research on the re-use of the existing weak soil of embankment with adhesives, namely fly-ash and cement have not yet been mentioned and have no specific study. Besides, fly ash is a waste from the electrical industry but it can be used in many material production industries.

In this paper, the authors focus on the study of solutions to strengthen the stability of the river embankment served as the upper base layer in the traffic load affected zone, and the study of foundation treatment solution in the river embankment sub-structure combined with transportation purpose. From which, proposing the appropriate river embankment sub-structure to guarantee flood prevention and transportation combination.

2. Materials and methods

2.1. Particle size distribution curve theory

Some basic theories about soil texture are mentioned such as Fuller Thomson Theory, Caquot and Faury, Papovic-Anderson, B.B. Okhotina and N.N. Ivanov, Talbot, Weymouth for discrete granular aggregate mixtures. Fuller's curve provides the optimum density due to the wide range of particle sizes but often leads to difficulty in compacting the mixture [6]. However, due to the "rolling" effect of fly ash, the friction between the particles will be reduced. Fuller's curve is a parabola or part of a parabola at its natural rate. Therefore, this study will use Fuller's curve to study and construct an appropriate gradation aggregate with size of small particles are from 1 μm to 12,500 μm being suitable for the current status of river dikes.

2.2. Materials and composition of reinforced river embankment layer

The soil samples used for experiments in the laboratory were taken from the existing river embankment of Huu Duong dike (ĐHĐ), Bac Ninh Province. The parameters of ĐHĐ soil sample are shown in table 1.

Table 1. Soil particle size distribution of the ĐHĐ soil samples.

Sieve size (mm)	Mass retained (g)	Percent retained (%)	Cumulative retained (%)	Percent passing (%)
10	504.45	19.432	19.432	80.568
5	406.88	15.673	35.105	64.895
2	450.63	17.359	52.464	47.536
1	448.12	17.262	69.726	30.274
0.5	300.04	11.558	81.284	18.716
0.25	165.74	6.384	87.668	12.332
0.1	217.23	8.368	96.036	3.964
< 0.1	102.91	3.964	100	0

Fly ashes used in combination with cement to reinforce the embankment soil are Dong Trieu fly ash (TBĐT) which is taken from Dong Trieu thermal power plant, and Cam Pha fly ash (TBCP) which is taken from Cam Pha thermal power plant. These two types of fly ashes have a total content of three major oxides ($SiO_2 + Al_2O_3 + Fe_2O_3$) greater than 70%, thus they are classified as Class F according to American Standard C618 [1]. The contents of SO_3 and Cl^- are small, therefore these fly ashes are suitable for ground reinforcement. The specification of Dong Trieu fly ash (TBĐT) and Cam Pha fly ash (TBCP) fly ash are shown in table 2.

Table 2. Specification of Dong Trieu and Cam Pha fly ash samples.

Parameters	Unit	Dong Trieu fly ash (TBĐT)	Cam Pha fly ash (TBCP)
$SiO_2 + Al_2O_3 + Fe_2O_3$	%	85.34	84.5
SO_3	%	1.52	0.59
Loss on Ignition	%	6.66	10.09
CaO	%	3.67	0
ion clo (Cl^-)	%	0.01	0.02
(Na_2O)	%	0.35	0.24

In this study the Nghi Son PCB40 cement is used. The chemical and mineral components of Nghi Son PCB40 cement are shown in table 3.

Table 3. The chemical and mineral components of Nghi Son PCB40 cement.

Chemical components (%)									Mineral components (%)			
SiO_2	Al_2O_3	Fe_2O_3	CaO	MgO	SO_3	Na_2O	K_2O	CaO	C_3S	C_2S	C_3A	C_4AF
20.32	4.98	3.24	62.2	0.10	1.12	0.25	0.72	0.150	51	25	8.16	10

After selecting the optimal relative values, the optimal designed Dong Trieu fly ash-soil ratio for Huu Duong dike is 15%, the same ratio is obtained for Cam Pha fly ash.

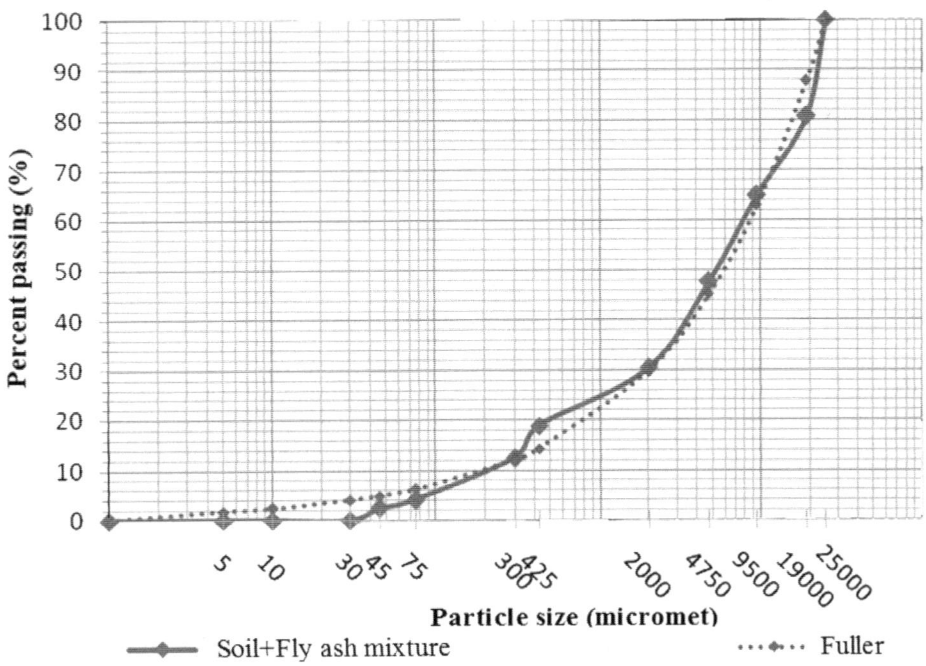

Figure 1. Particle size distribution of ĐHĐ soil and 15% fly ash mixture.

A total of 72 reinforced soil samples (the mixture of soil, cement and fly ash) were built, maintained and tested in a standardized laboratory to determine the sample properties at 14 days of age.

Table 4. Number of reinforced soil samples to be taken for each soil component.

Type of test	Soil condition	10% cement + 90% ĐHĐ	(15% TBCP + 85% ĐHĐ) + 10% cement	(15% TBĐT + 85% ĐHĐ) + 10% cement	Total of samples
Compressive strength	Saturated	6	6	6	18
Compressive strength	Watering	6	6	6	18
Split compressive strength	Saturated	6	6	6	18
Elastic modulus	Saturated	6	6	6	18

2.3. Materials and composition of reinforced crushed aggregate base course

Particle size distribution of crushed aggregate base course is shown in figure 2.

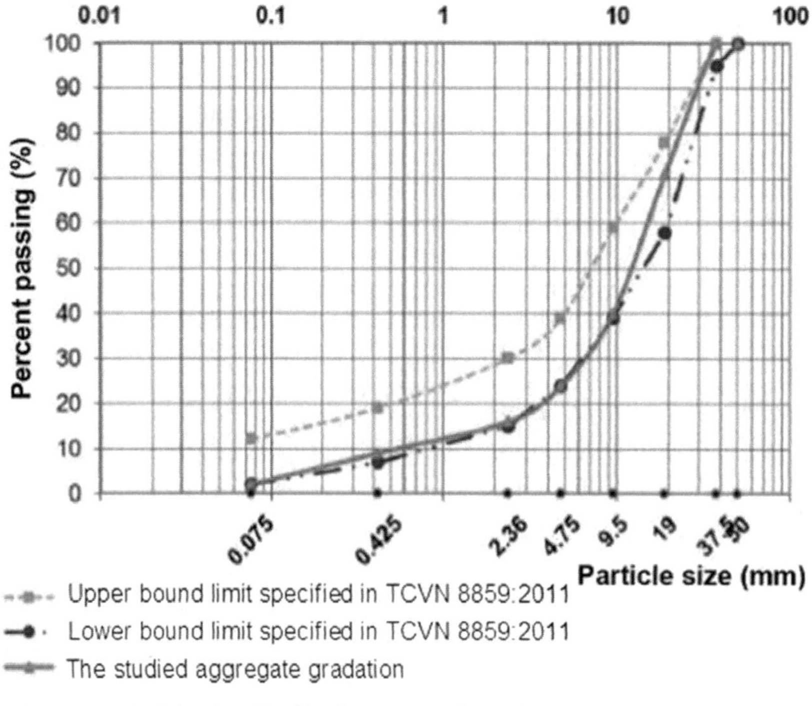

Figure 2. Particle size distribution curve of crushed aggregate base course.

Gradation aggregates are reinforced with cement and fly ash (TB), in which the ratio of replacing TB to cement referred as adhesives (CKD) are 0%, 10%, 20% and 30% respectively. The two types of TB used in the study were from Dong Trieu and Cam Pha thermal power plants.

3. Results and discussion

3.1. Laboratory test result and reinforced river embankment analysis

The following table shows the above experimental results:

Table 5. Experimental results of reinforced soil samples at 14 days of age.

Type of test	Unit	10% cement + 90% ĐHĐ	(15% TBCP + 85% ĐHĐ) + 10% cement	(15% TBĐT + 85% ĐHĐ) + 10% cement
Compressive strength (Saturated)	(MPa)	2.10	2.11	2.27
Compressive strength (Watering)	(MPa)	2.45	2.52	2.67
Split compressive strength (Saturated)	(MPa)	0.29	0.27	0.31
Elastic modulus (Saturated)	(MPa)	962	970	1048

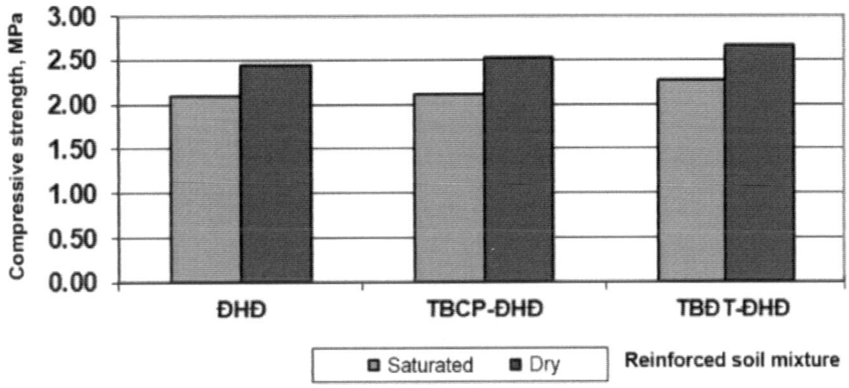

Figure 3. Comparison between saturated compressive strength and non-saturated after 14 days.

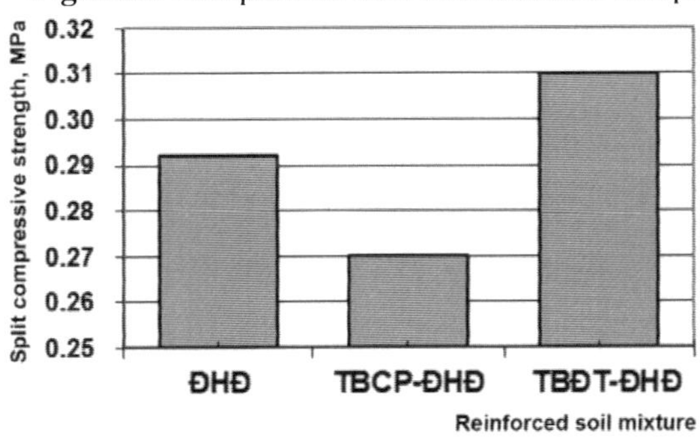

Figure 4. Comparison between split compressive strength of reinforced soil mixtures after 14 days.

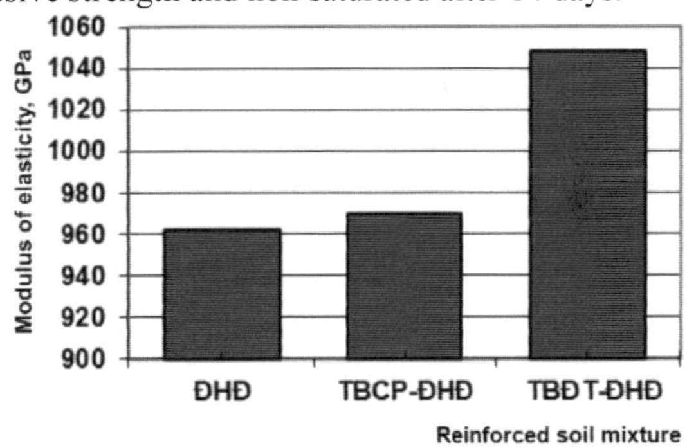

Figure 5. Comparison of elastic modulus among reinforced soil mixtures after 14 days.

The test results show that:

- The saturated compressive strength is varied between 83.0÷85.6% of unsaturated compressive strength. The saturated compressive strength of the reinforced mixture is higher than that of TCVN 8858:2011 [2] ($R_n \geq 1.5$MPa);
- Split compressive strength of the reinforced soil mixture varies from 0.27÷0.31MPa at age of 14 days, the ratio between the compressive strength and the split compressive strength of a saturated soil sample varies from 7.24 to 7.81 times;
- The elastic modulus of the reinforced soil layer with selected optimum ratios is varied from 962MPa to 1048MPa, the elastic modulus of this soil layer is quite high, which is a good basis for the material used in the road pavement.

IOP Conf. Series: Materials Science and Engineering **507** (2019) 012012 doi:10.1088/1757-899X/507/1/012012

- The proportion of 90% of reinforced soil (85% soil + 15% fly ash) + 10% of cement brought the highest values of Compressive strength, Split compressive strength and Elastic modulus. Hence, this proportion is the optimal ratio to reinforce the base of the dike to achieve the required combination of transportation.

3.2. Laboratory test result and reinforced crushed aggregate base course analysis

The test results show that:

- Reinforcement using fly ash to replace part of cement in the adhesives results in significantly increasing in the compressive strength (R_n), split compressive strength (R_{kc}) and elastic modulus (E_{dh});
- Compressive strength, split compressive strength, elastic modulus of gradation aggregate (CPĐD) used adhesives fly ash of Dong Trieu thermal power plant is higher than that of Cam Pha thermal power plant;
- The values of compressive strength, split compressive strength, elastic modulus increase when the volume of fly ash increases from 0%, 10%, 20% of CKD and decrease when the approach to 30%, thus using fly ash at the ratio of 20% obtains the best effect for the study.

3.3. The proposed river embankment sub-structure of the dike after studying

Through the experiments in the laboratory, the authors have determined the optimal proportion of gradation aggregate for reinforcement and the technical parameters achieved as follows:

- Determine the proportion of 90% of the reinforced soil (85% soil + 15% fly ash) + 10% of cement is the optimal ratio to reinforce the base of the dike to achieve the required combination of transportation.
- Determine the proportion of 96% of the crushed aggregate base course, reinforced by 4% adhesives (80% cement + 20% fly ash) is the optimal ratio to reinforce the surface of the dike.
- Gradation aggregate reinforced by adhesives (CKD) can be easily compacted, reducing the number of the compaction of aggregate.

From experimental results on the basis of technical specifications, the authors propose new river embankment sub-structure to ensure flood control and transportation combination including layers: concrete surface M300; crushed aggregate base course reinforced by 4% of adhesives (CKD); layer (soil + fly ash) + cement.

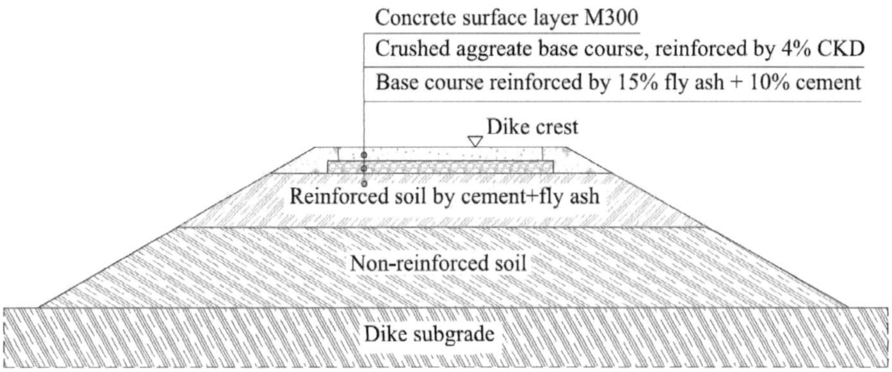

Figure 6. The proposed river embankment sub-structure of the dike after studying.

4. Conclusion

In this study, the authors propose a technical solution of using existing soil, fly ash and cement to reinforce the existing river embankment which meets the flood protection and transportation combination requirements. The optimal proportion of appropriate component is a mixture with 15% of fly ash, 85% of dry soil and 10% of concrete. Moreover, the authors also propose the crushed aggregate base course reinforced by fly ash and cement, in which the appropriate proportion, including 96% of crushed aggregate base course and 4% adhesives (80% cement and 20% fly ash).

References

[1] ASTM-C618 2005 ASTM International *Standard Specification for Coal Fly Ash and Raw or Calcined Natural Pozzolan for Use in Concrete.*

[2] TCVN 8858:2011 2011 Vietnam Ministry of Science and Technology International *Cement Treated Aggregate Bases for Road Pavement - Specification for Construction and Acceptance.*

[3] TCVN 8859:2011 2011 Vietnam Ministry of Science and Technology International *Graded Aggregate Bases and Subbases Pavement - Specification for Construction and Acceptance.*

[4] Mitchell J K and Freitag D R 1959 *Rev. Eval. Soil-Cem. Pavem.* **126** 1123.

[5] Wei H. and Z. ZHANG 2016 *1st Int., Conf., on Transportation Infrastructure and Materials (China)* vol 1 pp 513-520.

[6] Goltermann P, V. Johansen and L. Palbøl 1997 *ACI Mater. J.* **94** 435- 443.

[7] Johansson L 1979 *The effect of aggregate grading and mix proportions on the workability for concrete made with entirely crushed aggregate* (Sweden: Swedish Cement and Concrete Research Institute) pp 147-160.

IOP Conf. Series: Materials Science and Engineering **507** (2019) 012013 doi:10.1088/1757-899X/507/1/012013

Effects of Geometric Changes at the Fan Side of a TEFC on the Flow and Heat Transfer: A Numerical Study

Feng-Hsiang Lai[1*] and Huan-Sen Peng[1]

[1]TECO Group Research Institute, TECO Electric & Machinery Co., Ltd., Taipei, Taiwan

Abstract. CFD is performed to study the flow and heat transfer features of a TEFC induction motor. Geometry of an end shield and diameter of the fan cover are varied and examined. The modification of the end shield at the fan side can help increase the flow rate at the loading side by 10% due to a reduction in recirculation between the fan and the end shield. The change of cover diameter can lead to a variation of 40% in the external flow rate at the fan side and around 15% at the loading side. A combination of the re-designed end shield and the fan cover with large diameter is proposed in our study which can lower the average winding temperature by 4°C.

1. Introduction

Electric motor manufacturers have been pushing the limits of power density of motors due to the demand for higher efficiency, smaller and lighter electric machines. The same power output can be delivered with the use of less material or in a smaller size, which means the heat generation per unit volume will be higher as a result of being compact. The maximum temperature in the winding is of great concern since an electric machine will deteriorate in its performance or breaks down if the insulation material of the winding exceeds the limit and fails. Obviously heat dissipation plays an essential role in helping develop better performance motors, and more numerical and experimental studies regarding heat transfer performance are required in order to achieve a better design.

Totally Enclosed Fan Cooled (TEFC) induction motor is one of the motor types used most widely in a variety of industrial equipment. Heat resulted from the stator and the rotor is mainly dissipated via conduction to the parts including a finned frame along with end shields and then via convection to the ambient. Self-cooling is achieved by introducing an external fan with a fan cover at the non-drive side (or the fan side). What we can expect is the external flow features will be critical to the cooling performance which requires good designs for a fan, fan cover, end shield and finned-frame.

Micallef et al. [1] numerically and experimentally examined the flow field as well as the heat transfer in the end region of the winding and proposed an alternative configuration employing a fan mounted on the shaft near the end shield and the winding tip, and an increase by 19% in the windage loss along with a decrease by 33% in the thermal resistance was found. Chen et al. [2] adopted Response Surface Methodology with multi-quadratics as basic functions in order to optimize the fins on the frame under specific operating conditions. Results showed a more than 50% reduction in the thermal resistance in some cases. CFD was performed by Valenzuela and Tapia [3] to examine the air flow rate, air velocity and temperature of a motor, and the CFD results were well validated. Andres et al. [4] characterized radial fans used in electrical machines with the application of CFD. Flow-pressure characteristic curve of fans were obtained by CFD. More numerical and experimental works regarding air flow with fans [5] and cooling ducts [6] can be found in the literature.

From the literature survey we know how important the external flow is for cooling motors including TEFC. A lot of researchers have conducted studies to characterize fan, fan cover, finned frame and so on, although not many works regarding the overall effect of combining the components mentioned above on the air flow in axial direction and the winding temperature are available. In the present work we investigate the flow features at the fan side, re-design the geometry of the end shield, modify the fan cover and finally examine the effects of the end shield and the fan cover on the flow rate and the temperature in the winding in order to get a deeper insight into the overall effect of the external flow.

2. Mathematical model and simulation methodology

2.1. Geometric configuration

NEMA F#5810 series is selected in the present work. This motor is suited to wind/ hydraulic power facilities and can be applied in oil refinery, power plants and so no. We want to further improve the cooling performance and consequently CFD is conducted to study the flow and heat transfer features of a 60Hz, 4P-1000HP TEFC induction motor, as shown in figure 1.

2.2. Boundary and volume conditions

6 surfaces enclose the motor. The bottom surface simulates the ground on which the motor is placed and hence an adiabatic wall is introduced as the boundary condition. The rest of the 5 surfaces are set as pressure outlet, with the upstream and downstream surfaces 2.5 times and 8 times the length of the motor away from the motor, while the 3 lateral surfaces 2.5 times the width of the motor away from the motor. Ambient temperature is set 25°C. The motor runs at 1800rpm; a total loss of around 22kW is introduced accordingly to the stator, winding, rotor and rotor conductor as volume heat source.

2.3. Numerical method

The commercial software ANSYS FLUENT^{TM} is used to study the flow and thermal features of the full motor. This conjugate heat transfer problem is assumed to be steady-state and run with an implicit and segregated solver. The rotation of the fan is simulated with the multiple reference frame model (MRF). Unstructured mesh mainly composed of tetrahedral cells is applied for complicated geometries while structural mesh is applied to sweepable parts. Fine grids are deployed where high velocity or temperature gradient takes place, including air gap between the stator and rotor, blades, fins and so on. Y+ is checked to solve for laminar sub-layer.

3. Results and discussion

3.1. Flow features of the original design

From figure 2, a recirculation is clearly observed between the fan and the end shield at the fan side which will lead to a loss in energy and flow rate in the axial direction from the fan side to the loading side. In the present study the fan will be left unchanged due to the higher cost in modification while the fan cover and the end shield are allowed to be re-designed.

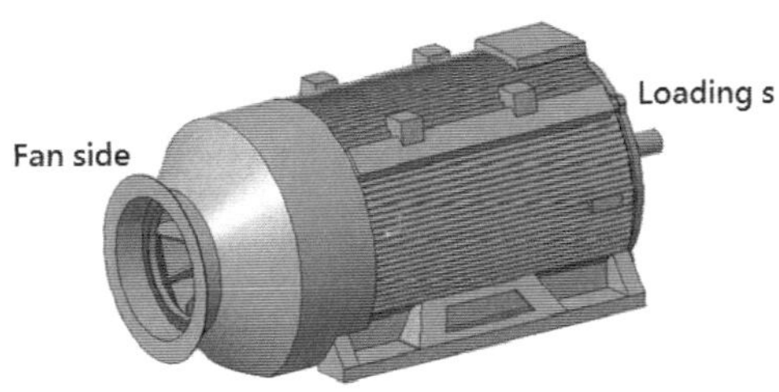

Figure 1. 3D CAD of NEMA 5810 induction motor.

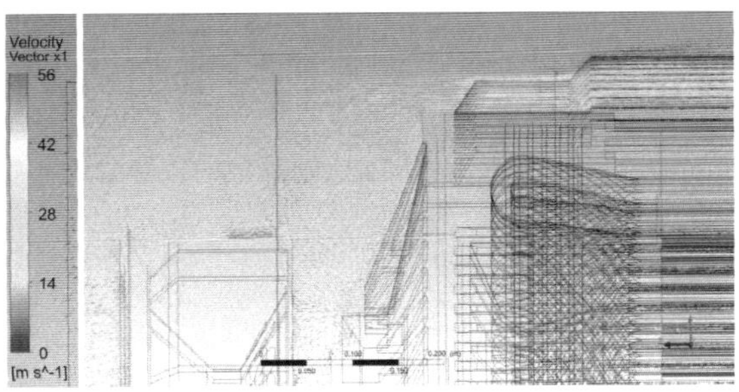

Figure 2. Vector distribution of the original version at the rear of the fan.

IOP Conf. Series: Materials Science and Engineering **507** (2019) 012013 doi:10.1088/1757-899X/507/1/012013

3.2. Reduction in recirculation

To eliminate this energy loss, we modify the end shield to be more streamlined, as shown in figure 3. Here we take a look at the flow rates in the inter-fin spacing at the 7 sections along the flow direction (z-dir.). From figure 4, an enhancement in flow rate is noticed with the addition of the streamlined end shield by 2 and 10% at the fan side and the loading side, respectively. The improvement is more announced at the downstream to make sure that the air flow is still sufficient to remove the heat at the loading side. Figure 5(a) demonstrates that with a more streamlined air-guiding end shield, the recirculation is reduced significantly and the air flow in the inter-fin spacing is improved as we can see the main flow is extended and hence the separation point at the loading side is shifted further to the downstream as shown in figure 5(b).

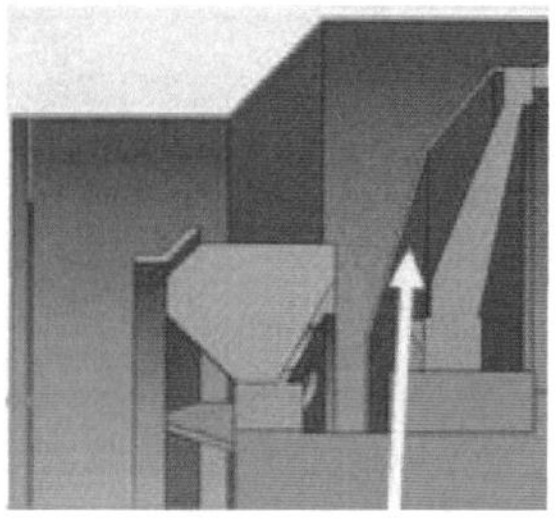

Figure 3. Sketch of the air-guiding end shield.

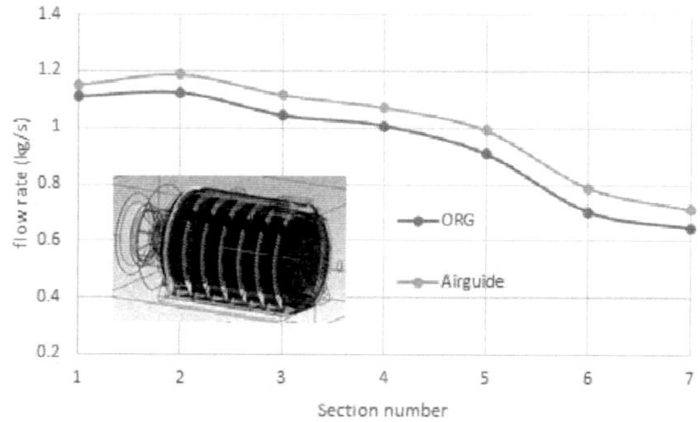

Figure 4. Local flow rate at sections along the flow direction.

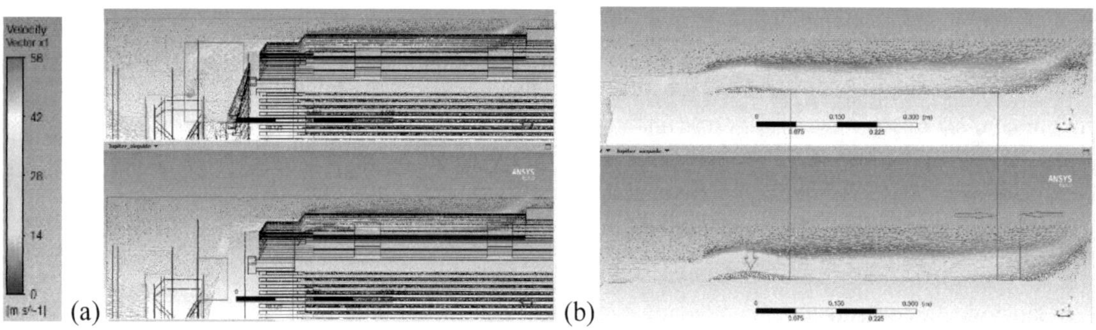

Figure 5. Vortex of the original design and the one with a more streamlined end shield at (a) the rear of the fan and (b) the inter-fin spacing.

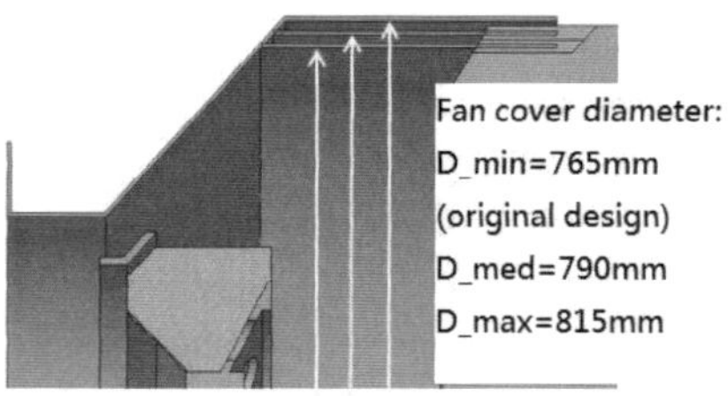

Figure 6. Three diameters of fan cover.

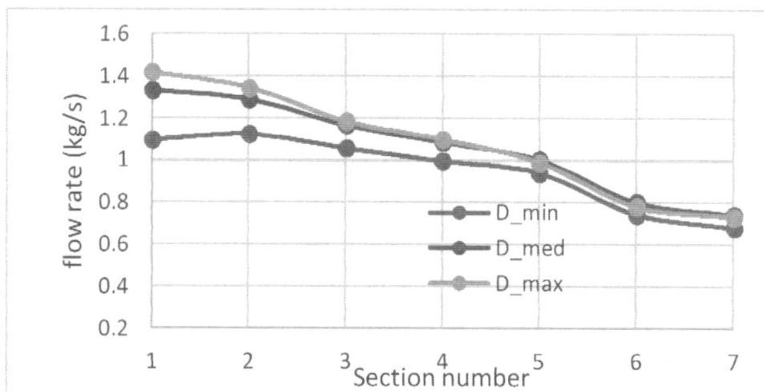

Figure 7. Flow rates using different fan cover diameters.

Figure 8. Vector using different diameters of fan cover, D=765, 790 and 815mm, from top to bottom.

3.3. Cover outlet diameter

Here the diameter of the fan cover is varied to investigate the effect on the flow rate. Three diameters are examined, D=765 (original design), 790 and 815mm, where a gap of 5mm between the fan cover and the fins on the frame is remained, as shown in figure 6. Flow rates using different fan cover diameters are demonstrated in figure 7. An increase in flow rate is noted as the diameter increases. One can notice a significant rise in flow rate at section 1 by 22% and 29% with D_med and D_max, respectively. However, the flow rate drops dramatically to a very unnoticeable difference at the loading side (section 7). Winding temperature at the loading side might be much higher than that at the fan side due to the huge flow rate drop. Moreover, 22% increase in flow rate is achieved with D_med compared to D_min while only 7% increase is revealed as the diameter is further increased to D_max, showing that there might be some optimal design in the fan cover diameter and will be studied in the future work. Figure 8 shows the flow vector using different fan cover diameters. Flow rate is increased in the fin region while a recirculation is observed in all three cases.

3.4. A combination of the air-guiding end shield and the fan cover with large diameter

From the above studies, the 'convex' design for the end shield at the fan side is noticed to help reduce recirculation at the rear of the fan. In addition, it is found that the flow rate increases as the diameter of the fan cover increases. In this part we combine these two features, that is, the modified end shield at the fan side along with the fan cover having D_max=815mm in our new design, as sketched in figure 9. Figure 10 shows the flow rate with the new design. An increase in the flow rate is noted by 40% and 15% at the fan side and the loading side, respectively, when comparing new to original design. Note in the figure that when comparing D_max+NewES with D_max, the flow rate can increase by 5-10% along the flow direction with the modification of the end shield at the fan side, resulting in a decrease in average winding temperature by 1°C as demonstrated in table 1. The winding temperature is lowered by almost 4°C when comparing new to original design. Figure 11 shows the vector on x-plane across one inter-fin spacing. From figure 11(a), a reduction in recirculation is observed as mentioned previously, while figure 11(b) shows an overall increase in velocity, even at the root of a fin.

Table 1. Temperature rise in the winding.

Case	Average temperature rise in winding, K
D_min (original)	103.87
D_med	102.62
D_max	101.30
D_max+NewES	99.98

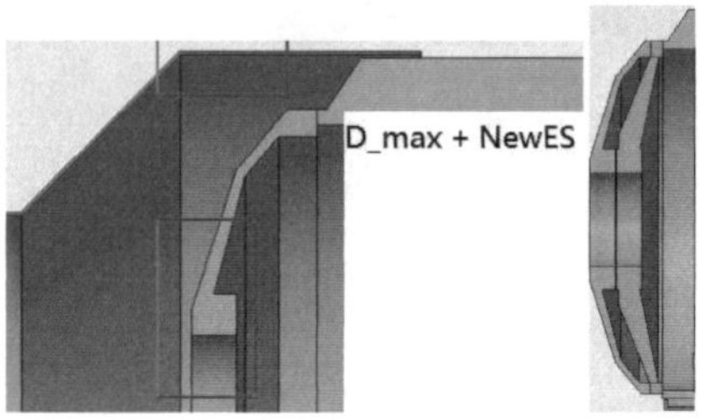

Figure 9. The new combination.

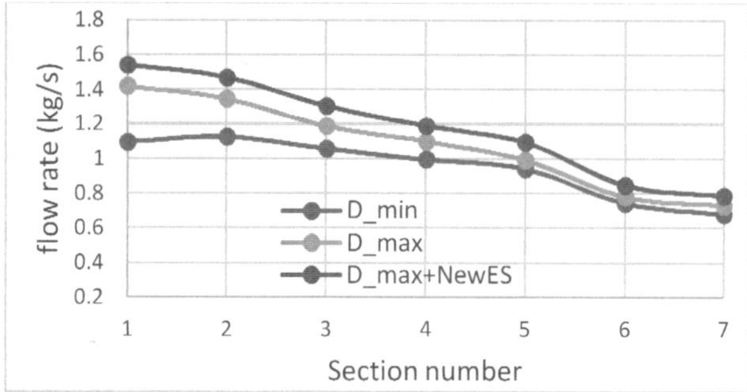

Figure 10. Flow rates using different fan cover diameters.

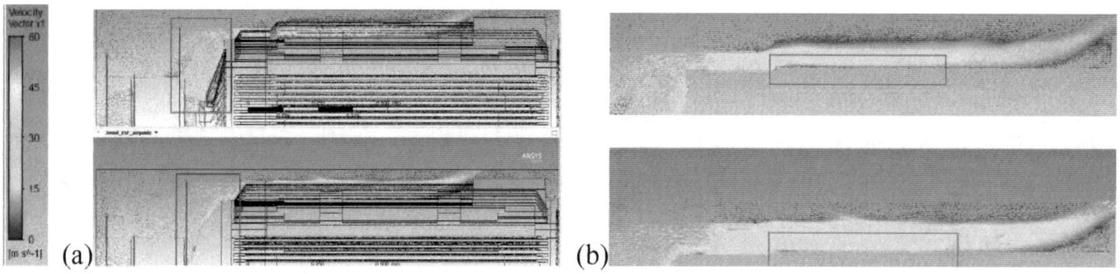

Figure 11. Vector at x-plane, (a) at the rear of the fan (b) at inter-fin spacing.

4. Conclusions

A 60Hz, 4P-1000HP F#5810 TEFC induction motor is studied with CFD approach. Results of the original design reveal recirculation at the rear of the external fan. In order to reduce the recirculation we modify the geometry of the end shield at the fan side. Making the end shield at the fan side more streamlined enhances the air flow even at the downstream side by 10%, showing less flow energy wasted in a recirculation and higher fan efficiency achieved. In addition, larger fan cover diameter is found to lead to higher air flow rate in the present study although the dramatic drop in flow rate from the fan side to the loading side is observed, showing some flow dissipated to outside of the fin region. The combination of the air-guided end shield at the fan side and the fan cover with larger diameter is proposed and an increase by 40% /15% in the flow rate at the fan/ loading side is noticed. Finally, a full thermo-fluid simulation is performed to confirm a reduction in the winding temperature by 4°C.

Acknowledgments

Authors wish to acknowledge the technical supports from colleagues at Group of Heavy Industrial Products and TGRI.

References

[1] Micallef C, Pickering S J, Simmons K A and Bradley K J 2008, Proc. 4th IET Conf. Power Electron. Mach. Drives pp 305–9.

[2] Chen Y, Chen B, Chen C and Dong J Q 2005, *Proceedings of the ASME Summer Heat Transfer Conference HT 2005* **3** 625–9.

[3] Valenzuela M A and Tapia J A 2006, *Proc. Conf. Rec. IEEE IECON* pp 4835–40.

[4] Andres U S, Almondoz G, Poza J, Ugalde G and Escalada A J 2014, *International Conference on Electrical Machine (ICEM) (Berlin, Germany)* 2179–85.

[5] Connor P H, Pickering S J, Gerada C, Eastwick C N and Micallef C 2012, *Power Electronics, Machines and Drives (PEMD 2012), 6th IET International Conference on.* pp 1–6.

[6] Schrittwieser M, Marn A, Farnleitner E and Kastner G 2012, Electrical Machines (ICEM) XXth International Conference on. 385–90.

Potential of Traditional Sago Starch: Life Cycle Assessment (LCA) Perspective

M A Yusuf[1,2*], M Romli[1], Suprihatin[1], E I Wiloso[3]

[1]Department of Agroindustrial Technology, Bogor Agricultural University, Bogor, Indonesia
[2]Department of Agricultural Engineering, Musamus University, Merauke, Indonesia
[3]Research Center of Chemistry, Indonesian Institute of Science (LIPI), Serpong, Indonesia

Abstract. Half of the world's sago production comes from Indonesia, and about 90% of Indonesia's sago originates from Papua, including West Papua. There are three types of technology used to produce sago starch in Indonesia, namely traditional, semi-mechanical and mechanical processes. In Papua generally the processing of sago starch is carried out with traditional processes. This study is intended to evaluate the environmental impacts caused by the traditionally processed sago starch products using the product life cycle (LCA) approach. Environmental impacts are estimated by analyzing the carbon footprint of GHG emissions resulting from 1 ton of dry sago starch. The scope of the product life cycle is limited to the stage of harvesting, transportation and extraction to produce dry sago starch. The results of the study show that the impact of GHG emissions generated to produce 1 ton of sago starch is 17.9 kgCO2eq. The biggest contribution came from the extraction process, especially from the drying process which was 12.25 kgCO2eq (68%). This value of impact is much lower when compared to corn starch (2700 kg CO2eq), potato starch (2402 kg CO2eq) and cassava starch (4310 kg CO2eq). This is mainly due to the low requirement of energy input of this traditional technology.

1. Introduction

Indonesia's sago production is around 50% of world sago production, and around 90% of Indonesia's sago potential is in Papua, including West Papua [1]. The production area of sago in Papua Province is around 4 700 000 ha and in West Papua around 510 000 ha [2]. The high potential of Indonesian sago can drive the development of Indonesia's sago industry. Data of sago area and production for 2004-2017 is presented in table 1.

Merauke regency is one of the regencies in Papua, which has a land area of 4.6 million hectares consisting of 3.1 million hectares in the form of forest cover including sago forest and 1.5 million hectares in the form of non-forest cover [3]. The processing of sago starch in this regency is carried out in traditional way, not yet supported by adequate sago starch processing technology. This is because the people still adhere to local customs which require that the processing of sago be carried out by manual processes without the use of machines.

Tabel 1. Area and sago production by farming category, 2004-2017.

Year	Area (Ha)				Production (Ton)			
	Smallholders	Government	Private	Total	Smallholders	Government	Private	Total
2004	87.113	-	-	87.113	14.544	-	-	14.544
2005	94.458	-	-	94.458	15.301	-	-	15.301
2006	94.528	-	-	94.528	14.202	-	-	14.202
2007	102.601	-	-	102.601	85.960	-	-	85.960
2008	48.115	-	-	48.115	31.767	-	-	31.767
2009	100.319	-	-	100.319	87.955	-	-	87.955
2010	102.174	-	-	102.174	89.629	-	-	89.629
2011	102.601	-	-	102.601	85.960	-	-	85.960
2012	106.957	-	20.200	127.157	93.265	-	39.044	132.309
2013	107.906	-	20.200	128.106	93.893	-	61.168	155.061
2014	115.284	-	20.200	135.484	249.488	-	61.168	310.656
2015	176.215	-	20.200	196.415	277.129	-	146.817	423.946
2016	193.080	-	20.200	213.280	283.511	-	157.005	440.516
2017	199.778	-	20.200	219.978	328.444	-	161.199	489.643

Source : Directorate General of Estate Crops

Traditional sago processing starts from the process of harvesting, transportation, and extraction. Each stage of these processes contributes to GHG emissions. GHG emissions are the release of greenhouse gases into the atmosphere in certain areas within a certain period of time which causes the earth's surface temperature to rise because of the ability of greenhouse gases to absorb solar radiation, which should be reflected outside the earth. According to the Intergovernmental Panel on Climate Change [4], the greater the total GHG emissions from an activity, the greater the value of potential warming generated (GWP) and the greater the activity has an impact on environmental damage. Total GHG emissions are calculated using a carbon footprint assessment. The carbon footprint is a measurement of GHG emissions that contribute to global warming which is expressed in units of kg CO2eq for each kg of products [5].

To find out the amount of GHG emissions released from sago starch production, a carbon footprint is assessed by using the Life Cycle Assessment method. LCA is an analytical method used to collect and evaluate inputs, outputs and potential environmental impacts of a product or service throughout its life cycle [6]. In addition to analyzing the impact of a product life cycle, the LCA method can be a means of developing improvement alternatives that can be used by policy makers in determining the more sustainable processes and products.

This research work was carried out in the local community village of Wasur, Merauke Regency, which aims to analyze the carbon footprint generated from each process stage in the traditional sago starch production. GHG emissions impact is represented in the form of kg CO2eq per one ton of dried sago starch produced. Thus, the output of this study is expected to serve as a database of environmental performance of one of Indonesia's sago processing methods and can provide recommendations for improvements related to energy use and waste treatment in sago starch production.

2. Methodology

This study uses the LCA methodology to evaluate the input, output, and potential environmental impacts of the life cycle of sago starch. This method consists of four stages, namely the determination of objectives and scope, inventory analysis, impact analysis and interpretation.

2.1 Purpose and scope

The purpose of this study is to analyze the input-output flow of materials throughout the life cycle of sago starch and to determine the impact to the environment. The variety of sago plants processed by the Wasur community in Merauke Regency is Metroxylon sago Rottbol (sago molat), which is not spiny. In general, the saplings produced by "sago molat" are much less than the other type of sago grown there (sago tuni). The scope of analysis is cradle to gate, which starts from harvesting natural sago trees in the forest, transportation, and extraction processes (figure 1).

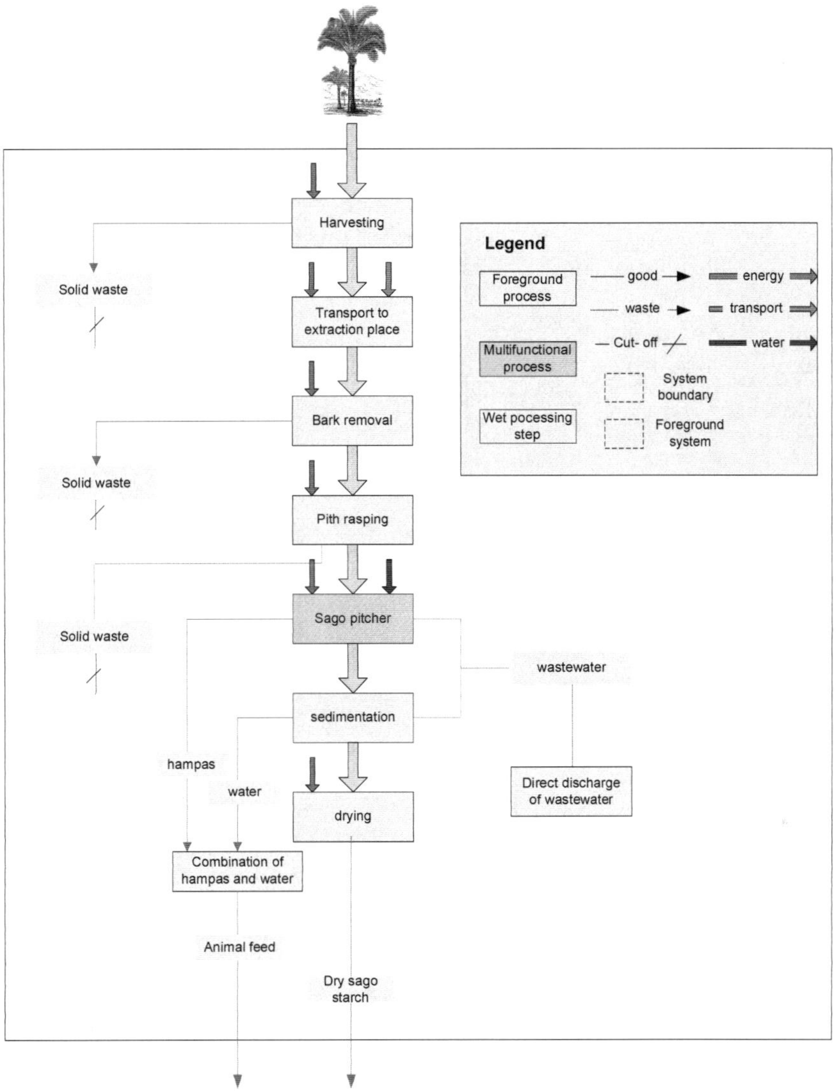

Figure 1. General LCA flowchart of traditional sago starch

Figure 1 describes the system boundaries of the sago starch life cycle used in this study, which consists of tree felling and cutting of sago stems, transportation of raw materials from the location to the extraction site, and extraction processes. The functional unit selected as a basis for determining the impact of GHG emissions is 1-ton dry sago starch. Inventory analysis was carried out by conducting an inventory of all input and output flows involved, in units of mass and energy per ton dried sago starch produced. Furthermore, each process/operation unit within the system boundary will be explained in detail starting from the harvesting process, stem cutting, transportation, skin stripping, pith destruction, squeeze, sedimentation, and drying to obtain dry sago starch.

2.2 Inventory analysis

This stage is carried out by collecting primary and secondary data related to the life cycle of sago starch to analyze the carbon footprint. SIMAPRO software (version 8.5.0.0) along with a database from ECOINVENT (version 3.3) is used to assist with this inventory analysis. Because sago is produced by cutting sago trees from the forest, sago trees are considered as the input of nature and therefore standing tree data is used (taken from SIMAPRO). Other input data on harvesting and transportation, such as weight and length of stem, weight of solid wastes, and distance from forest to extraction site are measured. In the extraction process, the weight of bark, residual pulp (extracted pith), volume of water used and waste generated (solid and liquid) are measured directly. The summary of all process input data from which the calculation of the impact of emissions made is given in Table 2.

Table 2. Summary of inventory analysis

Inventory	**SIMAPRO Analysis**			
	Input from nature	Input from technosphere	Output to technosphere	Final waste flow
Harvesting and transportation	Woods/unspecified, standing/kg	-	Sago stem	Waste, organic
Bark stripping	-	Sago stem	Sago pith	Waste, organic
Pith disintegration	-	Sago pith	Disintegrated pith	Waste, organic -
Squeezing	Water	Disintegrated pith	Starch solution	Waste pulp
Sedimentation	-	Starch solution	Wet sago starch	Waste water
drying	Wood and wood waste	Wet sago starch	Dry sago starch	-

Primary data directly measured from field are used as input category from technosphere and output category to technosphere. Secondary data taken from SIMAPRO library are used for input category from nature and final waste flow. These inventory data are then used to estimate the environmental impact of sago starch.

2.3 Impact assessment

The impact assessment stage is carried out by converting inventory data into an assessment of the environmental impacts that arise. The GHG emission indicator is carried out by analyzing the carbon footprint of the sago starch from the calculation of the estimated amount of CO2 emissions using SIMAPRO software which refers to the IPCC (2006) equation guide with the following formula:

$$E = AD \times EF$$

where,

E = Emissions
AD = Activity data
EF = Emission factor (kg CO2eq/AD)

2.4 Interpretation

The last stage of the LCA method is to interpret the results of the inventory analysis and environmental impact assessment by summarizing and analyzing conclusions in accordance with the objectives and scope that has been determined. The interpretation phase is carried out to conclude the final results of the assessment of the environmental impact of the life cycle of sago starch and provide advice to improve energy and water efficiency and reduce GHG emissions released in the traditional sago starch extraction process.

IOP Conf. Series: Materials Science and Engineering **507** (2019) 012014 doi:10.1088/1757-899X/507/1/012014

3. Results and discussion

3.1. Traditional Sago Starch Production

Sago starch production is traditionally carried out by the local population. The starch is used as a staple food for everyday consumption. Sago tree felling is carried out in mutual cooperation using simple equipment, such as machetes or axes. The harvested sago stems are cleaned and cut to size 1-2 m long, and are then transported to the extraction site by linking the sago stem pieces with sago leaf midribs and pulled manually by human power as far as 2 km. The extraction sites are primarily chosen based on the availability of the water supply.

At the extraction stage, activities carried out include skin stripping, pith disruption, squeezing, sedimentation and drying. Here the sago stems are removed from their bark, approximately 5 cm thick, using axes to get the starch-rich sago pith. The piths are disintegrated or disrupted by a device called "owyuh ombuk" (a kind of manual rasper) and the work of disintegrating the sago pith is called "undapa". The process is done in such a way that the pith is quite crushed and the starch is easily separated from the pith fibers. The disintegrated pith will be brown in colour when stored in open air in more than a day. Therefore, the rasped pith must be extracted on the same day. "Undapa" can be continued the next day until all stems are disintegrated. In this traditional way, cutting a whole sago tree can be completed within 1-3 weeks.

The disintegrated pith is then beaten by using wood rod to ease the starch extraction during squeezing. Extraction of sago flour is done by hand squeezing, and assisted with watering. In some areas, the water used comes from swamps in the location. In Merauke, the sago flour extraction facility is called "sahani", which is made from sago bark and equipped at the end with coconut fibers as a filter. The sago starch solution extracted is then flowed on the sago bark and got filtered. The sago starch is then deposited, and separated from the supernatant. Starch obtained from this traditional method is still wet. To dry it, the local tribe will wrap the sago starch with sago or banana leaves and burn it until the desired level of dryness.

3.2 Life Cycle Inventory

Results of inventory analysis are grouped according to the stages of the process started from the harvesting to drying. The mass balance of the traditional sago starch extraction process can be seen in Table 3.

Table 3. Mass balance of each stage of sago starch processing

Processes	Input		Output	
	Material	Quantity (kg)	Material	Quantity (kg)

Harvesting	Sago tree	1200	Sago stem	975
Bark stripping	Sago stem	975	Pith Bark	790 185
Pith disintegration	Pith	790	Disintegrated pith Pith waste	632 158
Squeezing	Disintegrated pith Water	632 470	Starch solution Wet pulp	679 423
Sedimentation	Starch solution	679	Wet strach Supernatant	252 427
Drying	Wet starch	252	Dry sago starch	127

* *in this study human energy is not taken into account*

Based on the mass balance above, the yield of dry sago starch obtained was 13%. Starch content in piths obtained from commercially harvested sago stems varied between 18.8% and 38.8% [7]. The manual handling and traditional extraction method results in relatively low yield of sago starch produced.

3.3 Carbon footprint analysis

The carbon footprint of traditional sago starch production is calculated using SIMAPRO software which uses the basic formula in the form of inventory data multiplied by material emission factors. Emission factor (EF) is the amount of GHG emitted from a material used in the production of a product in units of kg CO_2 equivalent / unit of material weight. The results of the carbon footprint assessment throughout the life cycle of dry sago starch products are presented in figure 2

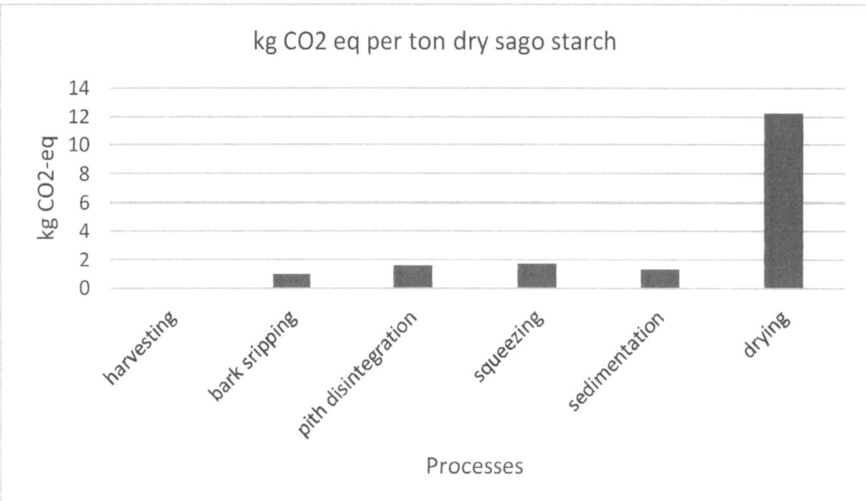

Figure 2. Results of assessment of carbon footprint on traditional sago starch production

From the results of the carbon footprint assessment, it is known that the total GHG emissions from 1 ton of dry sago starch is 17.9 kg CO_2eq. Each stage of the production process has different level of emissions impacts, and the biggest contribution is at the drying stage, which contributes to 12.25 kg CO_2eq / ton of sago starch (68%). This is because the drying process by burning wet sago starch with solid wastes of sago trees causes high carbon emissions. The second largest contribution comes from squeezing process that generates solid waste in the form of 423 kg of wet sago pulp, which contributes to a carbon footprint of 1.74 kg CO2eq / ton dry sago starch (10%). Overall, however, the value of this emission impact is still much lower when compared with the impact value on potato starch and corn starch, which are 2402 kg CO2eq and 2700 kg CO2eq [8], respectively, and cassava starch of 4310 kg CO2eq [9]. This low impact value is mainly due to the minimum use of energy input in traditional sago processing.

3.4 Interpretation
From the results of the assessment of the carbon footprint, there are various opportunities for developing process improvements to reduce GHG emissions in the production of traditional sago starch. One of the opportunities is by utilizing sago pulp for various purposes. Process development scenarios that can be applied to the sago starch extraction process can be seen in Table 4.

Table 4. Scenarios of process development

Process scenario	Activity
Utilization of sago pulp	1. Processing for animal feed 2. Processing for sago pulp flour 3. Composting 4. Processing of bioethanol
Other drying method	Use of sun drying for sago starch

4. Conclusion
The environmental impact caused by the traditional sago starch production process through a carbon footprint assessment has been calculated to be 97.9 kg CO2eq / ton dry sago starch. The largest carbon footprint contribution comes from the drying process, which is 29.4 kg CO2eq / ton dry sago starch (30%). In addition, the squeezing process generates solid waste in the form of 423 kg of wet sago pulp which contributes to a carbon footprint of 17.4 kg CO2eq / ton dry sago starch (18%). Therefore, it is necessary to develop several scenarios of production improvement, especially to further utilize sago pulp for other products and by applying alternative drying processes using sunlight.

References
[1] Jong FS, Widjono. 2007. The sustainability of sago palm (Metroxylon spp.) cultivation on deep peat in Sarawak, Sago Palm studies **3** 13-20
[2] Bintoro HMH, Asmona D, Erwin. 2009. Some efforts to rehabilitate sago palm plantation at meranti district riau province indonesia. 1st ASEAN Sago Symposium; 2009 Oct 29-30; Kuching, Malaysia. Kuching (May): 1-4
[3] Siti MMK, Siti NM, Siti AH Fakrul RA. IMPROVEMENT ON SAGO FLOUR PROCESSING. International Journal of Engineering and Technology, **4** (1), 2007, 8-14
[4] [IPCC] Intergovernmental Panel on Climate Change. 2007. IPCC Fourth Assessment Report: Climate Change 2007 [internet]. [downloaded 2016 December 3th]. Available at: http://www/ipcc.ch
[5] Ercin A, Hoekstra A. 2012. Carbon and Water Footprints Concepts, Methodologies and Policy Responses [ulasan]. Polandia (PL): United Nations Educational, Scientific and Cultural Organization

[6] Azapagic A, Lehtinen H, Saarentaus A, Rouhiainen J, Pitts M. 2011.A Review of LCA Methods and Tools and their Suitability for SMEs[review]. (GB): Pöyry Management Consulting Oy, Chemistry Innovation and Univ Manchester
[7] Singhal RS, Kennedy JF, Gopalakrishnan SM, Kaczmarek A, Knill CJ, and Amar PF. *Industrial production, processing, and utilization of sago-palm derivated products*. Carbohydrate Polymers 72 (2008) 1-20. Doi : 10.1016/j.carbpol.2007.07.043
[8] Wurdinger E, Roth U, Wegener A & Peche R. (2003) *Kunststoffe aus Nachwachsenden Rohstoffen: Vergleichende okobilanz fur Loose-fill-Packmittel aus Starke bzw. aus Polystyrol".* Final report. BIfA, IFEU, Flo-Pak (eds.), Projektforderung: Deutsche Bundesstiftung Umwelt; Augsburg, Marz 2003, 514 p.Andrea L, Fernando D, Gonzalo A 2016 *Int. J. LCA.*, **22**(4) 502-15.
[9] Cristiane M, Alex R, Luiz K, Carmen C. 2016. Environmental and energy analysis of biopolymer film based on cassava starch in Brail. J. Clepro. doi: 10.1016/j.jclepro.2016.12.14

IOP Conf. Series: Materials Science and Engineering **507** (2019) 012015 doi:10.1088/1757-899X/507/1/012015

Shear Block Test Performance of Melunak and Mengkulang Parallel to Glueline

Nor Jihan Abd Malek[1], Abdullah Abdulrahman Ahmed Al-Afif[1], Rohana Hassan[2]

[1]Faculty of Engineering and Built Environment, SEGi University, Kota Damansara, 47810 Petaling Jaya, Malaysia
[2]Engineered Wood Product (EWPro), Universiti Teknologi MARA, 40450 Shah Alam, Malaysia

Abstract. This experimental research is conducted on shear strength performance of glue laminated timber parallel to glue line using Malaysian tropical hardwood Melunak and Mengkulang. The objective of this research was to determine the shear strength values of Melunak and Mengkulang with the observation and comparison of the wood and glue failure percentage of both species. Total 30 numbers of wood specimens which; 15 numbers of wood for Melunak and 15 numbers of wood Mengkulang. The block specimen dimension was 50x50x50mm, then tested using a shear block fixtures and Universal Testing Machine (UTM). The Malaysian Standard, MS758:2001 was used in conducting shear block test and to obtain the shear strength values for both specimen. Both wood tested specimen results in 9.05% different in its shear strength values, where Melunak average shear strength was higher 0.92N/mm2 than Mengkulang. Test result of shear failure to grain direction in both specimen shows that failure occurred in wood was due to wood surface area. The Melunak average percentage failure of wood and glues area 93.7% and 6.3% respectively. While for Mengkulang average failure percentage for wood and glue area were 89.7% and 10.3% respectively. The difference in result of wood failure of both specimen area directly related to the density of specimens, in which Mengkulang wood has a higher density than Melunak. Both specimens reached to its minimum shear strength capacity that was about 6N/mm2 set by the Malaysian standard MS758:2001. Both Melunak and Mengkulang wood tested specimen results in 9.05% different in its shear strength values, where Melunak average shear strength is higher about 0.92N/mm2 than Mengkulang.

1. Introduction

Engineered woods such as glulam are manufactured and available around the world. Glue-laminated timbers are one of timber products whose ability to resist shear loading along glued surface need to be known through authenticated testing methods in order to be used. The evaluation of the shear strength of adhesive bonds between solid timber materials using standard test method outlined ASTM 4501-01 (2001) [1] shows that adhesive bond performs excellently against shear when the shear loading acts parallel to the timbers grain along the bonding surface. There are quite a number of researches done on woods products (engineered) in Malaysia, but more emphasis was placed on the woods strength and its bending leading to the scarcity of proven information on shear block test in determining the shear strength along bonded surface of local Malaysian timbers (Melunak and Mengkulang). This lack of information creates limitations on the use of Malaysian wood as it is paramount to be fully aware of the structural ability of combined wood in this case, glued wood against the most likely and frequent loading to face, (in-plane shear loading). This tease was opted by the shortage or inadequate

IOP Conf. Series: Materials Science and Engineering **507** (2019) 012015 doi:10.1088/1757-899X/507/1/012015

information on the shear block strength, lack of data with respect to percentage of bond failure, inadequate information on the comparison of behavior of bonded tropical Malaysia timbers which are specifically Melunak and Mengkulang. Hence, the shear block test was carried out in this report to help generate reliable information on some of the most abundant wood species in Malaysia (Melunak and Mengkulang). This study was slightly important because it used typical Malaysian (Melunak and Mengkulang) woods and using Malaysian standardized design method for structural timber as it is in MS758:2001 [2]. This study is designed to experimentally analyse the in-plane shear strength capacity of Melunak and Mengkulang wood block in accordance to structural use of timber under the Malaysian Standard MS758:2001 [2] using shear block test. It is specifically to determine and compare the shear strength for Melunak and Mengkulang parallel to the glue line thus to observe the failure percentage between glue and wood.

2. Research Methodology

2.1. Preparation of timber specimen
Two type Malaysian timbers, which are Melunak and Mengkulang. Both was conducted an experimental research in this project. The shear block test was conducted in UiTM structural laboratory and for the timber specimens were prepared in structural laboratory, SEGi University. The 15 block specimens for each species were dimensioned using size of (50mmx50mmx50mm) in correspondence to BS EN 392:1995 [3] and MS758:2001 [2].

2.2. Timber band saw machine
The wood used in this experiment were firstly acquired in large block form, hence, the need to cut it to dimension to fit the requirement of this experiment was a necessity. It was required that a specimen should have a dimension measuring (50mmx50mmx50mm). The band saw machine was used to execute this cutting with precision and with the availability of strong locking edges, the large timber block was able to fit properly during the cutting without disturbance.

2.3. Shear fixtures
In accordance to BS EN 392-1995 [3] requirements for a shear test device relating to adhesive bond quality assessment, the test apparatus was set up as shown in figure 1. The lapped block shear test method was used to specifically deduce the quality of adhesive on the bonded wood.

Figure 1. Fixtures of shear block test.

3. Result and Discussion

3.1. Melunak specimen
Figure 2 shows the Melunak specimen No11 graph of load (kN) versus deformation (mm), has its ultimate load applied (Fu) of 30kN and maximum deformation of the specimen 3.25mm. The specimens were loaded with 6kN at its initial stage until its reached the ultimate load of 30kN a sudden break is recorded at the ultimate load, results in 3.25 mm of deformation in the wood specimen. The load versus deformation increments also observed to be linear along the load increment, although little drop in deformation can be seen but still considered as consistent deformation at gradual load.

Figure 3 shows the ML14 (Melunak wood specimen) that had the maximum load (Fu) applied of 38.84kN and the maximum shear strength achieved is 15.39N/mm2. The specimens determine to be the highest outlier amongst all the 15 wood specimens as its split in two pieces under the highest load. The specimens failed 100% in wood because it was achieved its shear strength limit, determined by the Malaysian timber council which was 10.8N/mm2. Most of the specimens that fail 100% in wood achieved shear strength limit because it fails in wood also indicating that the adhesive bond performed excellently against shear loading allowing the load to cause the wood failure at its shear strength limit. Some of the specimens fail 100% in wood while it did not achieved its shear strength limit, could be cause by the undistributed loading from the shear block test or the moisture percentage contain in the specimen. Timbers tend to have higher strengths when the water content is low and lower strength when the water content is high.

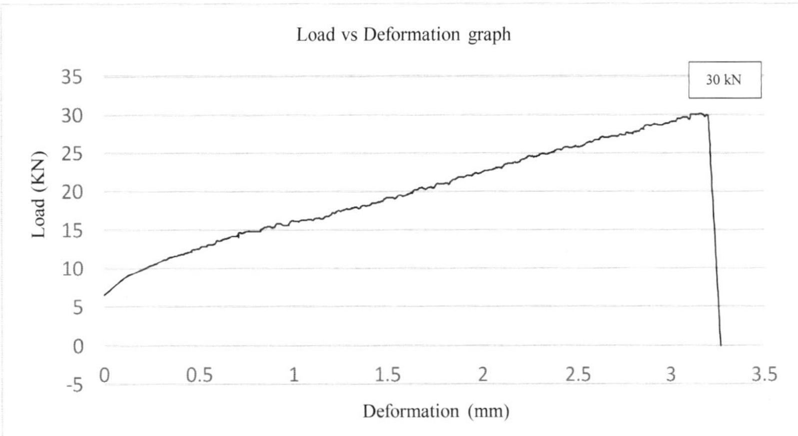

Figure 2. Load versus deformation for Melunak specimen.

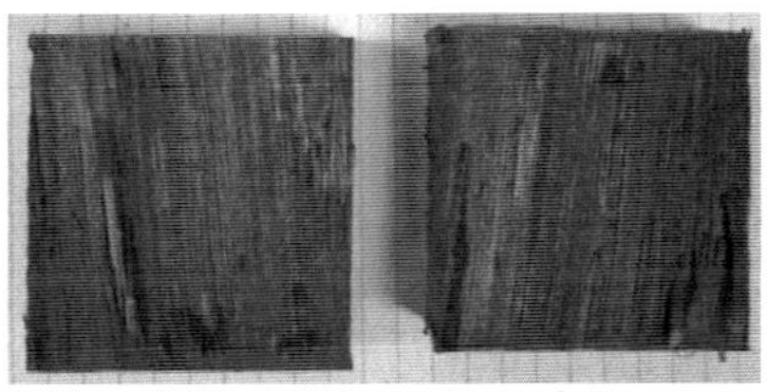

Figure 3. Melunak typical shear failure specimen.

3.2. Mengkulang specimen

Figure 4 shows the MK4 (Mengkulang wood specimen) graph of load (kN) versus deformation (mm). It has its ultimate load applied (Fu) of 28.16kN and maximum deformation of the specimen 2.7mm. The specimens were loaded with 0.75kN at its initial stage until it was reached the ultimate breaking load of 28.16kN results in 2.7mm of deformation in the wood specimen. The load versus deformation increments also observed to be linear along the load increment while a few small drops in deformation seen along different load interval but had a consistent deformation at gradual load. Figure 5 shows the MK13 (Mengkulang wood specimen) that had the maximum load (Fu) applied of 29.88kN and the maximum shear strength achieved was 11.95N/mm2. The specimens determine to be the highest outlier amongst all the 15 wood specimens as its split in two pieces under the highest load. The specimens failed 100% in wood because it was achieved its shear strength limit, determined by the Malaysian timber council which was 10.8N/mm2. Most of the specimens that fail 100% in wood achieved shear strength limit because it were failed in wood also indicating that the adhesive bond performed excellently against shear loading allowing the load to cause the wood failure at its shear strength limit. Some of the specimens failed 100% in wood while it did not achieved its shear strength limit, may cause by the undistributed loading from the shear block test or the moisture percentage containing by the specimen.

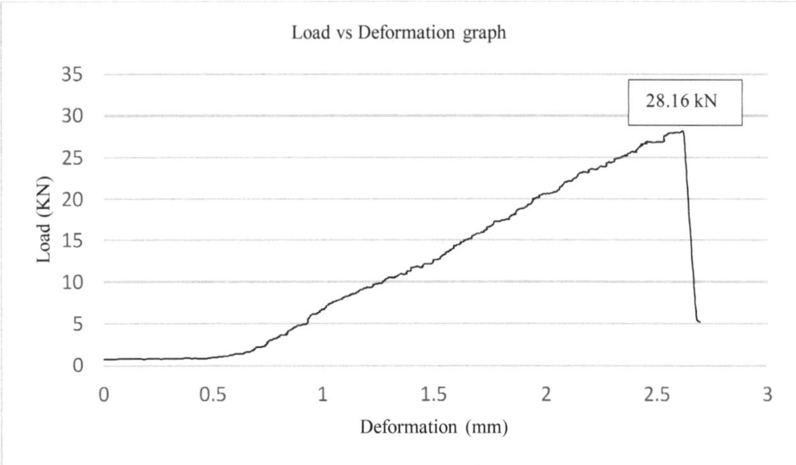

Figure 4. Load versus deformation for Mengkulang specimen.

Figure 5. Mengkulang typical shear failure specimen.

3.3. Comparison percentage failure between Melunak and Mengkulang specimen

Figure 6 shows the percentage failure that evaluated by the two wood specimen (Melunak and Mengkulang) experimental test conducted. The failure of the wood and its percentage differences between wood and glue in terms of shear failure in direction of surface grain of the wood specimen. In this experimental test result, where glue can resist strength and resulting in less percentage of failure occurs. The occurrence of failure in wood itself was because of its contact surface area to the force applied to shear the wood. It also shown the failure occurred in Melunak surface area and glued laminated are 93.7% and 6.3% respectively. While Mengkulang result in 89.7% and 10.3% failure in surface area for wood and laminated glue area respectively (Malek et. al., 2018) [4] found both specimen failed in wood. The difference in failure results for both Melunak and Mengkulang were entirely depends on the wood density, is that Mengkulang density is higher than Melunak result it in less failure than Melunak by 4% in wood and higher in glue failure. The higher the density of timber the less failure occur in wood area, and the lower the density the higher the failure in wood area (Vick and Okkonen, 1998) [5]. The structural adhesive was stronger than the timber in most of the joint. Thus, the failure in shear occurred in wood and not in the glue lines.

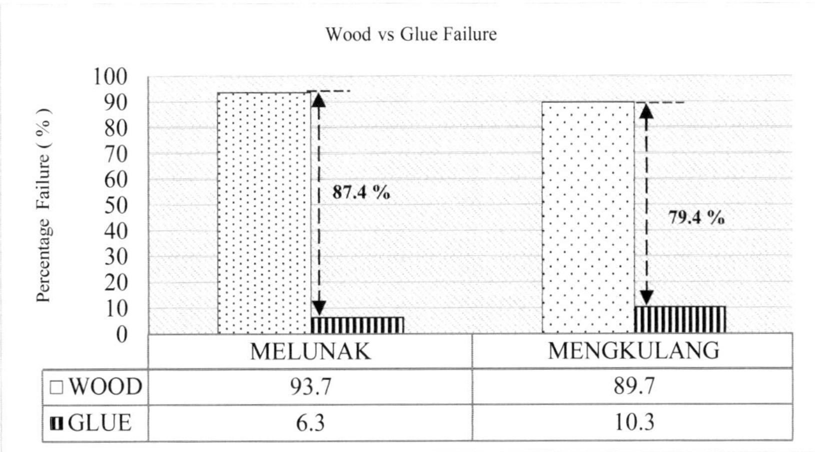

Figure 6. Percentage failure comparison of shear failure between wood and glue for Melunak and Mengkulang glulam.

4. Conclusion

Based on the result discussed, it could concluded that difference in result of wood failure of both specimen areas directly related to the holding density of specimens, in which Mengkulang timber holding a higher density then Melunak. Testing the specimens parallel to glue line, results indicating that the adhesive bond performed excellently against shear loading and allowed the wood to reach to its shear strength limit. However, Melunak showed higher average shear strength than Mengkulang average shear strength due to its differences in density and grain patterns. The higher the density the stronger the wood may result in less failure in wood and the lighter density the less stronger wood may resulted in high failure crack in wood. This can concluded that to utilize the full strength capacity of wood under the load and less failure occurs in wood. The glue bond should be higher in shear stress resistance than wood. Due to the good adhesive bonding the shear failure occurs mostly in wood area rather than in the glue line, showed good bonding performance of both Melunak and Mengkulang wood species with glue.

Acknowledgments

Authors would like to acknowledge the technical staff (UiTM, Shah Alam) and financial support (SEGi University, Kota Damansara).

References

[1] ASTM Standard D 4501-01 2001 1985 *Standard Test Method for Shear Strength of Adhesive Bonds Between Rigid Substrates by the Block-Shear Method*, American Society for Testing and Materials, West Conshohocken, Pennsylvania (first published in 1985).
[2] MS 758. 2001. *Glued Laminated Timber - Performance Requirements and Minimum Production Requirements (First Revision)*. Malaysia: Department of Standards Malaysia.
[3] British Standard Institution (1995). BS EN 392. London: British Standards Institution.
[4] Malek N J A, Hassan, R, Ibrahim, A, and Alhebshi M H 2018 *Int. J. Eng. Tech. (UAE)* **7(3)** 210-3.
[5] Vick C B and Okkonen E A 1998 *Forest Prod. J.* **48(11/12)** 71.

IOP Conf. Series: Materials Science and Engineering **507** (2019) 012016 doi:10.1088/1757-899X/507/1/012016

Ranking of Seaweed Agro-Industry in Indonesia Using Fuzzy OWA Method

L Chairani[1,2*], Sukardi[1], T Chandra[1]and F Udin[1]

[1]Department of Agroindustrial Technology, Bogor Agricultural University, Bogor, Indonesia
[2]Department of Industrial Engineering, Pancasila University, Jakarta, Indonesia

Abstract. Seaweed has an important role in efforts to increase Indonesia's fisheries production capacity because seaweed is one of the three main commodities of the fisheries revitalization program which is expected to improve community welfare. At 2016, Indonesian seaweed production had reached 11 million tons and has to increase in the coming year.Indonesia is one of the largest producers of Euchema and Gracilaria, and dominate 50% of market share in the world.Nowadays, 80% of the exported products are still in the form of raw material (dried seaweed). Therefore, the government plans to reduce the export of seaweed in the form of raw materials and process it into several potential commodities. To rank the most potential commodities, this study uses the Non-Numeric Fuzzy method and the OWA Method. The results indicate that there is a lack of development from the seaweed agro-industry in Indonesia, while the demand for processed products is high. From the three alternatives of the processed products, SRC was the first product to develop with a weight of 0.5436 and the lowest value.

1. Introduction

Seaweed, known in science as algae. Seaweed was first known in China around the year 2,700 BC. At that time, seaweed was used for medicines and self-consumed. In 65 BC, the Romans used seaweed as a cosmetic ingredient. At present, seaweed especially its processed products, is widely used as additional ingredients for the food industry, cosmetics industry, and other industries. The growth and distribution of seaweed are depended on oceanographic factors (physics, chemistry, and dynamics of sea water), as well as the type of substrate. Seaweed often found in shallow waters with sandy waters, a little mud, or a mixture of both [1].

Seaweed has an important role in efforts to increase Indonesia's fisheries production capacity because seaweed is one of the three main commodities of a fisheries revitalization program that plays an important role in improving the welfare of the community [2]. Van Bosse through the Siboga Sea expedition in 1899-1900 reported that Indonesia had approximately 555 types of 8,642 seaweed species found in the world [3]. Seaweed from the class of red algae (Rhodophyceae) is the highest in Indonesian marine waters, after those green algae (Chlorophyceae) and brown algae (Phaeophyceae) [3]. Behind its ecological and biological role in maintaining the stability of marine ecosystems and as a place to live as well as protection for other biotas, this macroalgae group has economic potential, namely as raw materials in industry and health [3]. Seaweed is a good source of nutrition because it contains carbohydrates, protein, a little fat, and ash which is mostly sodium and potassium salt compounds. Seaweed also contains vitamins, such as vitamins A, B1, B2, B6, B12, and C; Beta-carotene; and minerals, such as potassium, calcium, phosphorus, sodium, substances, and iodine.

At 2016, Indonesian seaweed production had reached 11 million tons and has to increase in the coming year. Indonesia is one of the largest producers of Euchema and Gracilaria, and dominate 50%

of market share in the world. Nowadays, 80% of the exported products are still raw material (dried seaweed) [4]. Indonesian seaweed is dominated by two types of seaweed; Eucheuma Cottonii and Gracilaria [5]. Both types of seaweed if processed will produce carrageenan and gelatin. There are two types of carrageenan; refined carrageenan and semi refined carragenan. Based on the information above, this research was conducted to find out which of the three existing commodities are potential commodities.

2. Literatur

2.1. Fuzzy

Many problems cannot be encoded in two values (existing and non-existent), but are more likely to be mapped by a logic that requires incomplete, inaccurate, fuzzy and uncertain processing of information. Therefore, a Fuzzy system and fuzzy groups can be modeled. With fuzzy logic, the input is characterized by linguistic terms, not by numbers [6].
There are several fuzzy group decision-making methods that have been developed to solve the multi-hierarchy problem, among others [7]:

- Semi non-numeric method
- Non-numeric method

Non-numerical methods are used to analyze the four commodities of seaweed. The characteristics of the non-numerical method, are followed:

- A Simple model and flexible.
- Consider the criteria explicitly.
- Can be used to manipulate satisfaction levels.

Factors that can use non-numeric fuzzy applications are [8]:

- The level of requirements for quality, quantity, and continuity of raw materials.
- Availability of human resources.
- Availability of processing technology.
- Market absorption of the products.
- Fulfillment of business capital.
- Professional of management.
- Etc

2.2. Ordered Weigth Average (OWA)

In a fuzzy environment, ranking fuzzy numbers is very important in decision making procedure. There are several methods that can be used to rank a number, OWA method is one of them [9]. OWA is a multicriteria procedure which depends on some parameters and can be specified by fuzzy [10]. OWA operators provide parameters from the aggregation that has been used in several applications. The OWA concept was introduced by Yager in 1985. This concept provides a unified framework for making uncertain decisions, where aggregation methods are different from decision criteria such as optimistic (maximization), pessimistic (maximal), most likely (Laplace) and Hurwicz criteria [11]. The application of OWA to decision makers is the determination of weight.

3. Methodology

The methodology is a systematic thinking in solving problems. The Methodology in this research are:

- Find out the product conditions and identify criteria based on literature. The criteria of this research are demanding (C1), return on investment (C2), market absorption (internal (C3) and external (C4)) and raw material (C5). The alternative commodities are; refined carrageenan (RC), semi-refined carrageenan (SRC) and gelatin.
- The frequency distribution approach has used to find the standard values, see table 1. The weight for standard value is based on 0 - 1, using the frequency distribution approach that number is divided into 7 classes.

IOP Conf. Series: Materials Science and Engineering **507** (2019) 012016 doi:10.1088/1757-899X/507/1/012016

Table 1. Standard Value.

No	Weight	Value	Score
1	0,0313 – 0,1603	Lowest (Lst)	1
2	0,1604 – 0,2894	Very low (VL)	2
3	0,2895 – 0,4185	Low (L)	3
4	0,4186 – 0,5476	Medium (M)	4
5	0,5477 – 0,6767	High (H)	5
6	0,6768 – 0,8058	Very high (VH)	6
7	0,8059 – 0,9349	Perfect (P)	7

- Determine the weights for each criterion. The weight of criteria based on the commodity probability of each criterion on literature data[5,12]. For example, the demand for refined carrageenan is 42,070 kg from 117,565 kg. The probability of refined carrageenan for demand criteria is 0.3578. The probabilities will be the weight of criteria, so the value of refined carrageenan of demand criteria is low. The value of other alternatives in each criterion can be seen in table 2.

Table 2. Value of Alternative in Criteria.

Alternative	C1		C2		C3		C4		C5	
	1	2	1	2	1	2	1	2	1	2
RC	L	L	M	VL	Lst	Lst	Lst	Lst	P	P
SRC	H	H	M	L	VL	L	VH	VH	P	P
Gelatin	Lst	Lst	VH	L	H	H	Lst	L	Lst	Lst

- The weight of each alternative can be known by using the geometric mean, see equation 1.

$$G = (X_1 x X_2 x X_3 x \dots x X_n)^{1/n} \quad (1)$$

- The values for each alternative (i) from each decision maker (j) can be known by using equation 2.

$$V_{ij} = \text{Min}\ [Neg\ (W_{ak})\ v\ V_{ij}(ak)] \quad (2)$$

The negation of value for each alternative as is follow [9]

Neg (P) = Lst   Neg (M) = M   Neg (Lst) = P
Neg (VH) = VL   Neg (L) = H
Neg (H) = L   Neg (VL) = VH

- The value of each alternative can be known by using the OWA method, see equation 3.

$$V_i = f(V_i) = \max\ [W_j \wedge b_j] \quad (3)$$

4. Result and Discussion

Before carrying out further analysis, the weight of each alternative is calculated using equation 1. All data related to each commodity (alternatives) are averaged using geometric mean, the result of the weight calculation, then adjusted to the standard values in table 1. Weight and Value for each alternative can be seen in table 3.

Table 3. Weight and Value of Alternatives.

Alternatives	Weight	Value

RC	0,2694	Very Low
SRC	0,5436	Medium
Gelatin	0,2064	Very Low

Based on data analysis in table 2 and by using equations 2, the aggregation results for each alternative shown in table 4.

Table 4. The Aggregation result.

DM	Alternatives		
	RC	SRC	Gelatin
1	Min[H,M,P,P,Lst] = Lst	Min[L,M,VH,VL,Lst] = Lst	Min[P,VL,L,P,P] =VL
2	Min[H,VH,P,P,Lst] = Lst	Min[L,H,H,VL,Lst]= Lst	Min[P,H,L,H,P]= L

The weight of each decision maker is considered the same. This decision is based on their expertise related to seaweed commodity. The weight of each decision maker, then adjusted to the standard value in table 1. Weight and value of decision maker can be seen in table 5.

Table 5. Weigth and Value of Decision Maker

DM	Weight	Value
1	0,5000	Medium
2	0,5000	Medium

After obtaining the value of decision makers in table 5 and aggregation results in table 4. An alternative value can be known using the OWA method. Equation 3 is used to find the value of each alternative. The value of each alternative is:

- RC : Max[M^Lst, M^Lst]
  Max[Lst,Lst] =Lst
- SRC : Max[M^Lst, M^Lst]
  Max[Lst,Lst] = Lst
- Gelatin : Max[M^VL, M^L]
  Max[VL,L] = L

The results of the calculation for the three commodities indicate that the three commodities (alternatives) have a low value, especially for RC and SRC. The output obtained from the five input parameters (criteria) that used in this study. These results indicate a lack of seaweed agro-industry in Indonesia that is not comparable with the amount of seaweed produced. The high market demand both domestically and abroad has not been able to be met by seaweed agro-industry in Indonesia. Of the three alternatives (commodities), SRC was the first product to develop. This is based on the weight of SRC higher than the other two commodities (see table 3) and the value of SRC is lowest (Lst). This result also shows that the SRC was ranked first in the seaweed agro-industry in Indonesia, followed by RC and Gelatin.

5. Conclusion

Based on the results and discussion above, the conclusions from this research are:

- All alternatives have low values.
- From the five existing criteria, SRC has the highest weight if we compared to RC and gelatin

From the result above, carrageenan especially SRC needs to be developed in the future time. The aggregation results show that SRC has the lowest Value. The lowest value of SRC is caused by an imbalance between the amount of seaweed production and the number of SRC production in Indonesia, while the demand for this product is high.

Acknowledgement

This paper is funded by Indonesia Endowment Fund for Education (LPDP).

References

[1] Priono B 2013 *Media Akuakultur* **8(1)**1-8.
[2] Susilowati T *et al* 2012 *Jurnal Saintek Perikanan* **8(1)** 7-12.
[3] Suparmi and Sahri A 2009 *Sultan Agung* **XLIV(118)**92–116.
[4] Hikmah 2015 *J. Kebijakan Sosek KP vol5 no1.*
[5] Carolina RA 2015 *Info Komoditi Rumput Laut* (Jakarta: Badan Pengkajian dan Pengembangan Kebijakan Perdagangan-Indonesian Ministry of Trade) pp 25 - 42.
[6] Marimin, et al 2013 *Teknik dan Analisis Pengambilan Keputusan Fuzzy dalam Manajemen Rantai Pasok* (Bogor: IPB Press)
[7] Rukmayadi D and Marimin 2000 *J.II.Pert Indon,* **9(2)**
[8] Santoso I and Marimin. 2001 *Jurnal Teknologi dan Industri Pangan*.**XII(2)** 163–70.
[9] Lamata MT 2004 *Int. J. Intell. Syst.***19** 473-82 *DOI 10.1002/int.20002*
[10] Mokarram M and Hojati M 2016 *J. Sol. Earth DOI 10.5194/se-2016-17.*
[11] Cheng C H*et al* 2012 *African Journal of Business Management vol 6(21) pp 6358-6368. DOI 10.5897/AJBM11.534*
[12] Hendrawati TY 2016 *Pengolahan Rumput Laut dan Kelayakan Industrinya* (Jakarta: UMJ Press)

IOP Conf. Series: Materials Science and Engineering **507** (2019) 012017 doi:10.1088/1757-899X/507/1/012017

Roadway embankment stabilization on permafrost using thermosyphons with Y-shaped evaporators

Vamsi Kukkapalli and Sunwoo Kim

Department of Mechanical Engineering, University of Alaska, Fairbanks, AK, 99775-5905, USA

Abstract. Constructing surface facilities on permafrost soils is a substantial engineering challenge. Moreover, evading thaw-settlement on roadway embankments is not an easy task due to the shifting thermal regime of underground soil because of seasonal temperature variations. Of all engineered cooling solutions available to stabilize permafrost, thermosyphons have attracted many researchers because of the benefits of using this cooling method compared with other methods. In this paper, we describe the optimization of thermosyphons with a Y-shaped evaporator. The cooling effects were compared with those of a conventional parallel design. The optimal spacing between thermosyphons for both the parallel design and the proposed Y-shaped design are documented.

1. Introduction

Permafrost is soil that is at or below the freezing point of water (0°C) for two or more years. Permafrost is a major problem in high-latitude regions such as the Arctic and Antarctic. This soil will thaw in a shifting thermal regime. Stabilizing thaw-settlement on roadway embankments in permafrost zones is a highly technical engineering challenge. Roadway embankments are one of various engineering structures affected by thawing permafrost. Thermosyphon cooling is a passive cooling method widely used in cold regions to avoid thaw-settlement in permafrost [1–5]. Our main interest in this research is the evaporator, which is buried in the soil at a certain known depth. The evaporator absorbs heat from the surrounding soil and delivers it to the condenser where the heat is ejected to the atmosphere. There is growing interest in designs that improve the performance of the evaporator section in thermosyphons. The classic design of the evaporator section in a thermosyphon consists of pipes of simple parallel shape, and this design has been used widely [2, 6, 7]. In our research, we employ a Y-shaped design to improve the parallel evaporator.

The Y-shaped tree design is depicted in figure 1. We have compared the evaporator performance of the classic design (straight evaporators placed parallel) with the Y-shaped design. The net volume occupied by the pipes in the evaporator is fixed in all designs for a fair comparison.

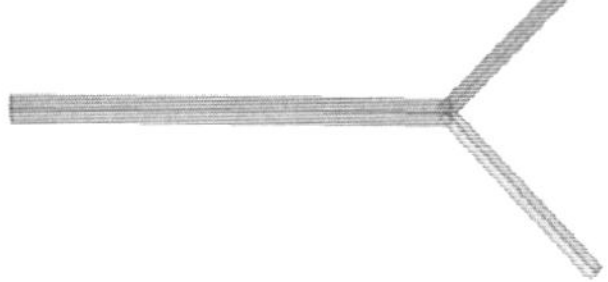

Figure 1. Y-shaped architecture

IOP Conf. Series: Materials Science and Engineering **507** (2019) 012017 doi:10.1088/1757-899X/507/1/012017

2. Mathematical model

The geometry of the roadway embankment model is adopted from a highly cited paper [8]. The roadway embankment dimensions are shown in figure 2. This model consists of a highway embankment with a driving surface width of 6 m and height of 2.5 m. The lower boundary is 9 m beneath the pavement surface, and the outer vertical boundary is 17 m from the centerline. Due to the symmetry, we can only consider the half portion of the model. We are interested in finding the optimal spacing between two neighboring thermosyphons by minimizing the volume-averaged temperature. We will consider that the road is of infinite length and that the thermosyphons are installed equidistantly in the direction perpendicular to the plane. By varying the width of the model (the spacing between two neighboring thermosyphons), the minimum volume-averaged temperature is sought. The width resulting in the minimum volume-averaged temperature is the optimal spacing. The outer surfaces of the evaporator, which are buried in the solid, are modeled as isothermal. This assumption is reasonable because the flow in the pipes is intense and the convective heat transfer coefficient in the pipes is very high. The boundary of the entire volume is modeled as adiabatic. The solid is initially at temperature (T_s), which is higher than the temperature of the pipes (T_p). In time, due to the temperature difference, heat transfer occurs between the ground soil and the pipes, and the cooled zone grows around the evaporator.

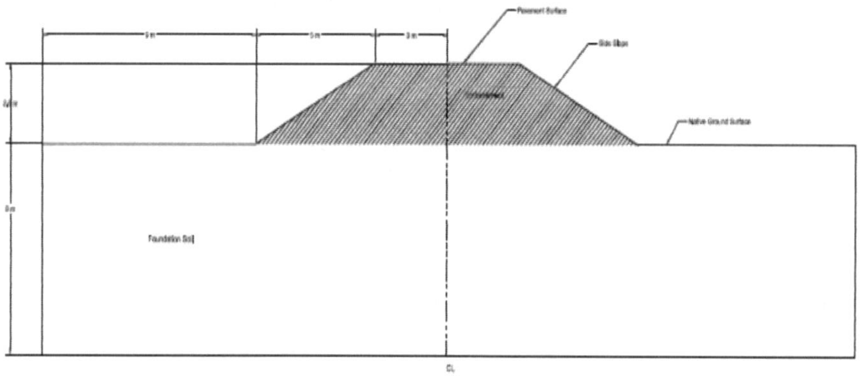

Figure 2. Model embankment geometry

The Y-shaped design (figure 1) has the following dimensions:

$$L_1 = 3m, \text{ and } L_2 = \frac{L_1}{3} \quad (1)$$

All of the pipe cross sections are round. The minimization of fluid flow resistance calls for a particular sequence of diameter sizes, and the round shape holds for both laminar and turbulent flow. The diameters of the pipes in Fig. 1 are sized relative to one another, according to the Hess-Murray rule [9],

$$\frac{D_i}{D_{i+1}} = 2^{1/3} \quad (i = 1, 2, 3\ldots) \quad (2)$$

The conservation of energy in the solid that surrounds the branch structures is governed by

$$\frac{\partial T}{\partial t} = \alpha \nabla^2 T \quad (3)$$

where α is the thermal diffusivity of the solid, and $\nabla^2 = \frac{\partial^2}{\partial x^2} + \frac{\partial^2}{\partial y^2} + \frac{\partial^2}{\partial z^2}$, where x, y, z are the coordinates of the model. For greater generality, we determine the temperature field in terms of the dimensionless variables.

$$(x_*, y_*, z_*) = \frac{x,y,z}{V^{1/3}} \quad (4)$$

$$t_* = \frac{\alpha t}{V^{2/3}} \tag{5}$$

$$\theta = \frac{T_p - T}{T_p - T_s} \tag{6}$$

where t and V denote the time and volume, respectively. Eq. (3) can be written in terms of the dimensionless variables as

$$\frac{\partial \theta}{\partial t_*} = \frac{\partial^2 \theta}{\partial x_*^2} + \frac{\partial^2 \theta}{\partial y_*^2} + \frac{\partial^2 \theta}{\partial z_*^2} \tag{7}$$

The initial boundary conditions are

$$\frac{\partial \theta}{\partial x_*} = 0 \; at \; x_* = 0 \text{ and } 17; \; \frac{\partial \theta}{\partial y_*} = 0 \; at \; y_* = 2 \text{ and } -6;$$

$$\text{and } \frac{\partial \theta}{\partial z_*} = 0 \; at \; z_* = 0 \text{ and } W \tag{8}$$

$$\theta = 1 \quad \text{at } t_* = 0 \tag{9}$$

The simulation is initiated from $t_*= 0$, where the heat is transferred from the solid to the pipes. The objective is to find the effective volume-averaged temperature with different designs. As the t_* value increases, the volume-averaged temperature of the embankment approaches that of the pipe temperature (T_p). At a fixed non-dimensionalized time at $t_* = 1$, we compute the effective volume-averaged temperature:

$$\theta_{avg} = \frac{T_p - T_{avg}}{T_p - T_s} \tag{10}$$

3. Simulations: Y-shaped vs. conventional design

The heat transfer process was simulated as transient heat conduction by using a computational package, COMSOL Multiphysics [10]. Figure 3 depicts examples of the simulations at $t_* = 1$. The value in the color temperature bar indicators is the non-dimensionalized time. As can be seen in figure 3, the left panel shows the straight evaporator (conventional design); the right panel shows a Y-shaped evaporator. In the vicinity of the evaporator, temperatures are low due to the cooling provided by the thermosyphons. Therefore, the embankment and the soil under it are well protected from the possibility of thawing permafrost.

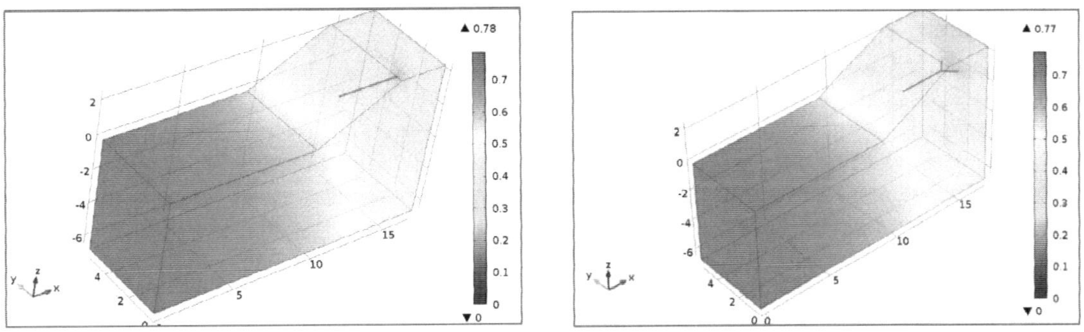

Figure 3. Simulation results for the conventional (left) and Y-shaped (right) designs at an optimum width of 5.5 m

Figure 4 shows the computed volume-averaged temperature with respect to the width of the embankment structure. Both Y-shaped and conventional parallel designs make a U-shaped curve, which implies that there is an optimal width for maximum cooling performance from the equal-volume occupied evaporators. The optimal width is formed at approximately 5.5 m for both designs. In other words, a thermosyphon performs best when it is installed an equal distance of 5.5 m apart from neighboring thermosyphons. When the distance (W) is reduced to 2 m, the average temperature rises to 0.604 and 0.594 for the parallel and Y-shaped design, respectively. That value also becomes greater than the minima when the distance between two neighboring thermosyphons is greater than 5.5 m.

In a comparison of the Y-shaped and conventional designs, the proposed Y-shaped structures transfer the heat much better than the classic evaporator model. Improvements are made by changing the classic design to the proposed Y-shaped design: the minimum averaged temperatures are 0.577 for the conventional design and 0.556 for the Y-shaped design. The Y-shaped design results in a 3.6% improvement.

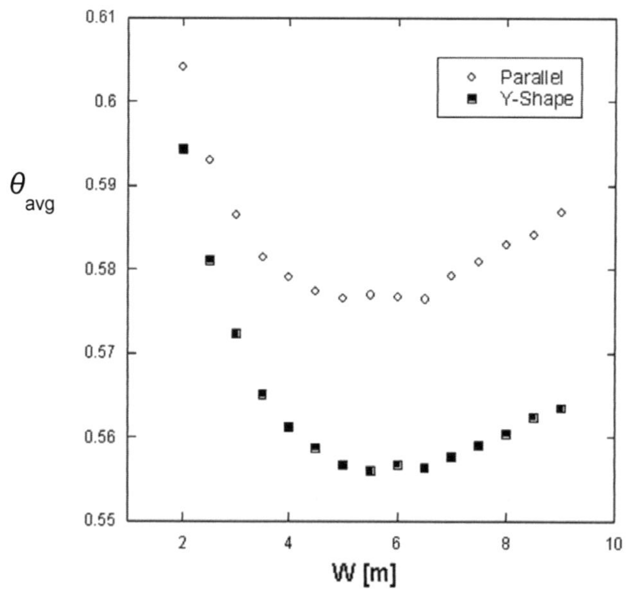

Figure 4. Comparison of volume-averaged temperature with respect to width for Y-shaped and parallel (classic) shaped evaporators

4. Conclusion and future work

In this paper, we propose a Y-shaped configuration to improve the heat transfer performance of thermosyphons used for roadway embankment stabilization on permafrost. We successfully analysed the thermal performance of the different architectures located in a conducting medium. The simulations computed the volume-averaged temperature for various cases. For fair comparison, the volume fraction of the evaporator was maintained at a fixed value for all cases studied. Our findings show that there is an optimal distance (width) between two neighboring thermosyphons.

We found that the Y-shaped configuration performs better than the classic design, as shown in figure 4. The optimum spacing was determined to be 5.5 m. The volume-averaged temperature (θ_{avg}) for Y-shaped structures produced an improvement over the conventional ones with about a 3.6% increase.

The improvement stems from the Y-shaped bifurcation. This design provides more effective heat transfer between the evaporator and ground soil. In future work, we will study higher bifurcation level designs (figure 5 shows an example of a two-bifurcation level design), as well as the effect of the branch length at each level. The angles of bifurcations will be numerically evaluated.

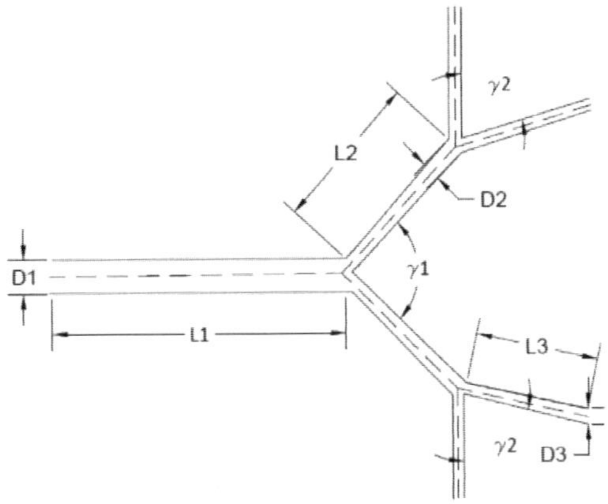

Figure 5. Tree structure lengths and branch angles

References

[1] Zhi W, Yu S, Wei M, Jilin Q, Wu J, 2005 *Cold Reg. Sci. Technol.* **43** 150–163.
[2] Wu J, Ma W, Sun Z, Wen Z, 2010 *Cold Reg. Sci. Technol.* **60** 234–244.
[3] Cheng G, Sun Z, Niu F, 2008 *Cold Reg. Sci. Technol.* **53** 241–258.
[4] Song Y, Jin L, Zhang J 2013 *Cold Reg. Sci. Technol.* **93** 12–19.
[5] Xu J, Goering D J, 2008 *Cold Reg. Sci. Technol.* **53** 283–297.
[6] Haynes F D, Zarling J P, Gooch G E, 1992 *Cold Reg. Sci. Technol.* **20** 261–269.
[7] Bayasan R M, Korotchenko A G, Volkov N G, Pustovoit G P, Lobanov A D 2008, *Appl. Therm. Eng.* **28** 274–277.
[8] Goering D J, Kumar P, 1996 *Winter-time convection in open-graded embankments,* **24** 57–74.
[9] Comsol Multiphysics, http://www.comsol.com/.
[10] Bejan A, Lorente S, *Design With Constructal Theory,* 1st ed., John Wiley & Sons, Inc., Hoboken, New Jersey, 2008.

IOP Conf. Series: Materials Science and Engineering **507** (2019) 012018 doi:10.1088/1757-899X/507/1/012018

Performances Investigation of Filter Membrane Coated

Ling Shi[1]*, Yufeng Chang[1], Ruihua Fan, Hui[1] Zhang[1]

[1]Hubei Key Laboratory of Industrial Fume & Dust Pollution Control (Jianghan University), Hubei Wuhan 430056

Abstract. Filtration membrane coated is a key kind of aerosol pollution control technology. The mechanism of depth filtration and surface filtration is analysed, and several SEM structure also given and basic compared. Furthermore, the efficiency of filtration e-PTFE membrane coated is tested according to the GB6719-2009 and ISO 11057-2011used the VDI3296 experimental apparatus by the process of the standard test. The pressure is also tested in the same experimental apparatus. The investigation is useful to guide the choice of the filter membrane coated to reached the emission regulation of 10mg/m^3.

1. Introduction

In recent years, fine aerosol particles pollution is played widely attention, due to serious fog and haze pollution have happen in many cities all over the world. So a serial of strict regulation has been released to restrain the fine particulate matter emission, and many new technology and new materials are utilized to removal of fine aerosol particles flowing from pollutant source. The filtration is one of these technology. Filtration technology is considered as an effective method for meeting policy stipulation, because fiber media has distinctive feature: (1) high efficiency; (2) simple application device.

Filtration is used in diverse applications, such as respiratory protection, hazardous materials [1], removal of asbestos fibers [2]. Fibrous filtration has been extensively studied experimentally and theoretically, the most population theory of filtration is that inertia impact, interception, Brown diffusion and electrostatic of single fiber, and that of the exponential power theory of the packing fiber. Ryan Mead-Hunter [3] gave a review on filter from the aerosol-mist cleaning viewpoint, and another review on the filter was reported by C.Y.Chen [4] from the overall aerosol-particulates cleaning viewpoint. Based on the single-fiber efficiency were well developed and systematically documented [1, 3, 4]. Pao-Kuan Wang [5] gave a modified efficiency definition on particles collided an infinity long single fiber, where a constant that was the geometrical volume swept out by the collector per unit time was used. K.W.Lee and B.Y.H.Liu [6] has conducted a theory study of aerosol filtration used the Kuwabara flow field. A theory review is conducted on the filtration by S.K.Friledlander [7].

Based on the fiber filtration, some new feature is added, such as Centrifugal Filter [8], filtration of fiber array [9], electrostatic enhancement of fabric filter, filter membrane filter [10]. A type membrane filter material has been developed by W. L. Gore & Associates Inc. a viewpoint is consider that membrane filter material has transformed the filter method, i.e. depth filtration to surface filtration, due to captured particles adhering the surface filter, it makes the deashing easier. Jingxian Liu [11] reported also the results of efficiency of three membrane filter, and John F. McCarthy reported [12] the results of fractional penetration of particles on GS Membrane Cartridge Filters. These study demonstrations that filter media membrane coated can meet with the emission 10mg/m^3 regulation even lower.

First, basic law of filtration on fiber filtration are introduced, then, several membrane microstructure and the surface filtration process are demonstrated and are compared, last the experimental performance of a e-PTFE filtration media membrane coated is reported.

2. Structure and mechanism of membrane filter material

Two type of filtration are usually utilized: one is depth filter media, the other is surface filter media. Depth filter media, which has high porosity and high air permeability, should be made to have high dust holding capacity, since the cartridge will not be cleaned during the operation period [13, 14]. However, surface filter media, which is that of classic fiber media surface coated a layer membrane, such as e-PTFE membrane, PPS membrane, has smooth surface and micrometer or submicron meter pore. It is obvious that the efficiency of removal of particles is more high and particles captured on surface can easily be cleaned. Surface filter media are made used meltblown, hot press methods. Usually, the meltblown technology, which is based on mechanical dispersion, is a one step process.

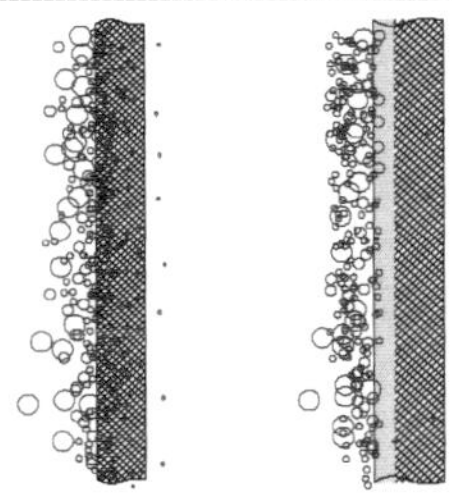

a b

Figure 1. Schematic of depth filtration and surface filtration

(a: depth filtration; b: surface filtration)

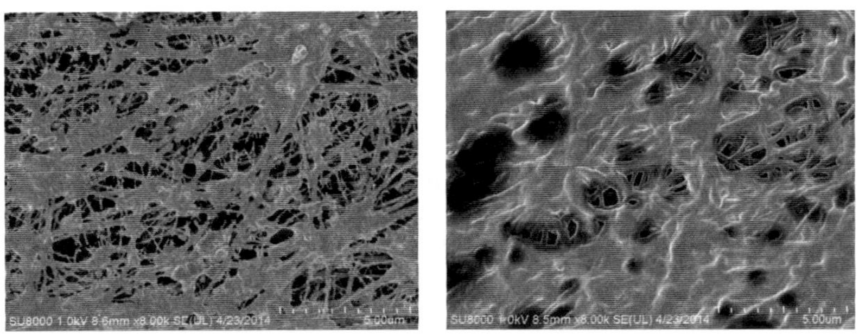

a. membrane b. membrane

Figure 2. Two type surface microstructure of e-PTFE membrane.

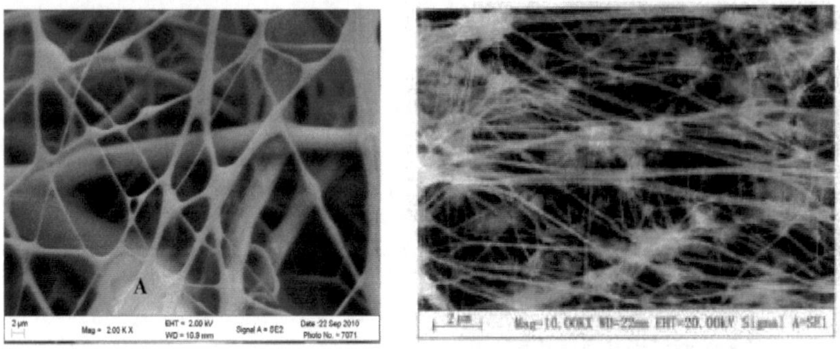

a b

Figure 3. Membranes SEM structure by meltbrown and hot press [14]
(a meltbrown membrane SEM structure b stretch and hot press SEM membrane structure)

Particles are removed, when gas stream particle- loaded pass through depth filter media, at first stage, the depth filter media is made function, after the dust cake formed, the particle are filtrated by the dust cake [15, 16]. However, sieving becomes main mechanism removal of particles for the surface filter media, since the membrane has very small pore at surface of filter media, and the dust cake is not necessary. The differential of two can be demonstrated in figure 1. The porosity is more than 85%. Microstructure of the several membrane filter material SEM image are shown in figure 2 and figure 3, respectively.

From those SEM images, there are obvious difference the base material and the micron structure of membrane, this may bring to perfect filtration performance of the filter membrane coated, and affects the pressure drop.

3. Experimental of filtration membrane coated

The efficiency of filter membrane coated experimental apparatus is shown in figure 4. The experimental apparatus is according with ASTM D6830-02, ISO11057, VDI/DIN3926 and GB/T1264 standard. It mainly consists of a rectangular, vertical raw-gas duct, a cylinder, horizontal suction unit incorporating the sample of the fabric to be tested and cleaning-system consisting of a 2.5 liter pressure tank, a quick acting diaphragm value and a blow-tube. Table 1 is particles distribution for the filtration efficiency tested.

Table 1. diameter distribution of particles.

Particles diameter/μm	0~5	5~10	10~15	15~20	20~25	25~40	40~45
distribution/%	53.9	14	13.5	7.9	4.1	5.2	0.7
D50=4.2μm		D10=0.89μm		D90=20.7μm			

According to the technology regulation and requirements of GB6719-2009, several experiments are conducted. Two filters membrane coated were used in this study. Their SEM image is shown in figure 2. They were made for industry application. The concentration at entrance is 5g/m^3. The dust which made by Sasol Germany GmbH company is Pural NF. The mass of sample filter tested and absolute filter is weighed with analytical balance with resolution ratio of 0.01mg.

The pressure drop experiment is conducted in another rig made by myself.

IOP Conf. Series: Materials Science and Engineering **507** (2019) 012018 doi:10.1088/1757-899X/507/1/012018

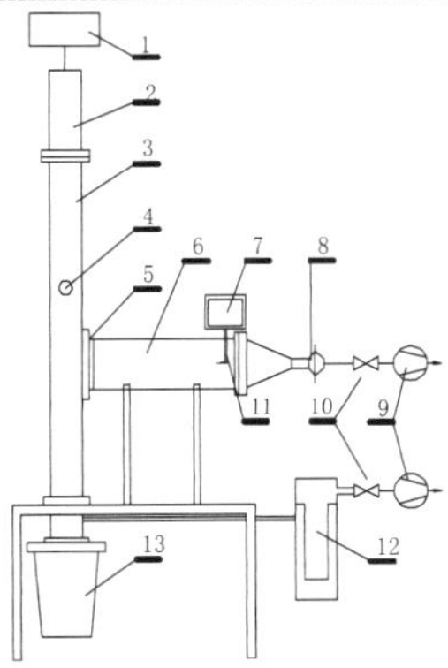

Figure 4. Schematic diagram of experimental apparatus

1. dust feeder 2. inlet mixing tube 3. vertical dirty-gas tube 4. photometric measuring pore 5. holder for filter sample 6. horizontal clean-gas dust 7. Compressed air tank 8. absolute filter 9. pump 10. mass-flow-meter 11. blow tube 12. dirty-gas back-up filter 13. dust container

3.1. Fraction efficiency

Two kinds of filtration membrane coated efficiency are acquired: number fraction efficiency and dust-laden mass fraction efficiency. Under the filter velocity 1.2m/min condition, the number fraction efficiency is conducted, and the result is shown in figure 5. Under the face velocity 2m/min condition, the mass fraction efficiency is conducted, figure 6 shows the experimental curve. As figure 5 shown, the number efficiency is more than 94 percent even through the diameter of particles is less than 0.5 μm, and shows than there is remarkable different between two kinds of filter membrane media. Contrast to the SEM image two kinds of filter e-PTFE membrane media, it is easily found that the pore quality of A filter membrane media is superior to B filter membrane. A filter membrane media can be applying to filtering PM2.5. This can be proved by the pressure loss of two kinds of the membrane.

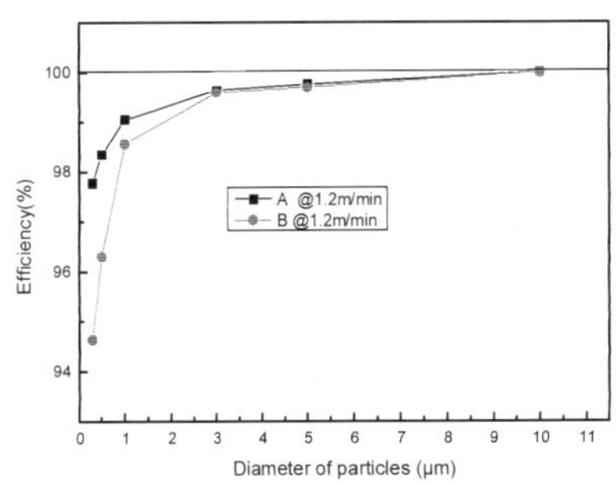

Figure 5. Counter fractional efficiency of A and B filter membrane coated.

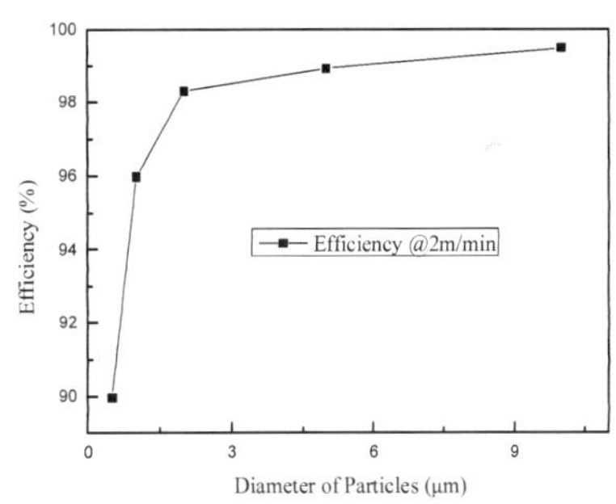

Figure 6. Mass fractional efficiency.

Figure 6 shows that the simulated dust-laden mass efficiency is more than 90 percent, this results of dust-laden simulated experiment is different from that of number efficiency, the cause may be the velocity is larger than that of the number efficiency measuring.

3.2. Pressure drop and ageing

To evaluate the performance of the filter membrane coated, it is necessary not only on the filtration efficiency but also on the pressure drop and ageing.

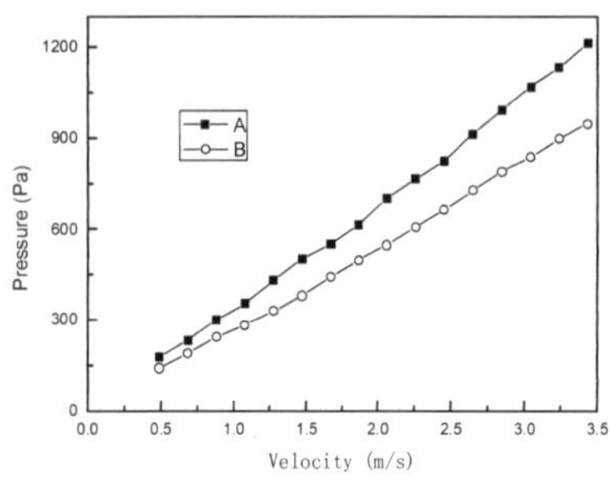

Figure 7. The pressure drop versus velocity for filter.

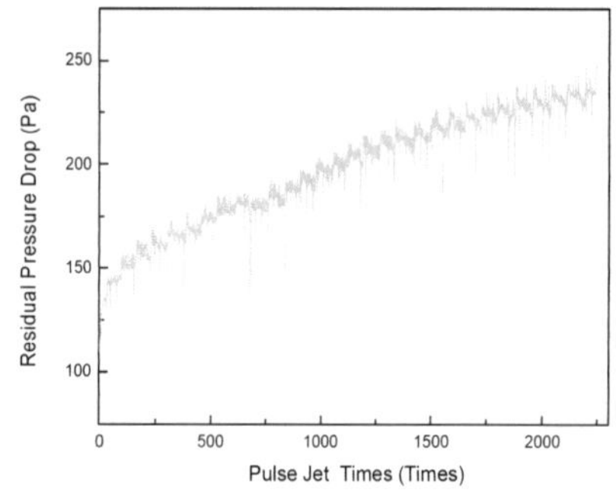

Figure 8. Residual Pressure Drop changing Curve cleaning with Pulse Jet.

Figure 7 shows that the relationship of pressure drop versus face velocity under cleanable gas flow. The pressure drop increase with the face velocity increase for two filter, but the pressure drop of B filter is less than that of A filter. This is consistent with the efficiency curve, the pressure drop data and SEM structure.

Figure 8 shows that the change of Residual pressure drop is with pulse jet in first 2300 cycles. The data is under the condition that Cleaning pressure is 1000pa, the face velocity is 2m/min, and the pulse jet pressure is 5 bar. The residual pressure drop is increase at first 2300 cycles. The residual pressure drop is less than 250 Pa, it is good for the peel performance of the filter membrane coated.

To evaluate the life of filter, the aging experiment also are conducted, the result is shown in figure 9 under the constant pressure pulse jet 10000 condition with dust- laden gas. There are different about two filter. The pressure drop has a decrease for A membrane the last fifteen jet, it might be that the pore become large result to the pressure drop small.

However, the pressure drop does not decrease, it demonstrates that the pore of B membrane hasn't changed basiclly, and performance is better for tolerating gas flush. So the life period is more long for B membrane, and the efficiency is stable.

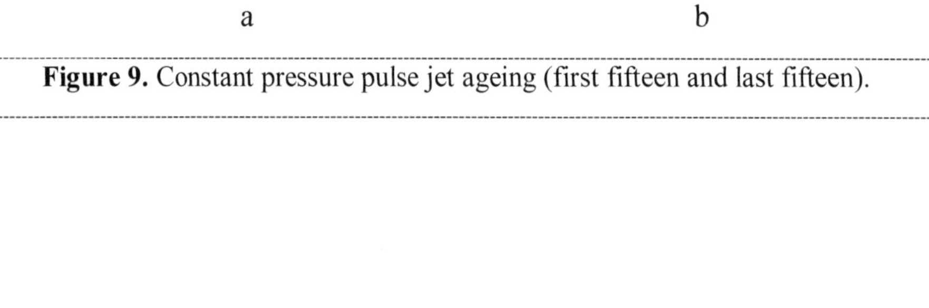

Figure 9. Constant pressure pulse jet ageing (first fifteen and last fifteen).

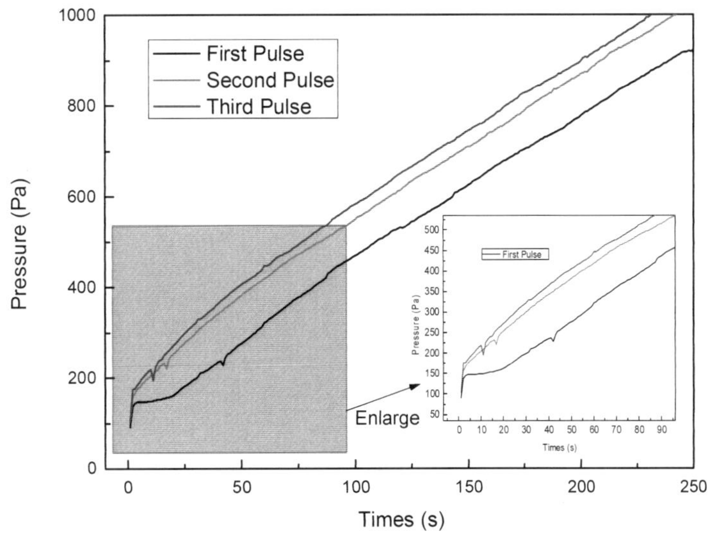

Figure 10. Pressure drop versus times.

The Curve of pressure drop excursion versus the filtering times is shown in figure 10 under first three pulse jet. As figure 10 shown, the pressure drop excursion curve versus specific aerosol loading is slow increase at initial short time span, following it is linear increase. This may be the transition from the dust cake form to dust accumulate. There are a distinct different among first three times pulse jet under 0.5bar constant pressure, this may be cause that the dust cake doesn't form for cleanable filter firstly used.

4. Conclusions

Several membrane microstructures are given out, and the collection efficiency of two e-PTFE membrane filter media may reach 99 percent for 1μm, and 98 percent for 0.5μm.

Filtration coated membrane is a high collection efficiency, and pressure drop increase with increasing face velocity.

B filter is more life period than A filter according to pressure drop experiment.

Acknowledgments

The authors would like to thank Hubei province natural science foundation directive project(2016CFA093) and Hubei Key Laboratory of Industrial Fume & Dust Pollution Control open project (HBIK2017-05).

References

[1] Hinds W C, Harrington M S and Pui D Y H 2007 *J. Nanopart Res.* **9** 117-1125.
[2] Spurny K R 1986 Sci. *Total Enviorn.* **52** 189-99.
[3] Ryan Mead-Hunter et al 2014 *Separat. Purificat. Tech.* **133** 484-506.
[4] Chen C Y 1955 *Chem. Rev.* **55**(3) 595-623.
[5] Wang P K 1983 *J. Atmosph. Sci.* **40** 1051-052.
[6] Lee K W and Liu B Y H 1982 *Aeros. Sci. Tech.* 1 147-61.
[7] Lee K W and Liu B Y H 1982 *Aeros. Sci. Tech.* 1 35-46.
[8] Friledlander S K 1958 *Ind. Eng. Chem.* **50**(8) 1161-4.
[9] Nakajima S et al 2015 *Aeros. Sci. Tech.* **49** 959-65.
[10] Mostafa S et al 2016 *Microsyst. Techmol.* **22** 965-77.
[11] Frederik E R 1980 *J. Air Pollut. Contr. Assoc.* **31**(5) 426-31.
[12] Liu J X et al 2011 *Sci. Total Envir.* **409** 4868-74.

[13] McCarthy J F 1989 Capture efficiency of Balston Type DS membrane filters for submicron aerosol particles. Technology reported of Parker Hannifin Corporation.
[14] Hasoli N and Park Y O et al 2013 *Power Tech.* **237** 24-31.
[15] Lalagiri M et al 2013 *Ind. Eng. Chem. Res.* **52** 16513-8.
[16] Shi L, Han S Y, Yu X M 2014 *Ind. Saf. Envir. Protect.* **11** 64-6.
[17] Brown R C 1993 *Air filtration.* London: Pergamon Prss.

IOP Conf. Series: Materials Science and Engineering **507** (2019) 012019 doi:10.1088/1757-899X/507/1/012019

Design of L1 Adaptive Control for PMSG Wind Turbine System

Duc Ha Vu[1, 2], Shoudao Huang[1, b], Thi Diep Tran[1, 2], Duy Vinh Nguyen[3, 4*]
[1]College of Electrical and Information Engineering, Hunan University, Hunan, P.R. China
[2]Faculty of Electrical Engineering, Sao Do University, Hai Duong, Vietnam
[3]Faculty of Vehicle and Energy Engineering, Phenikaa University, Yen Nghia, Ha-Dong District, Hanoi, Viet Nam
[4]Faculty of Mechanical Engineering and Mechatronics, Phenikaa University, Yen Nghia, Ha-Dong District, Hanoi, Viet Nam

Abstract. This paper focuses on the design of an L1 adaptive controller for variable speed wind turbines on a permanent magnet synchronous generator (PMSG) to maximize the wind power exploited in the presence of parameters uncertainties and wind speed turbulences. This control method based on the class of state feedback state for multiinput multi-output (MIMO) system can be applied to the current control and speed control of the system. The proposed controller successfully deals with the essential nonlinear problems of wind turbines and the effect of unknown parameters. Moreover, the L1 adaptive controller with fast adaptation is not only beneficial for both performance and robustness but also effective for the appearance of the time-delay state variables of wind turbine. Finally, with comparison to the indirect model reference adaptive control (MRAC), the performances of the system are verified by mean of several simulation results.

1. Introduction

The wind turbine is one of the fastest growing renewable energy technologies available today. The Global Wind Energy Council (GWEC) released its Global Wind Report: Annual Market Update today, showing a maturing industry successfully competing in the marketplace, even against heavily subsidized traditional power generation technologies. More than 52GW of clean, emissions-free wind power was added in 2017, bringing total installations to 539 GW globally. With new records set in Europe, India and in the offshore sector, annual markets will resume rapid growth after 2018 [1]. Among currently available wind energy conversion systems (WECSs), variable speed wind turbines are preferable to fixed speed ones because of their ability to maximize power extraction energy and increase efficiency [2]. In variable speed wind turbines, those based on permanent magnet synchronous generator (PMSG) offer several benefits in the wind power applications, such as their high-power density, high efficiency, better reliability and easy maintenance. These features make it an attractive choice in the wind turbine systems [3, 4]. It is widely known that the main control objective for the wind turbine in the low speed region is power efficiency maximization. To achieve this goal, the turbine tip speed ratio should be maintained at its optimum value despite the variation of wind speed.

For such pressing purpose, many different control strategies can be approached. First, proportional-integral (PI) based control schemes were proposed to deal with the problem of the maximum power point tracking [5, 6]. Li et al represented the control methods using the conventional and direct-current vector configuration to implement the maximum power extraction. It should be realized that PI control is a linear regulator which is really suitable for linear time invariant systems. However, the electrical and mechanical components in wind turbines work globally as nonlinear systems where electromechanical parameters vary considerably. Additionally, wind turbines are expected to operate effectively under a wide range of wind speeds, therefore the control designs get more difficult. To overcome this drawback, there are many different intelligent and nonlinear control strategies such as, gain scheduling [7], fuzzy logic control [8], neural network [9], feedback linearization [10] and backstepping [11]. Especially, methods based on sliding mode control (SMC) seem to be an interesting approach because they make plants more robust and guarantee high tracking of output signals following their reference in the presence of system uncertainties and disturbances. In this paper, L1 control strategy applied to the PMSGWECS has been proposed to guarantee the high tracking rotor speed of wind turbine following their reference to ensure the maximum power extraction in the presence of nonlinear uncertainties parameters and wind speed turbulences. Compared with the L1 controller in [11] represented by L1 adaptive output feedback for the single input and single output (SISO) system, L1 adaptive state feedback for multi input and multi output (MIMO), which applies to the speed control and the current control, has been presented to ensure the performance of the electromagnetic torque and active power. It reveals the good performance and robustness in the presence of high adaptation gain of L1 controller compared with the indirect MRAC in the simulation section.

2. Wind turbine system description

Wind energy is known to be first transformed into mechanical energy through the wind turbine blades and then into electrical energy through the PMSG. The PMSG is connected into the grid through two-level back to back converter. The electrical power is completely delivered to the grid by regulating the controller of the two-level back to back converter which consists of machine side converter (MSC), grid side converter (GSC) and DC link. The generator side control comprised of the current control loop and speed control loop is shown in figure 1.

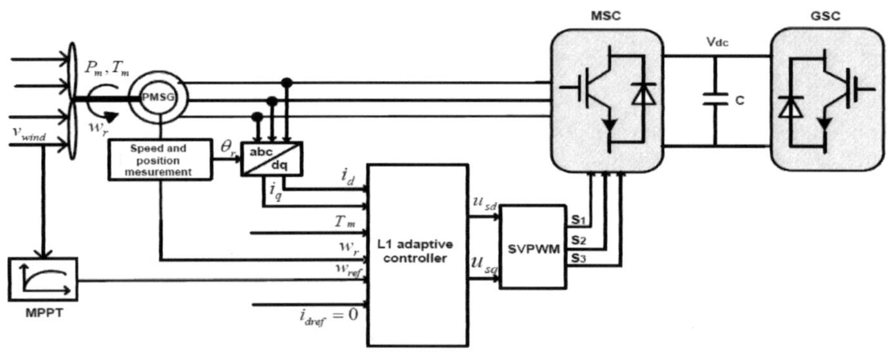

Figure 1. Control block diagram of the wind turbine system.

2.1. Wind turbine modeling

The aerodynamic power extracted from the wind is expressed [8].

$$P_m = 0.5\rho\pi R^2 C_p(\lambda,\beta)v^3 \tag{1}$$

where ρ is the air density, R is the wind turbine rotor radius, v is the wind speed, and $C_p(\lambda,\beta)$ is the turbine power coefficient. This coefficient can be calculated as follows:

$$C_p(\lambda,\beta) = 0.73\left(\frac{151}{\lambda_i} - 0.58\beta - 0.002\beta^{2.14} - 13.2\right)e^{\frac{-18.4}{\lambda_i}} \tag{2}$$

$$\lambda_i = \frac{1}{\dfrac{1}{\lambda + 0.002\beta} - \dfrac{0.003}{\beta^3 + 1}} \tag{3}$$

where β is the blade pitch angle, and λ is the tip speed ratio which is defined in [10].

2.2. PMSG modeling

Dynamic modeling of the PMSG can be described in (d, q) synchronous rotating reference frame. Thus, it is given by the following equations

$$\frac{dw_r}{dt} = \frac{T_m}{j_t} - \frac{k_t}{j_t} w_r - \frac{k_g}{j_t} i_{sq} \tag{4}$$

$$\frac{di_{sq}}{dt} = -\frac{R_s}{L_s} i_{sq} - p i_{sd} w_r - \frac{\psi_m}{L_s} \rho w_r + \frac{1}{L_s} u_{sq} \tag{5}$$

$$\frac{di_{sd}}{dt} = -\frac{R_s}{L_s} i_{sd} + p i_{sq} w_r + \frac{1}{L_s} u_{sd} \tag{6}$$

where T_m denotes the aerodynamic torque, i_{sd} and i_{sq} are d-axis, q-axis stator current, respectively; u_{sd} and u_{sq} are d-axis, q-axis stator voltage, respectively; R_s is the resistance of stator windings, L_s is the inductance of stator windings, ψ_r is the permanent magnetic flux, p is the number of pole pairs of the PMSG; $k_g = 1.5 p \psi_r$.

3. Simulation results

In order to verify the validity of the proposed L_1 adaptive control for the maximum power extraction of the wind turbine system, Matlab-simulink has been performed to illustrate the simulation results of the wind turbine system. The nominal values of the parameters employed in the simulations process. 50% uncertain parameters of wind turbine have been considered in this paper.

Time – delay state variables of wind turbine are defined as:

$$z_d(\mathrm{t}) = \begin{cases} z(\mathrm{t} - \tau) & \mathrm{t} \geq \tau \\ 0 & t < \tau \end{cases} \tag{7}$$

where τ is the time - delay margin.

In these simulations, the wind speed is assumed working under realistic condition which consists of the mean wind speed and its turbulence. The wind speed profile in this paper is divided into two scenarios based on the different turbulence intensity.

IOP Conf. Series: Materials Science and Engineering **507** (2019) 012019 doi:10.1088/1757-899X/507/1/012019

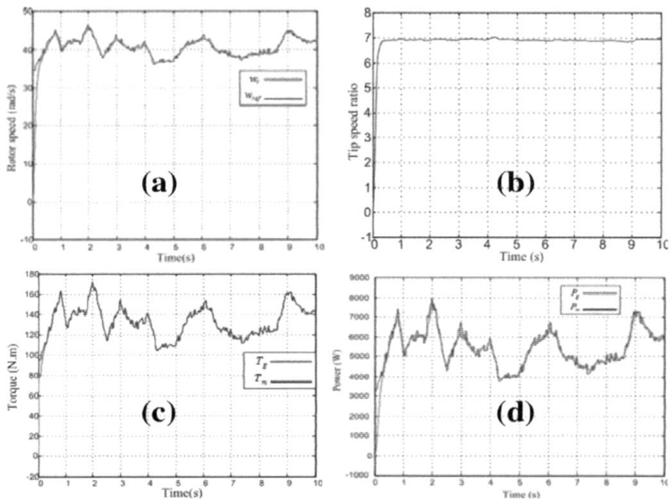

Figure 2. (a) Rotor speed and rotor speed reference; (b) Tip speed ratio; (c) Electromagnetic torque and aerodynamic torque; (d) Active power and mechanical power.

Figure 2 shows the tracking performance of actual variables tracking their reference variables in the scenario 1 by utilizing the L_1 adaptive controller. In response to the wind speed profile, the actual rotor speed W_r and its reference value w_{ref} are shown in figure 2a. It is clear that the rotor speed tracks more closely its reference value with good transient (the transient time 0.5 second and no overshoot phenomenon). Figure 2b shows that the uncertain tip speed ratio $\left(\lambda_{opt} = 6.91\right)$ with small oscillation. Figure 2c and 2d reveal that despite the wind speed turbulences and parameters uncertainties, the electromagnetic torque and the active power of the PMSG accurately follow their reference values (P_m is the maximum power of wind turbine, T_m denotes the aerodynamic torque). For the scenario 2, the turbulence intensity is much larger than it in the scenario 1. Figure 3 also shows the high tracking performances of rotor speed, electromagnetic torque and active power. It can be noted that the oscillations of wind turbine variables are bigger than them in the scenario 1, thus the tracking performances of this controller are influenced and become worse compared with scenario 1. However, these oscillations are still small and considered acceptable.

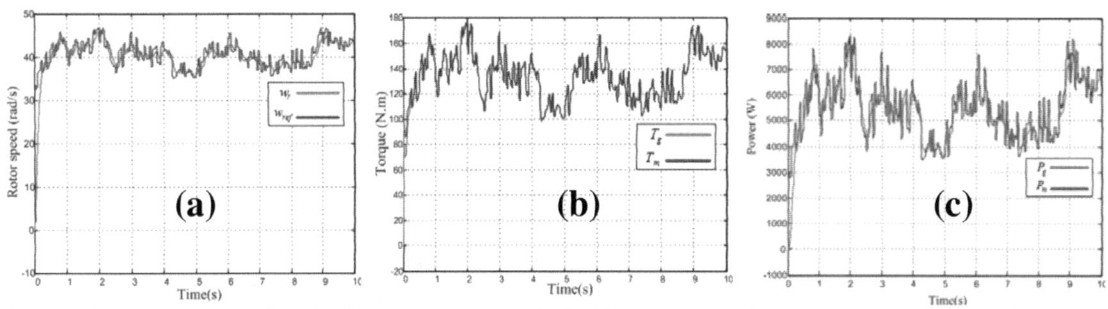

Figure 3. (a) Rotor speed and reference rotor speed; (b) Electromagnetic torque and aerodynamic torque; (c) Active power and mechanical power.

Figure 4 shows the expression of the tracking performance of the rotor speed regulated by L_1 adaptive controller and indirect MRAC controller under different step change of wind speed. Figure 4a (L_1 adaptive controller) reveals that increasing adaptive gain improves the tracking performance. When Γ is large value ($\Gamma \geq 10$), the distinction of their tracking performance is not remarkable. Moreover, the fast adaptive gain ($\Gamma \geq 10$), does not influence the robustness of the system. The improvement of system performance by increasing Γ is also exposed in figure 4b (indirect MRAC).

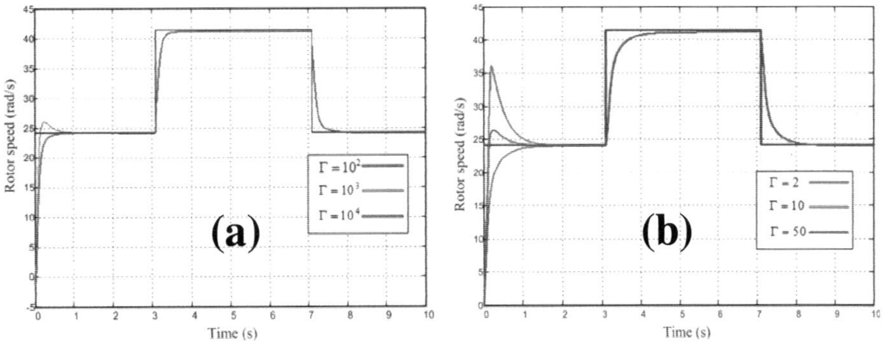

Figure 4. (a) Step respond of rotor speed (L_1 adaptive controller);
(b) Step respond of rotor speed (indirect MRAC controller).

Figure 5 illustrates the tracking performance of the rotor speed regulated by L_1 adaptive controller and indirect MRAC controller under different step change of wind speed with time-delay margin $\tau = 0.005s$. In figure 5a, it is clear that the tracking performance of the proposed control with time-delay state variables is unchanged compare with the proposed control described in figure 5a. Meanwhile, in figure 5b, with the small adaptive gain $\Gamma, \Gamma \leq 10$ the tracking performance of the indirect MRAC under time-delay margin $\tau = 0.005s$ is also unchanged compare with the performance presented in figure 4b. However, when high adaptive gain Γ the system controlled by the indirect MRAC controller may loses robustness. So, it can be seen that increase of the adaptive gain Γ affects the negative robustness of system. The loss of robustness happens regardless of the small adaptive gain $(\Gamma = 12)$ under appearance of delay-time state variables.

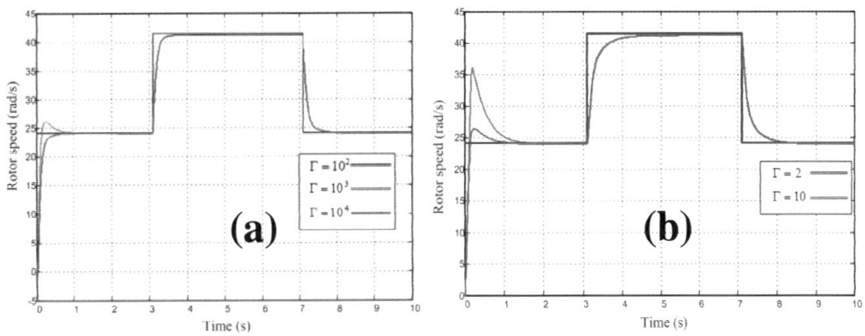

Figure 5. (a) Step respond of rotor speed (L 1 adaptive controller) with time-delay margin $\tau = 0.005s$;
(b) Step respond of rotor speed (indirect MRAC controller).

4. Conclusions

L_1 adaptive controller was proposed here for the PMSG wind energy conversion system to ensure the maximum power extraction from the wind. The control design using a low-pass filter in the definition of the control signal to appropriately limit the frequencies of the signal; therefore, it guarantees the robustness regardless of the fast adaptation. Consequently, the proposed controller showed a high tracking performance of those values such as the speed, the torque and the power toward their references and robustness with fast adaptation under the parameter uncertainties and wind turbine turbulences as shown in the detailed simulation section. Compared with the indirect MRAC, L_1 adaptive control is more robust while the indirect MRAC loses the robustness because of the fast adaptation. L_1 adaptive control also reached good performance and robustness under time-delay effects on the system.

References

[1] GWEC global wind energy council. (2018, February). Global wind statistics 2018. Available: http://www.gwec.net
[2] Wu Y L B, Zargari N, Kouro S 2011 *Power conversion and control of wind energy system,* vol. chap.1: John Wiley&Son.
[3] Chinchilla M, Arnaltes S, Burgos J C 2006 *IEEE Trans. Energy Conv.* **21(1)** 130-5.
[4] Polinder H, Pijl F F A v d, Vilder G d, et al. 2006 *IEEE Trans. Energy Conv.* **21(3)** 725-33.
[5] Li S, Haskew T A, Xu L 2010 *Electr. Power Syst. Res.* **80(3)** 328-38.
[6] Li S, Haskew T A, Swatloski R P et al. 2012 *IEEE Trans. Energy Conv.* **27(5)** 2325-37.
[7] Bianchi F D, Battista H N D, Mantz R J, 2007 *Wind turbine control systems: Principles, Modeling and Gain Scheduling Design,* Springer-Verlag.
[8] Muyeen S M, Al-Durra A 2013 *IEEE Syst. J.* **7(4)** 817-24.
[9] Cadenas E, Rivera W 2009 *Renew. Energy*, **34(1)** 274-8.
[10] Kim K, Jeung Y, Lee D et al. 2012 *IEEE Trans. Power Electr.* **27(5)** 2376-84.
[11] El Magri A, Giri F, Besançon G Et al. *Contr. Eng. Pract.* **21(4)** 530-43.

IOP Conf. Series: Materials Science and Engineering **507** (2019) 012020 doi:10.1088/1757-899X/507/1/012020

Design of the System for Automatic Grain Size Analysis at Ore Mining and Processing Enterprises

V N Kruglov[1], A V Kruglov[1, 2]

[1]Ural Federal University, Yekaterinburg, Russia
[2]Innopolis University, Innopolis, Russia

Abstract. The paper is devoted to the vision system for determining the grain-size distribution of finely crushed particles (~ 0.1mm in size) of bulk material moving on a conveyor belt. A block diagram of the system and data exchange protocol are presented. The commands for high-speed LED pulse backlight control are provided. The functional diagram and principle of the LED matrix control are described. The results of a full-scale testing of the system at the mining and processing plant are presented. As a result of the full-scale testing of the system at the mining and processing plant, the absolute error varies from 1.2% for the class "+ 4.0mm" to 5.2% for the class "-0.1mm". The absolute error for the target "+ 1.0mm" is 2.5%.

1. Introduction

Mineral ore is passed through several processing stages to obtain the desired mineral. At each of this stage, the crushing of a source material is performed. Beginning with blasting in the open-cast, the ore mass is comminuted until it becomes possible to concentrate it in one way or another in order to obtain the desired mineral.

For operational control of the mineral ore processing, devices with high reliability, performance, advanced functionality and extended accuracy are needed. The most all these requirements are met by devices based on the systems of machine vision [1-5]. An advantage of such systems is the non-contact way of the target objects measurement.

In machine vision systems, a radar [6], laser [7-9], or video camera are used to obtain the image of a bulk layer of a crushed or granulated material. Radar or laser allows you to form a 3D image, while a video camera gives a 2D image in common. However, in the analysis of the ore mass, the operative control of its geological composition is of great importance. This task can be solved with a video camera only by isolating and analyzing the color and texture features of ore particles [3,10].

By now, several systems for the operational grain-size analysis of crushed or granulated material have been developed and implemented. A brief description and comparative analysis of such systems as SPLIT, WIPFRAG, FRAGSCAN, CIAS, IPACS, TUCIPS is given in [5]. These systems use cameras with a minimum exposure time of 1/10,000 sec for the purpose of determining the grain-size composition of ore particles with dimensions of ~ 5 ÷ 7 mm. For the case of fine-crushed particles with a size of ~ 0.1 mm while the conveyor speed is about 1.0 m/sec and margin of sample error is 1%, the minimum exposure time of the video camera should be 100 times smaller. To fulfill such conditions, it is necessary either to use expensive high-speed video cameras or to control the intensifier pulse both by the duration and by the formation moment. This article is devoted to the design of the machine vision system that provides for the fulfillment of technical requirements for the obtaining of appropriate images of fine-crushed particles.

IOP Conf. Series: Materials Science and Engineering **507** (2019) 012020 doi:10.1088/1757-899X/507/1/012020

2. Structural configuration of the system

Structural scheme of the system is given in figure 1. The system consists of the following components:

- Adapter unit, which is connected to PC via COM port. Controller exchanges signals with adapter unit via symmetrical bus BUSU (twisted pair wire) using modified protocol m-bus.
- Switch unit CSU protect the power line from an overvoltage and the control line from a lighting discharge.
- Camera is powered by a constant voltage +12VDC from the controller card and exchanged control signals with controller using its own galvanically isolated interface.
- LED projectors are powered by voltage 50VDC from power supply unit. Signal exchange with controller is implemented using single-line asymmetrical bus BUSD.

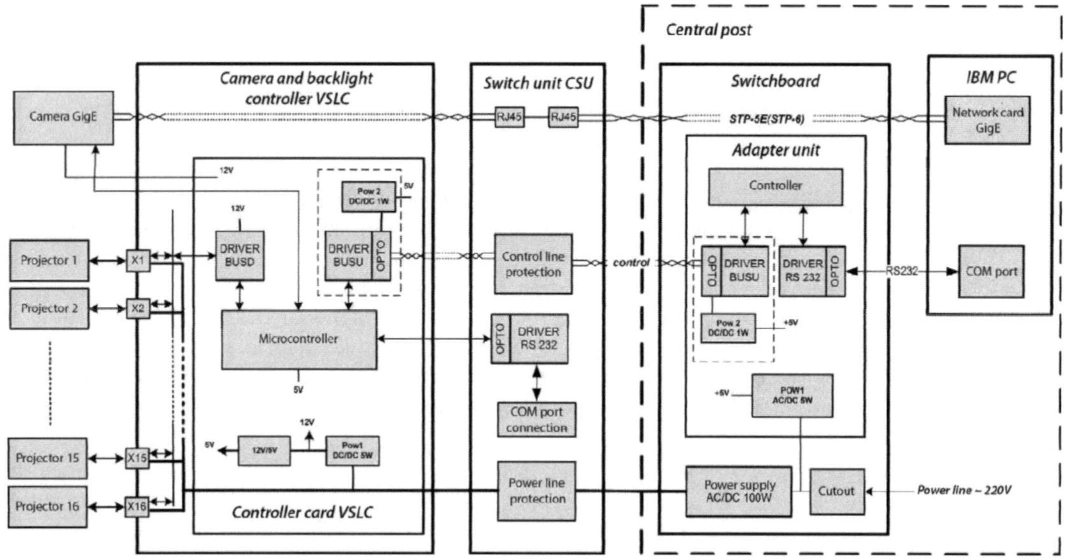

Figure 1. Structural configuration of the system.

3. Data exchange

As follows from the structural scheme in figure 1, the connection between the operator's workplace and the camera and backlight controller, as well as connected projectors, is made through a composite heterogeneous channel (figure 2).

Bottom-upwards: the common bus BUS_DOWN (BUS_D) that connects the LED projectors to the controller, bus BUS_UP (BUS_U) from the controller to the adapter unit, and the RS232 bus from the adapter unit to the PC.

At the transport level, all communications are carried out through a pass-through protocol such as modbus-ASCII.

The system devices exchange information packets with the source and destination addresses and contain commands and (optionally) data. In response to the received packet, any device in the system sends an acknowledgment (ticket) - ASK, which is a pulse of a certain duration for the BUS bus, and a short packet of two special symbols '$' and LF (0Ah) for the USART bus.

The system modules exchange data in the order shown in figure 2. The initiator of the exchange is the PC (control console program), all devices are only responding.

IOP Conf. Series: Materials Science and Engineering **507** (2019) 012020 doi:10.1088/1757-899X/507/1/012020

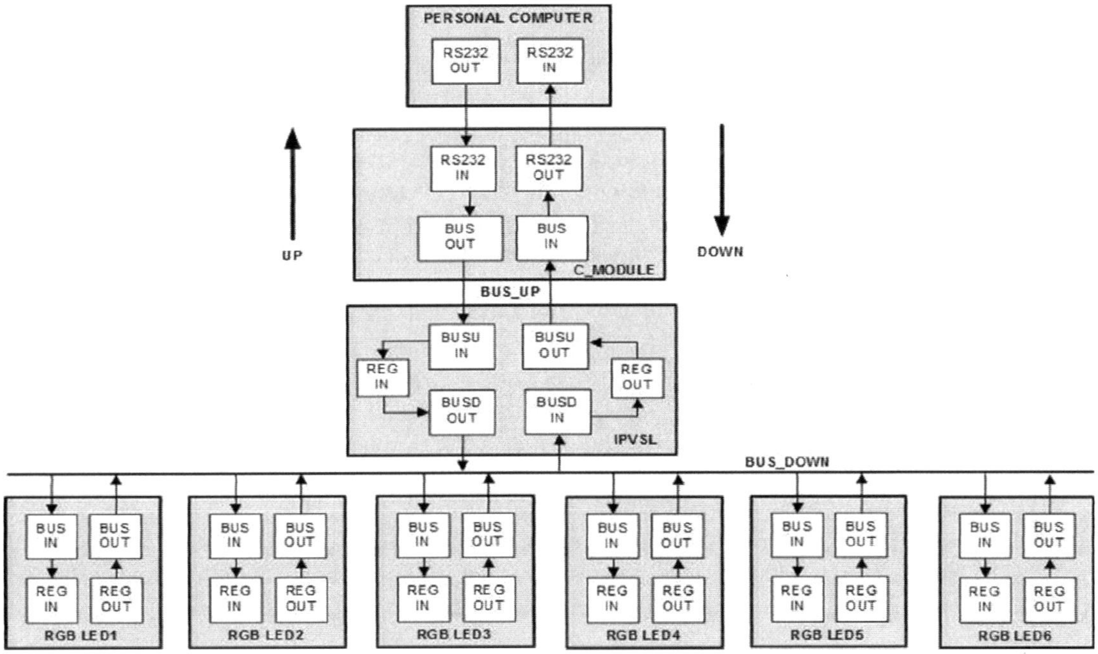

Figure 2. Data exchange in the system.

In the downward direction from the console, polling commands to the system units and control commands are received. In the upward direction, there are packets with the status data of the system units and responses to the console commands.

The camera and backlight controller accepts commands addressed to him and to projectors. Projectors' control commands are transmitted by the controller either to the specific projector or to a broadcast address, which allows them to be applied to all projectors at once. Upon polling command, the controller sends to the console address a packet with data on the state of the projectors and its own. The VSLC controller collects statistics on the state of the projectors by periodically generating polling commands via BUSD bus.

Table 1. HEX packet for the upper-layer protocol.

Byte num.	Data field num.	Data field name	System units				
			Operator console	Adapter unit	Receiver-corrector	Video sensor	Projector
0	1	Packet heading terminator	'$' (36h)				
1	2	Destination address	20h	40h	60h	80h	A0h – AFh, BFh
2	3	Source address	20h	40h	60h	80h	A0h – AFh
3	4	Command and data block code	0bCCCCKKKK, where C..C – command code, KKKK– data block code				
4 - 13	5	Data field	HEX - 10 bytes				
14	6	Hash total	XOR of bytes from 0 till 13				
15	7	Packet ending terminator	LF (0Ah)				

The structure of the information packet consisting of 7 fields is given in table 1.

The following fields are highlighted in the packet of 16 bytes: Destination address (1 byte), Source address (1 byte), Command and data block code (1 byte), Data field (10 bytes) Hash total. The package is framed by terminators: the heading and ending of the packet.

1. The device type (name) field (3 high-ordered bytes):
   - DEV = 1 – Operator console,
   - DEV = 2 – Adapter unit,

- DEV = 3 – Receiver –corrector (not used),
- DEV = 4 – Video sensor (camera and backlight controller),
- DEV = 5 – Projector.

2. Number field (5 low-ordered bytes) NUMB=0..31 – which is used for projectors only. Full addresses of projectors: A0h, A1h,... AFh. BFh – broadcasting address.

System management includes setting the operating mode of the projectors, the parameters of the intensifier pulse, the current of the LED array (through setting the array voltage).

The operating modes are controlled through the CONTRL_REG register (data block code -1). The matrix current is controlled via two-byte registers MX_CSET_REG, MX_VOLT_REG (data block code -2). The intensifier pulse parameters are controlled via 2 single-byte registers for pulse duration and delay WIDTH_PLS, DELAY_PLS (data block code -3). If the data block code is 0, then all control registers are transmitted. The status registers store information about the current values of the voltages and currents of the units, their statuses, as well as the version number, etc.

The command is transmitted in the third byte of the packet (the highest tetrad), coded by a nibble, so their maximum number is 15.

1. Device polling command - LIVE. Code, 3h. Multifunctional command:
   - It is periodically sent to the address of the controller, which, in turn, sends this command sequentially to the addresses of the projectors for the purpose of monitoring the console operability. The packet with the measured parameters should be sent back to the source address.
   - It is generated by the controller in a cycle of projectors polling.
2. Register bank writing command - REGS_WR, Code - 4h. This command is modified, the heading of the block and the number of bytes are excluded from the command, instead of them the code of the register group (the lowest tetrad of field 4) is used. The data for writing to the controllers' registers are transferred starting from the zero byte of the data field, no more than 10 bytes can be written per one command. The response packet is a copy initial one with inverted source and destination addresses.
3. Register bank reading command - REGS_RD, Code - 5h. The command is always sent by the master to the slave of any level. This command is modified, the heading of the block and the number of bytes are excluded from the command, instead of them the code of the register group (the lowest tetrad of field 4) is used. In the response packet, the slave device transmits the values of all control registers starting with a zero byte (you can read no more than 10 bytes per command).
4. Program version request command - VERSION, Code - 6h. The command is sent by the master to the slave device of any level. In the response packet, the version number of the software in the Ver.Rel format is transmitted in the data field, where Ver is the version number (the first byte of the data field), Rel is the release number (the second byte of the data field). (The version number is also sent in the response packet of the LIVE command).
5. Packet transfer error command - TXD_ERR, Code - 7h. The command is intended for the console and can be transferred by any device of the system.
6. Initialization mode switching command - SYS_INIT Code - 8h. The content of the zero byte of the data block defines the function of the SYS_INIT command: Byte 0 = RESET_INIT (0xC) is a system reset.

4. Design of the video sensor

The device integrates a video camera in its own protective case, a VSLC controller and 16 projectors.

The video sensor design (figure 3) is a tapered frame formed by two horizontal panels (top cover and mounting plate) and six inclined ribs. The lower part of the frame, for additional rigidity, is connected by spacer tubes. Removable protective cover plates (not shown in the figure) are mounted on the fins to prevent splashes and dust from getting inside the sensor.

The internal volume of the sensor is divided vertically into two parts: the upper, protected by the IP44 class, in which the video camera and the controller are located, and the lower - the open one, in which the projectors of the backlight are arranged in two circles of 8 items each.

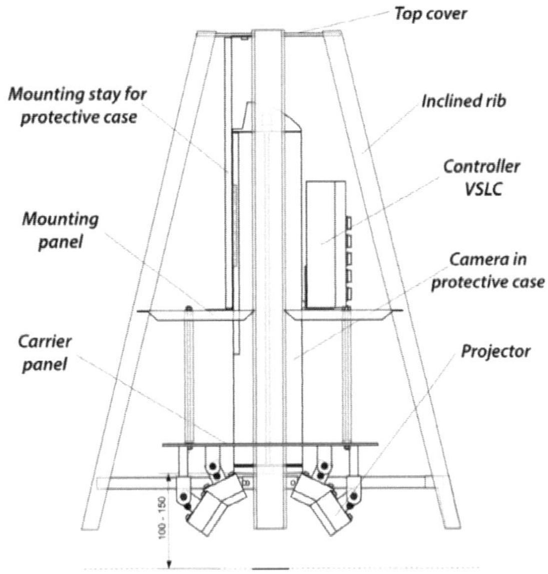

Figure 3. Video sensor.

The position of the carrier panel in vertical is selected by the criterion of maximum approach to the illuminated video object, in which the mechanical safety of the lower lamps is still performed. The supporting panel of the projectors 4 is fixed with vertically threaded ties that run inside the aluminum tubes. The upper ends of the couplers are fixed with nuts to the frame mounting plate.

Projectors are electrically connected to the controller by cables (not shown in the figure 3), which are passed into special holes made in the mounting panel. The camera is also connected to the controller with a power and control cable, through the connector on the casing.

Table 2. The results of the industrial tests.

№	Data source	Grain-size class output, %					
		+5.0	+4.0	+1.0	+0.5	+0.1	-0.1
1	Laboratory	0.8	1.0	46.5	15.4	10.5	25.8
	System's software	0	0	46	17	16	21
	Absolute error	0.8	1	0.5	1.6	5.5	4.8
2	Laboratory	1.1	1.7	46.3	15.6	14.3	21.0
	System's software	0	2	42.0	18	16	22.0
	Absolute error	1.1	0.3	4.3	2.4	1.7	1
3	Laboratory	1.5	1.6	45.6	14.9	12.4	24.0
	System's software	0	2	45	17	15	21
	Absolute error	1.5	0.4	0.6	2.1	2.6	3
4	Laboratory	20.5	4.1	37.7	14.1	18.1	5.5
	System's software	18	2	33	15	14	18
	Absolute error	2.5	2.1	4.7	0.9	4.1	12.5
5	Laboratory	17.7	4.3	37.5	14.5	18.4	7.6
	System's software	19	2	40	14	12	13
	Absolute error	1.3	2.3	2.5	0.5	6.4	5.4
Average error		**1,4**	**1,2**	**2,5**	**1,5**	**4,1**	**5,3**

IOP Conf. Series: Materials Science and Engineering **507** (2019) 012020 doi:10.1088/1757-899X/507/1/012020

5. Conclusion

The described system is developed and passed industrial tests at the enterprise of JSC "EVRAZ-KGOK". According to the program of industrial tests, the laboratory of product quality management carried out the sampling of fine-crushed particles for the purpose of grain-size distribution analysis for the classes: "-0.1mm", "+ 0.1mm", "+ 0.5mm", "+ 1.0mm", "4.0mm" and "+ 5.0mm". At the same time, the same parameters were recorded according to the output of the system software. The results of experiments on the evaluation of the grain-size distribution of fine-crushed particles are presented in Table 2.

The obtained results show that the developed system accurately determines the grain-size distribution of the fine-crushed material. The increasing of the error for the classes "-0.1mm" and "+ 0.1mm" is due to the fact that the controlled material was moist. In this state, small particles either stick together with each other and form larger pieces, or stick to large pieces, which reducing their quantity. However, according to experts of "EVRAZ-KGOK", in the production of sinter the class "+1.0 mm" is of the most interest. The average absolute error in determining this class was only 2.5%.

Also, the hardware failures were not detected during the industrial tests of the system. Thus, the proposed design of the hardware part of the system showed its high efficiency and reliability in the mining and processing industry.

References

[1] Mkwelo S, De Jager G, Nicolls F. 2003 *Watershed-based segmentation of rock scenes and proximity-based classification of watershed regions under uncontrolled lighting conditions,* Proceedings of the 14th Annual Symposium PRASA, 2003 Langebaan, South Africa, 107-112.
[2] Tessier J, Duchesne C, Bartolacci G 2007 *Min. Eng.* **20(12)** 1129-44.
[3] Perez C A, et al. 2015 *Int. J. Min. Proc.* **144** 56-64.
[4] Chatterjee S 2013 *Appl. Intell.* **39(1)** 14–27.
[5] Thurley M J, Ng K C 2005 *J. Comp. Vision Image Understanding,* **98/2** 239-70.
[6] Benter A 2014 *Characterising the Internal Composition of Rock Piles Using Ground Penetrating Radar,* Australia: Charles Sturt University. Doctoral/Master's Thesis, pp. 139.
[7] Thurley M, Andersson T 2008 *Min. Eng.* **21** 405-15.
[8] Thurley M J, Ng K C 2008 *Comp. Vision Image Understanding,* **111(2)** 170–8.
[9] Thurley M 2006 *On-line 3D surface measurement of iron ore green pellets*, International Conference on Computational Intelligence for Modelling, Control and Automation: CIMCA 2006; Sydney, 2006 IEEE. pp.229-35.
[10] Perez C A et al. 2011 *Int. J. Min. Proc.* **101(1–4)** 28-36.

IOP Conf. Series: Materials Science and Engineering **507** (2019) 012021 doi:10.1088/1757-899X/507/1/012021

LED Projector of Pulse Backlight for Machine Vision System

V.N. Kruglov[1], A.V. Kruglov[1,2], A.L.Karev[1], A. G. Gruh[3], I. Khurelchuluun[4]

[1]Ural Federal University, Yekaterinburg, Russia
[2]Innopolis University, Innopolis, Russia
[3]AO «EVRAZ KGOK», Kachkanar, Russia
[4]Erdenet Mining Corporation, Mongolia

Abstract. The paper presents technical description of pulse LED backlight. This system is implemented in machine vision systems designed for high-speed processes control. Pulse projectors managed on the basis of starting time and duration of the pulse. The system provides control of electrical parameters of projectors, such as input voltage and current of the LED array, which are measured during backlight pulse operate time. This information gives an indirect measurement of the LED arrays' state. The design of the system and implementation example are given.

1. Introduction

At present, machine vision systems are widely used for operational control of technological processes. An important part of such systems is the artificial illumination of the scene. Solution to this problem is given a lot of attention, according to the literary sources [1-7]. Correctly designed backlight provides more contrast image of the controlled objects. This leads to a significant reduction in computational operations in the construction of object contours and, subsequently, effective segmentation of the image [8].

Modern technologies make possible development of the high-performance backlight projectors based on LED. Such projectors are used by WipWare in machine vision systems Momentum and Solo, which are designed to control the grain-size distribution of crushed material moving on a conveyor belt [9]. Similar projectors are used in the VisioRock system manufactured by Metso Cisa [10]. These systems used LED projectors in a continuous mode. However, when tracking the high-speed processes, such as the motion of ore on the conveyor, a video camera operates in electronic shutter mode. In this mode, the video frame forming time ($\sim 0.1 \div 1$ msec) is significantly shorter than the time between the formations of consecutive video frames: at a shooting frequency of 60 frames per second, the interval between frames is 16.6 msec. Thus, there is no need for continuous operation of projectors in this mode. Synchronizing the video frame forming time with the pulse starting time can significantly increase the light flux. For example, according to the technical documentation for LEDs LEMWA33X70JX [11], it is possible to form a light flux almost 3 times greater in the pulsed mode than in the continuous one. Moreover, there are high chances that the operation of LEDs in pulse mode will increase their operation time. This article is devoted to the description of a LED projector operating in a pulsed mode.

2. LED projector of pulse backlight

LED projector is a dual-board design in a sealed case (IP65) with a glass-covered lid. The boards are stacked on each other and are interconnected by two single-row PLS (PBS) connectors. On the top board the LED array of 10 LEDs LEMWA33X70JX, with monitoring circuits of the current array, is unsoldered. LEDs are equipped with focusing lenses LA-11, which increases the light flux by concentrating it on the object. On the lower board the project control circuit is mounted.

LED projector ready fitted with mounting bracket provides possibility to change the angle of the projector case in two directions. Through the hermetic cable entry, the projector is connected to a camera and backlight controller (VSLC), which provides power and control.

2.1. Functional circuit of the LED projector

The functional circuit of LED projector (figure 1) consists of the following functional units: input voltage sensor, controlled voltage regulator, LED array, backlight pulse key, current array control circuit, control microcontroller, bus driver and power supply regulator.

The LED array of the projector should be switched on for a certain time interval synchronously with the exposure moment of the video camera, which is provided by the microcontroller by the pulse on the BUSD bus. BUSD bus receives the control commands for the backlight pulse parameters and the service commands from the controller. In response, the information about the input voltage and the current of the array, measured during the backlight pulse operate time, is transmitted from the projectors. This information indirectly characterizes the state of the LED arrays.

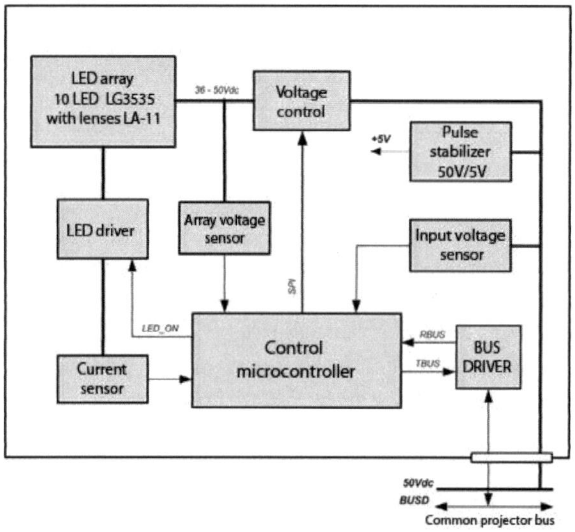

Figure 1. The functional circuit of LED projector.

The control of the luminous intensity of the LEDs is not carried out, since the maximum possible luminous flux is required from the projectors. A current sensor is used to monitor the channel current.

2.2. Forming of the backlight pulse

The LED projectors bus is used both for information exchange and for synchronization of backlight processes, i.e., transmission of sync pulses.

Figure 2 illustrates the relationship of cyclic processes on BUSD bus. At the beginning of the cycle, the VSLC controller generates a 4 ms pause, keeping the bus at zero level. Then SYNC_PULSE sync pulse of the video camera exposure with a duration of 20 μs is formed at the output of SYNC. After that the FLASH_DELAY pause for the ending of the transient processes of the video camera is started. At the end of the Flash Delay pause, BUSD bus is set to a high level for EXPOS_TIME exposure time, and a low level after its completion, thereby creating a trigger pulse with the duration of EXPOS_TIME.

Projector controllers on the rising edge of the trigger pulse include the keys of their arrays, thus triggering a work flash (the key is turned off after the FLASHWIDTH pulse expires).

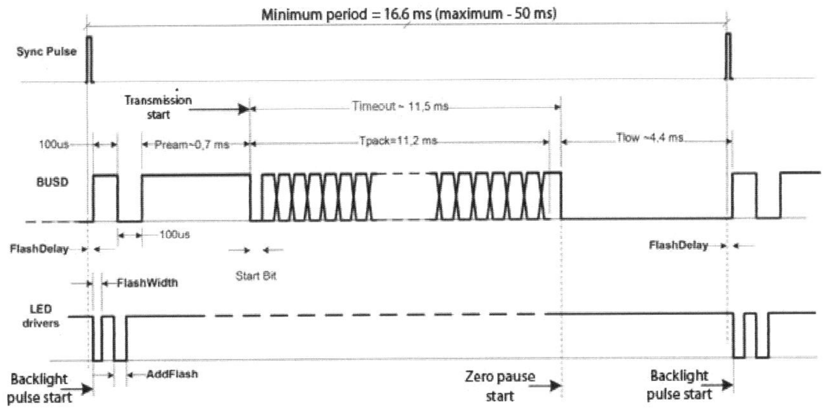

Figure 2. Signal envelops, which illustrate the interconnection of processes on the BUSD bus.

By the decay of the pulse on BUSD bus, the projector controllers turn on the array again to form a 20 μs ADDFLASH measurement pulse. Only the current is measured during the pulse. The input voltage and voltage of the LED array are measured when the array is turned off.

Then the logic-1 level pause before the exchange is established and maintained for 700 μs. In case if the VSLC controller does not start transmitting a packet for projectors during this time (time instant "Start transfer"), then after another 50 μs the receiving of the response packet can start.

In any case, the timeout starts for 11.5 ms, which is reliably more than transmit-receive time for the data packet. At the end of the timeout any exchange is terminated, and on the bus a zero level is generated for the time remaining until the end of the cycle (16.6 ~ 4.4 ms).

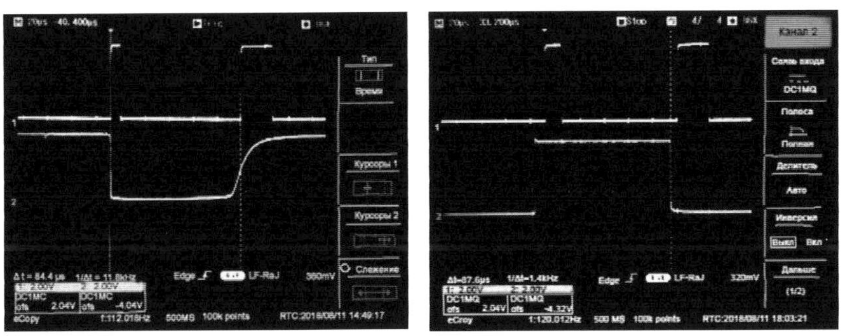

Figure 3. Oscillograms of the formation of backlight pulses. The upper (yellow) indicator shows the backlight pulses (left pulse is working, 7 μs, right one is measuring, 20 μs); the lower (blue) indicator shows FLASH pulse from the video camera on the left oscillogram and a trigger pulse in the BUSD channel on the right one.

The oscillograms in figure 3 should be interpreted as follows. Synchronization of the oscillograph is carried out by an external channel from the front of the video camera trigger pulse (a pulse of 20 μs, which image is absent in the oscillograms). At the end of this pulse, a delay is triggered, which takes into account all the transient processes in the video camera, and after that the exposure is started. This time point on the oscillogram corresponds to the Flash Window pulse droop. If the beginning of the working pulse is coincided with the specified moment or located to the right of it on the time axis, then the target object will be highlighted with maximum efficiency.

The working backlight pulse is formed by all projectors on the rising edge of the trigger pulse on the bus, while the measuring pulse – by its the droop (right oscillogram at figure 3), which is eliminated an ambient light. The backlight pulses are formed with a delay about 8 μs from the front and the decay of the trigger pulse on BUSD bus.

The width of the trigger pulse on the bus is 84 μs, which corresponds to the set exposure time of the video camera - 80 μs, plus the program response time to the timer interruption. On the left oscillogram (figure 3) you can see that the camera "removes" the Flash Window signal also after 80 μs.

2.3. BUSD bus. Data exchange with LED projectors

The data exchange between the controller and the LED projectors includes the transfer of commands from the controller, either broadcast or an individual address, and the transmission of data from each projector about its current state. It is intended to use up to 16 LED projectors in the backlight system. While exchange, the following commands can be used:

- survey "LIVE" (individual address);
- request for version "VERSION" (individual address);
- write registers "REGS_WR" (broadcast address);

2.3.1. Link level. At the link layer, the exchange protocol works as follows. During the exchange the controller can generate either address packets for each of the 16 projectors, which should receive answers in the next cycle, or broadcast packets, which require no response.

Figure 4 shows that while exchange each transmission cycle follows by the reception cycle over the BUSD bus, and if a broadcast packet is transmitted, then the transmission from the projectors is absent in the next cycle.

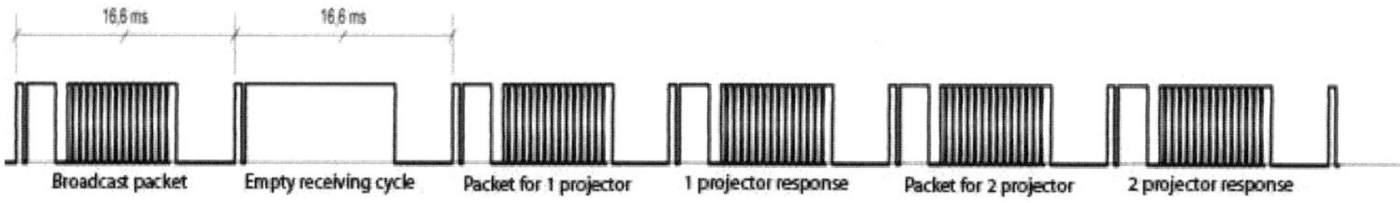

Figure 4. Exchange envelopes over the BUSD bus.

2.3.2. Physical level. Exchange over the bus at the physical level is similar to the physics of the USART protocol (RS-232). An 8-bit packet has start and stop bits, so the total number of bits per byte is 10, and the total number of bits per packet is 10·16=160. Due to this circumstance, the bit interval is reduced to 70 μs.

The transmission time of a 16-bytes packet with a duration of a bit interval of 70 μs, will be $70 \cdot 10 \cdot 16 = 11.2$ ms. For the minimum repetition period of the sync pulses equal to 16.6 ms (the shooting speed is 60 fps) and the packet preamble time is greater / equal to 10 bit intervals – 0.7 ms, it is remains about $16.6 - 0.2 - 11.5 = 4.9$ ms for pauses. The preamble before the exchange is selected equal to 0.7 ms, the pause before the sync pulse is 4.2 ms.

2.4. Assigning addresses to LED projectors

When manufacturing LED projectors, each device is assigned an individual number by means of jumpers. This number is indicated both on the board and on the case of the projector. During an installation the LED projectors can be placed in any order.

3. Local console

The VSLC controller provides the local service management using any laptop equipped with a COM port. For this purpose, the controller includes an optically isolated COM port driver.

Using a pre-installed terminal emulator program (for example, PCommLite by Moxa Inc), you can control the switching on of LED projectors, change the contents of the control registers, view the contents of the current and voltage measurement registers of both the controller and projectors.

To work with the local console, it is necessary to remove the upper cover of the switching box body and connect the RS232 cable to the COM port connector (DB9), run the console program on the computer and turn on the controller power. The information about controller registers will be displayed in the window of the console program. Using the command line, it is possible to perform certain actions with the controller.

The program supports 2 screens: the main screen and the help one, which are shown in figure 5. The help screen displays the syntax of the commands. To enter the help screen, input the command *help*, to return to the main screen - *work*.

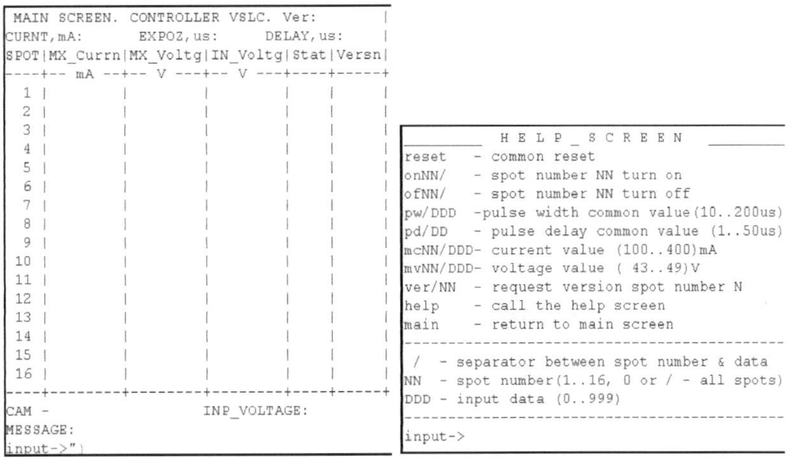

a) b)

Figure 5. Local console program: a) main screen, b) help screen.

The main screen displays the following information (top to bottom, left to right):

- Screen name. Name of the program. Software version (Ver)
- General settings of the backlight pulse parameters: current (CURNT), duration (EXPOZ), delay (DELAY).
- Projector parameters in a table of 6 columns and 16 lines: number (SPOT), array current (MX_Currn), array voltage (MX_Voltg), input voltage (IN_Voltg), projector status (Stat - ON/OFF) and software version (Versn).
- Video camera status (CAM).
- The input voltage of the controller (INP_VOLTAGE).
- Message line of the program (MESSAGE).
- Command input string (input ->).

The following commands are available to the user for input.

- General reset command - *reset.*
- Commands for switching on (*onNN/*) and off (*ofNN/*) of the projector with the *NN* number. If the number is not specified or equal to 0, then the command is valid for all projectors.
- Pulse width setting command (*pw/DDD*), where *DDD* is a number from 10 to 200 μs.
- Pulse delay set command (*pd/DD*), where *DD* is a number from 1 to 50 μs.
- Current setting command (*mcNN/DDD*) with current value *DDD* of the projector with number *NN*. If the number is omitted or equal to 0, then all projectors are executed.
- Voltage setting command (*mvNN/DDD*) with voltage value *DDD* of projector with number *NN*. If the number is omitted or equal to 0, then all projectors are executed.
- Projector software version request (*ver/NN*). Here the indication of the number is required.

The values of the measured parameters of the projectors: the current and voltage of the array, as well as the input voltage of the projector are displayed sequentially in the table.

4. Conclusion

The article describes the control system of LED backlight. The backlight consists of projectors and controlled by a microcontroller. The projector receives the parameters of the backlight pulse (beginning and duration of the pulse) and the service commands. In response, information about the input voltage and matrix current is transmitted from the projector to the controller. Such a system

IOP Conf. Series: Materials Science and Engineering **507** (2019) 012021 doi:10.1088/1757-899X/507/1/012021

provides generation of light pulses ranging in duration from a few microseconds to 1 ms. Feedback from LED projectors gives a control of their performance. In this system, there is no need to control the temperature of the LEDs. With an increase in the luminous flux (~ 2-3 times) and, consequently, the current of the LEDs, the time between adjacent pulses is large enough to cool the LEDs by natural means.

The described system was implemented in software and hardware complex, designed to assess the particle size distribution of pieces of crushed ore, moving on a conveyor belt. The complex is installed in the large-scale crushing workshop at EVRAZ-KGOK. LED backlighting ensures the formation of high-contrast images of pieces of crushed ore in conditions of high dust content of the workshop. For all the time of operation, the system demonstrates high reliability and excellent performance [12, 13].

References

[1] Guzhov V, Ilinykh S, Vagizov A, Kuznetsov R 2011 *New principle of the shaping the nonline illumination in optical measuring systems*, The 6th International Forum on Strategic Technology, Novosibirsk, Russia, pp. 652-4.
[2] Guan C, Hassebrook L, Lau D 2003 *Opt. Expr.* **11(5)** 406–17.
[3] Chen F, Brown G, Song M 2000 *Opt. Eng.* **39** 10–22.
[4] Mertz C, Koppal S, Sia S, Narasimhan S 2012 *A low-power structured light sensor for outdoor scene reconstruction and dominant material identification*, Proc. 9th IEEE International Workshop on ProjectorCamera Systems.
[5] Wang K, Ren G, ZhuFeng S 2012 *Adv. Mater. Res.* **411** 203-7.
[6] Zuo C, Chen Q, Gu G, Feng S, Feng F, 2012 *Opt. Expr.* **20(17)** 2012.
[7] Salvi J, Fernandez S, Pribanic T, Llado X, *Patt. Recognit.* **43(8)** 2666-80.
[8] Koh T K, Miles N J, Morgan S P, Hayes-Gill B R 2009 *Min. Eng.* **22** 537–43.
[9] www.wipware.com.
[10] www.metso-cisa.com
[11] led.lginnotek.com/front/product/productView.do?seq=198
[12] Krouglov V, 2015 *Industrial Tests of Programmed Equipment Complex "Granics" Performed in The Facilities of Erdenet Mining Corporation*, Mongolia, Proceedings of the 2015 International Conference on Electrical, Automation And Mechanical Engineering (EAME 2015), **13** 542-4.
[13] Kruglov V, Gruh A, Kruglov A, 2017 *System GRANICS for particle-size distribution determination by photoanalysis*, 4th International Young Scientists Conference on Information Technologies, Telecommunications and Control Systems. CEUR Workshop Proceedings, **2035** 96-101.

IOP Conf. Series: Materials Science and Engineering **507** (2019) 012022 doi:10.1088/1757-899X/507/1/012022

Non-identical shaped vortex generators with triangular and cylindrical fins to control in the heat transfer

Suresh Kumar Muthusamy[1], N.Sankar Ganesh[2]

[1]Lecturer, Madha Institute of Engineering & Technology, Sadhanathapuram, Thandalam Post, Chennai – 600 122.
[2]Research Engineer, Skill TeQ, Adambakkam, Chennai – 600 088.

Abstract. In this paper, a novel heat transfer enhancement mechanism is deployed, which improves the performance of heat transfer using longitudinal vortices generator in weak regions. Instead of plain fine, the fins embedded with vortex generator are used to reduce the fin area. This mechanism generates less drop in pressure with increasing heat transfer fluid flow. Further, this research aims to study the heat transfer over three row plate fin. Further, tube heat exchangers without and without vortex generators of pair different shape mounted behind a tube is studied through computational fluid dynamics model in a wake re-circulation zone. The proposed mechanism is tested through numerical simulation with different span angles for triangular and cylindrical shaped vortex generator for Reynolds number ranging from 400 to 3000.

1. Introduction

The increasing prices of energy in industries started to focus on cost effective solutions to reduce the cost of energy using efficient designs. The cost effective solution is a better way to trade-off between sustainability and efficiency using the upgraded systems. One such way is the intelligent system design with the existing system, which helps to improve the energy utilization. Such miniaturized devices improve the usage of energy with reduced system weight, cost and volume. In order to attain optimal rate of efficiency, the heat in system should be controlled with optimal cooling rate. However, if the system gets too much heated up, performance of the system gets affected. Hence, it is very essential to consider reduced size components in a system that should support well effective cooling rate in the system. The demand on these systems mainly aims to reduce the overall size of cooling equipment with optimal rate of heat transfer.

Some of the earlier works includes various shapes of vortex generators to improve the rate of heat transfer, this includes: rectangular [1, 4, 6], rectangular trapezius [1], angular rectangular [1], wishbone [1], intended [1], wavy [1], delta winglet [2], rectangular winglet [2], spanwise-varying passive [3], lateral sweep [5], half-cylinder [7], flapping [8], low-profile [9] and quadrilateral [10]. These studies provide necessary information about enhancing the heat transfer rate in vortex generation through slight modifications in its overall geometry. This work aims at modifying the shape of vortex generator cross section through convex and concave textures. Here, two or more textures are considered in trailing and leading faces of vortex generators. The domain is placed with vortex generator over heated plate and this is studied using computational fluid dynamics. The volume of the vortex generator is made constant. The major parameters considered for the proposed study involves the position, span angles and size of vortex generator, which is not considered in conventional literatures. In spite of this, the present study aims at the evaluation of modification in thermal effects due to span angle effects.

The present study considers three span angles α = 30°, 45° and 60° for various vortex generation geometries like aerofoil, elliptical and semi-circular. The investigations are carried out by setting the range of Reynolds number between 400 and 3000. A three row tube is used for performing the numerical simulation and the turbulence effect is simulated using CFD. Further, the conjugated convective heat transfers in the flow field and heat conduction in the fins are also considered. This paper mainly aims at maintaining reduced pressure drop to attain high rate of heat transfer using aerofoil, elliptical and semi-circular shaped vortex generator.

2. Materials and methods

To tackle the fluid flow problem in vortex generator, the structuring of CFD codes is carried out around numerical algorithm. The codes in CFD include its main elements like solver, post and pre-processor. These codes aim to reduce the time for solving the problem using its urbane user interfaces to its parameters for determining the results.

3. Geometry and formulations

The characteristic of heat transfer over the surface is influenced by the configuration of extended surface. This study focuses on surface pressure when the heat is transferred from plate to fluid. The vortex generator produces different pressured area zones, where there is a drop in pressure flow and it induces circular flow along the direction of flow to form vortices.

This study involves the analysis of thermal characteristics of vortex generation with various shapes like convex and concave surface with trailing and leading surface in vortex generator. An energy equation is used in heat flow of vortex generator, where thermal radiation is eliminated. Here, the convection dominates the heat transfer but this is neglected easily. The movement of thermal energy is prompted by thermal potential difference in vortex generator. This is realised in terms of temperature difference, which helps to study the thermal behaviour and temperature distribution by solving the energy equation. By eliminating the volumetric heat generation rate, the energy equation in coordinate independent form [29] is given by

$$\frac{\partial T}{\partial t}+\left(\vec{V}\cdot\nabla\right)T=k\nabla^2 T \tag{1}$$

Where, t is the time of vortex generator medium, k is the thermal conductivity of vortex generator medium, T is the temperature of vortex generator medium and is the velocity vector, which depends on coordinate axes. From the above equation, the computation of temperature involves the velocity vector and this involves momentum and mass for its perfect estimation, which is given respectively in following equation.

$$\nabla\cdot\vec{V}=0 \tag{2}$$

$$\frac{\partial \vec{V}}{\partial t}+\left(\vec{V}\cdot\nabla\right)\vec{V}=\frac{\nabla P}{\rho}+v\nabla^2\vec{V} \tag{3}$$

where, ρ is density, v is kinematic viscosity and P is pressure.

A set of boundary condition is used for solving these equations [9]. This is used to validate the present work and the main aim is to find the thermal structure of vortex generator. To solve the thermal convection, heat transfer co-efficient (h) and Nusselt number (Nu) are considered. The relative strength of convection to conduction is expressed as follows,

$$Nu=\frac{hD_w}{k} \tag{4}$$

where, Dw defines the length and it is estimated as volumetric ratio of computational domain to the heat transfer area and it is set as 0.04 m. k is the fluid thermal conductivity and the heat transfer coefficient (h) is given by,

$$h = \frac{q''}{T_w - T_m} \quad (5)$$

where, q″ defines the heat flux, Tw is the temperature of wall and Tm is the temperature of mean inlet.

Finally, the Reynolds number is used to find the water's flow velocity, which is given by,

$$\mathrm{Re} = \frac{\rho u D_w}{\mu} \quad (6)$$

where, μ is the fluid's dynamic viscosity and u is the fluid velocity.

4. Physical model

The vortex generator performance is estimated based on its size, position and span angles. The present study aims at evaluation of effects of span angles on thermal characteristics. Here, the Reynolds number ranging from 400 to 3000 is used to investigate the three span angles, namely 300, 450and 600. The numerical simulation is performed in a turbulent manner for heat transfer and fluid flow over three row tubes and the turbulence effect is simulated through CFD. This study further takes into account the heat conduction and heat transfer of the flow in fins. The three different span angles are shown in figure 1 to figure 3.

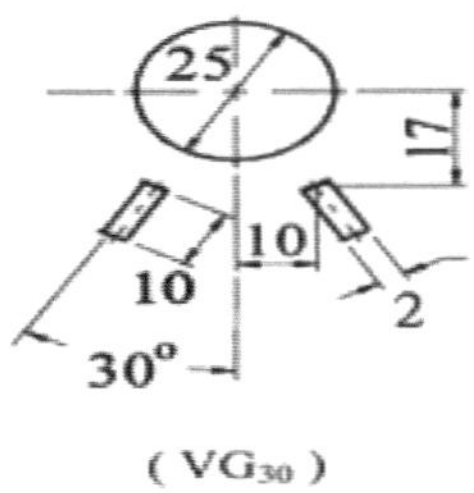

Figure 1. Span angle α = 30°.

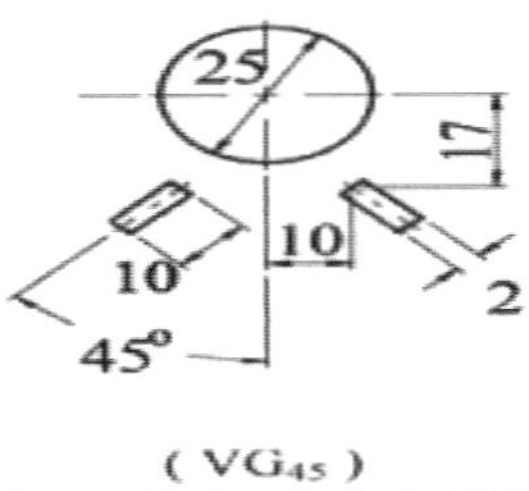

Figure 2. Span angle 45°.

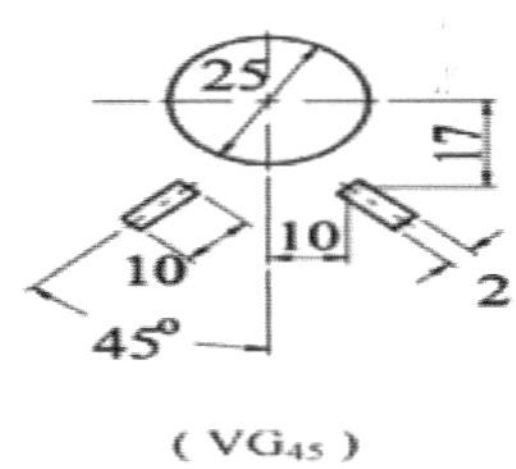

Figure 3. Span angle 60°.

5. Boundary condition

The fluid taken for study is incompressible with perpetual properties and the flow of fluid is turbulent, no viscous dissipation and steady. The velocity of fluid flow is assumed uniform at the boundary with inlet temperature of 200C. The turbulence intensity at inlet is set as 3%.The Neumann boundary condition or stream wise gradient for variables is zero. At the downstream end, the tube diameter is seven times the last downstream row tube. The temperature of constant tube wall, no slip condition is specified at solid surface. Between the boundary planes of XY and XZ, the cyclic match is found. The device for pressure drop has three tube row placed inside an inline array. Finally, the comparison of pressure drop and heat transfer over elliptical, aero foil and semi-circular generators is studied with and without wings.

6. Heat exchanger solving procedure

Gambit 2.0 solves the problem in pre-processor and Fluent 6.2 is used to solve the problem in solver and post-processor. Gambit is used to form geometries and meshes are generated to solve the post-processor problem. The heat exchanger model and test section is shown in figure 4 and 5, respectively.

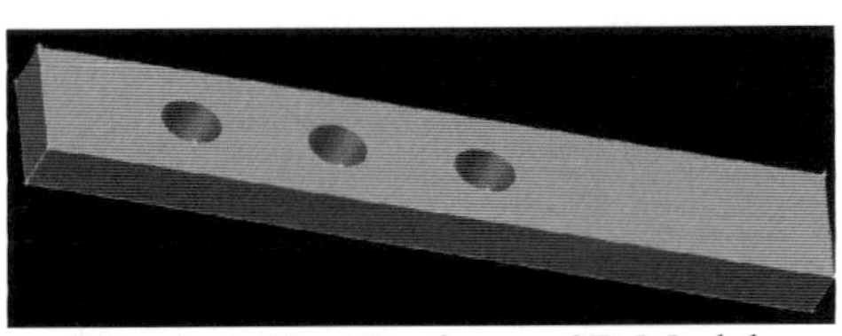

Figure 4. Heat Exchanger3D Model.

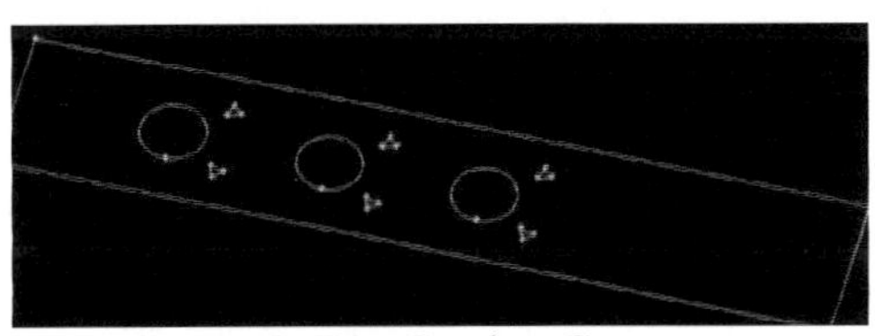

Figure 5. Test Section Geometry.

7. Results

To evaluate the proposed model, the system is numerically modelled and independency test is conducted using boundary conditions and governing equations before formal analysis. A rectangular shaped vortex generator is considered in the study with a base plate. The vortex generator is placed at three different angles. The system is placed with rectangular parallelepiped to find the flow conditions over base plate and vortex generator. The water is allowed to flow at different Reynolds number. At the outlet, the pressure is maintained at standard temperate and pressure. Finally, the base plate has constant heat flux to perform the numerical simulation.

A good quality mesh is used to capture the required information from CFD. At vortex generator, a better accuracy is maintained to determine the pressure, velocity and temperature. To maintain the grids to attain better solution, grid dependency test is performed. It is found that surface average Nusselt number and surface skin friction coefficient varies as the grid changes.

8. Comparative result

8.1. Tringale

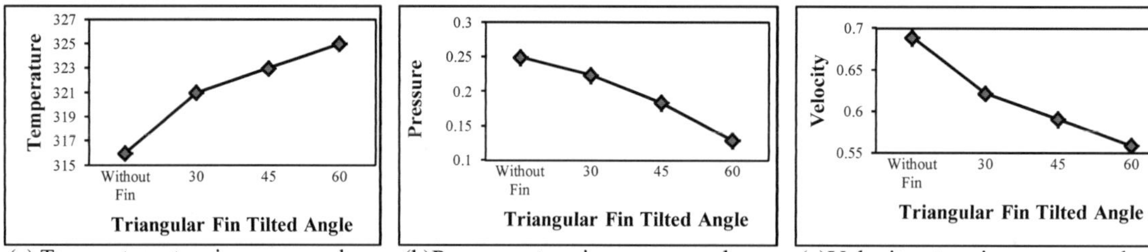

(a) Temperature at various span angles (b)Pressure at various span angles (c)Velocity at various span angles

Figure 6. Cylindrical votex generator.

8.2 Cylindrical

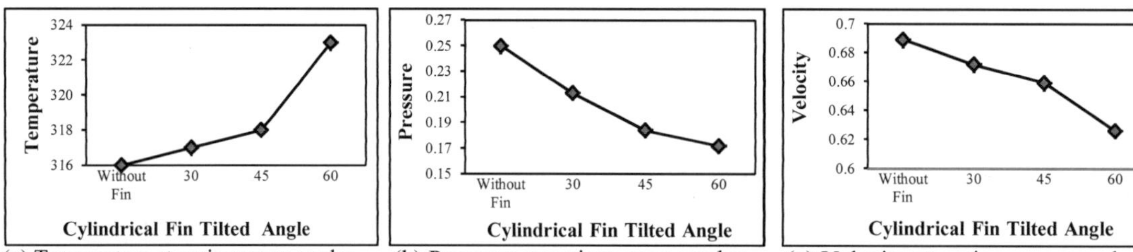

(a) Temperature at various span angles (b) Pressure at various span angles (c) Velocity at various span angles

Figure 7. Cylindrical votex generator.

Temperature, Pressure and Velocity of triangular vortex generator at various span angles is shown in figure 6. Similarly, Temperature, Pressure and Velocity of cylindrical vortex generator at various span angles are shown in figure 7. The result shows that the temperature of vortex generator without fin in triangular shape is less than cylindrical shape. However, as the span angles increases, the triangular shaped vortex generator increases than cylindrical shaped. The pressure of both shapes are same without fin and the span angle with fin is lesser for triangular at 30° and it is same at 45° and further it is higher for cylindrical at 60°. Finally, the velocity of triangular without fin and cylindrical shapes are same. However, the pressure in cylindrical shaped fin vortex generator with fin increases as the span angle increases. The pressure with fin at varying angles is higher than triangular shaped vortex generator.

9. Conclusion

In this paper, the triangular and cylindrical vortex generator is simulated using CFD model under three span angles, which varies between α =30°, 45° and 60°. The entire system is tested with the Reynolds number ranging from 400 to 3000. The optimised heat transfer over three row tube with and without vortex generator is attained using the proposed design. Numerical simulation proves that proposed triangular model is better in its design that circular model. Here, the use of rectangular vortex generator improves the rate of heat transfer from plate to fluid. The utilisation of convex and concave modifications on trailing and leading surface of vortex generator improves the accuracy. The presence of high pressure at the walls alters the thickness resulting in decaying of vortex generator. Finally, it is seen that numerous convex profiles on trailing and leading faces leads to degradation in local skin friction. Hence, it is avoided in the present design and this is true in case of circular design than rectangular design.

References

[1] Samadifar M, & Toghraie D 2018 *Appl. Therm. Eng.* **133** 671-81.
[2] Awais M & Bhuiyan A A 2018 *Therm. Sci. Eng. Progr.* **5** 524-45.
[3] Xin D, Zhang H, Ou J 2018 *J. Wind Eng. Ind. Aerodyn.* **175** 100-10.
[4] Kashyap U, Das K, Debnath B K 2018 *Int. J. Therm. Sci.* **127** 61-78.
[5] Aravind G P, Deepu M 2017 *Int. J. Heat Mass Transf.* **115** 809-25.
[6] Xu Z, Han Z, Wang J, Liu Z 2018 *Int. J. Heat Mass Transf.* **116** 61-72.
[7] Han Z, Xu Z, Wang J 2018 *Appl. Therm. Eng.* **128** 1456-63.
[8] Li Z, Xu X, Li K, Chen Y, Huang G, Chen C L, Chen C H 2018 *Int. J. Heat Mass Transf.* **118** 1340-56.
[9] Martínez-Filgueira P, Fernandez-Gamiz U, Zulueta E, Errasti I, Fernandez-Gauna B 2017 *Int. J. Hydr. Energy*, **42(28)** 17700-12.
[10] Dang W, Nugud J, Lin Z M, Zhang Y H, Liu S, Wang L B 2018 *Appl. Therm. Eng.* **134** 437-49.

Microstructure Analysis and Mechanical Properties of LM25 alloy with AC4B Nano-Composite using Gravity Die Casting Method

M Pasupathi[1], S Berlin Raj[2]

[1]Associate Prof., Department of Mechanical Engineering, Madha Institute of Engineering & Technology, Sadhanathapuram, Thandalam Post, Chennai – 600 122.
[2]CAE Analyst, Skill TeQ, Adambakkam, Chennai – 600 088.

Abstract. Casting of Aluminium alloys with zero defects would be the ultimate aim of any manufacturer. In the present work, the Alloy LM25 which was being used for the manufacturing of intricate Automobile components is compared to Alloy AC4B Nano-composite using Gravity Die Casting method. In order to investigate the mechanical properties of the concerned materials, Tensile followed by Hardness, and Compressive tests was implemented. Microstructure investigation is examined using Scanning Electron Microscope. The outcome from micro structural study confesses the uniform distribution, grain refinement, and low porosity in Nano-composite specimen. Mechanical results affirm that addition of B4C Nano-composite leads to enhancement of yield strength, ultimate strength and hardness. The uplift in mechanical properties manifested that the type of fabrication process and particle size involved in it.

1. Introduction

In present scenario, metal matrix composites are under column consideration for outsized range of structural application. The mechanical properties and also relatively less cost create them as a gorgeous choice [David L.et.al 1,25 1985]. A laye variety of fabrication techniques that manufacture MMC materials consistent with stir casting.

Dinesh kumar et. al [1] fabricated aluminum 6063 MMC one reinforced with silicon carbide and other reinforced with boron carbide by using stir casting technique various mechanical tests like tensile, hardness test, were conducted and it concluded that there is an increased in values of tensile strength hardness value and flexural strength of composite SIC and B&C particulates. N Subramani et .al fabricated aluminum 2042 metal matrix composites reinforced with carbide sulphite by stie casting techniques and their mechanical properties are evaluated.

The stir casting process is most effective in terms of productivity and cost efficiency, which is implemented to prepare the particular MMCs. The motivation for this work was to compare and study the mechanical and micro structural properties of M25 alloy with AL4 nanocomposite at different volume fraction.

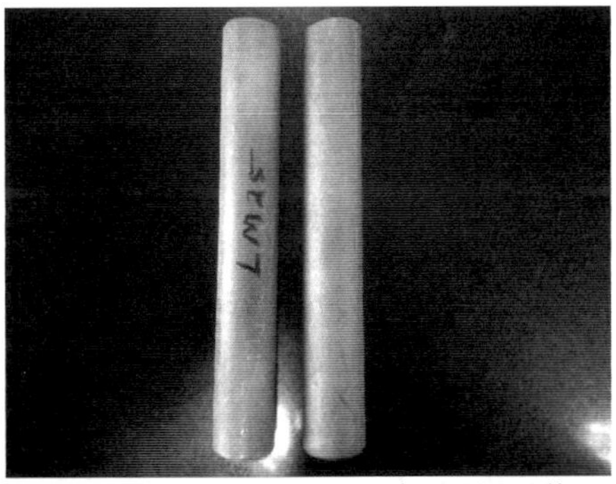

Figure 1. Micro Structural LM25 Alloy.

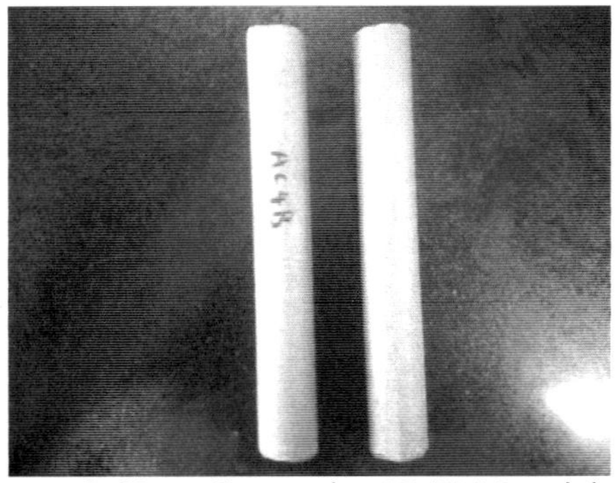

Figure 2. Nano Composite AL4B Materials.

2. Experimental procedure

2.1 Composition

The specific material alloy used in this study where LM25 and AC4B nanocomposite. The chemical composition of these elementary materials has been exhibited in table 1.

Table 1. Chemical composition of elementary materials.

Elements	Si	Cu	Fe	Mn	Mg	Ni	Zn	Sn	Ti	Pb	Cr
LM25	7-10	2-4	0.8	0.5	0.35	1	0.1	0.1	0.2	0.2	0.2
AC4B	6.5-7.5	0.2	0.5	0.3	0.2-0.6	0.1	0.1	0.1	0.05	0.1	

The refinement in terms of grain size is affirmed to be transverse, by Linear Intercept method.

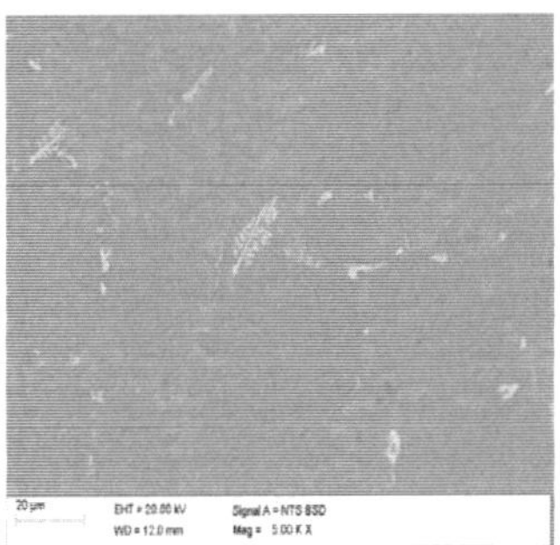

Figure 3. AC4B.

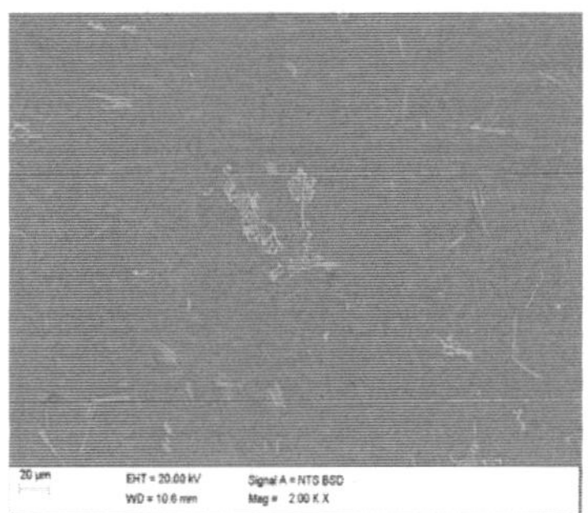

Figure 4. LM25.

2.2. Tensile strength and density

The ultimate tensile strength, yield strength, percentage elongation are shown in figure 5 and figure 6.

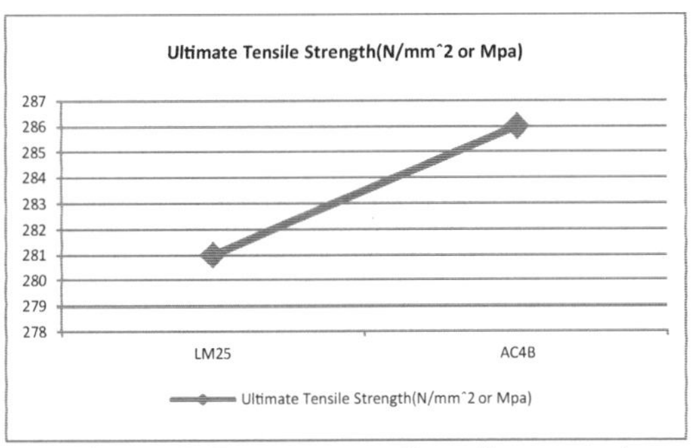

Figure 5. Percentage of tensile strength.

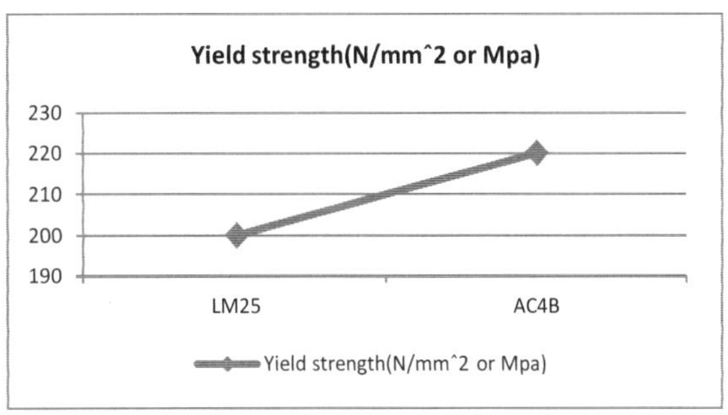

Figure 6. Percentage elongation of yield strength.

As the density level of the alloy increases, it gives the direct proportionality towards its increasing strength from the Gravity die-casting method.

3. Results and Discussion

3.1 Metallurgy Structure

Grain size plays a vital role in microstructural analysis. Figure (7) shows the microstructure of a gravity die casting sample of LM25 alloy. Figure (8) shows the microstructure of gravity die casting sample of AC4B a nanocomposite alloy. Good grain structures or size are desirable, leading to the enhancements of mechanical properties, tear resistance, pressure tightness. From above consideration the LM25 alloy, the grains are loosely packed which is verified from the Scanning Electron Microscope. Loosely packed grains leads to porosity and shrinkage defects.

Figure 7. Ac4B.

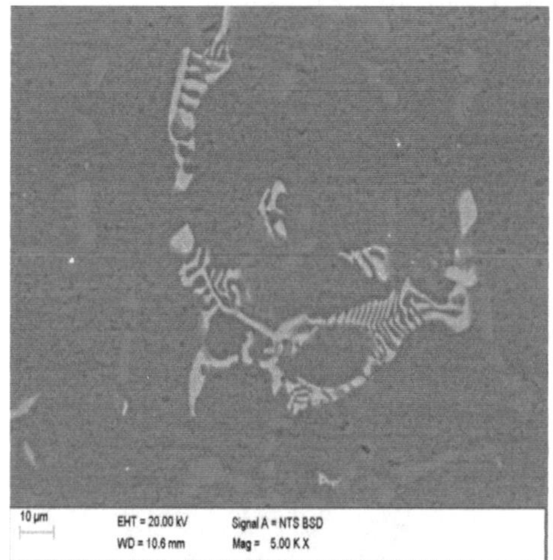

Figure 8. LM25.

In AC4B aluminium alloy, the C4B acts as grain refiners up to 3-5%, Grain refiners acts as a nuclei for grains to grow. More nucleation of likely eight sites may appear on every individual particle. In AC4B alloy, C4B make the grains to occur closely with each other as a refiner. This is formed in the form of equation.

$$Al + C4B \rightarrow \alpha - \text{Al alloy solid solution} \quad (1)$$

As a result of closed packed grains and good grain structure AC4B alloy does not have any defects and the mechanical properties are comparatively good than LM25 alloy (1). In gravity die casting sample of AC4B alloy, porosity within the microstructure was not seen on the surface or located in very small amount. AC4B shows good enhancement in mechanical properties which are resulted below.

3.2. Tensile Strength

The values of the ultimate tensile strength, yield strength, and percentage elongation are shown in table 2. In addition of a nanocomposite material, the yield strength of the material increases which had been examined by the Scanning Electron Microscope. Followed, the ultimate strength also increases by adding C4B a nanocomposite material was resulted through microstructure analysis. The testing machine and specimen is shown in figure 9 and figure 10.

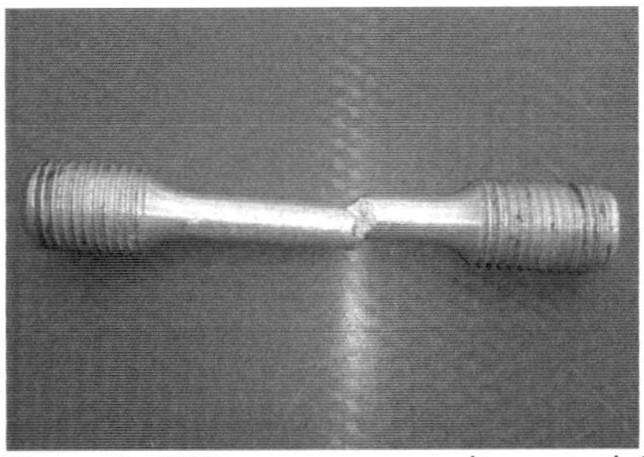

Figure 9. C4B a nanocomposite material.

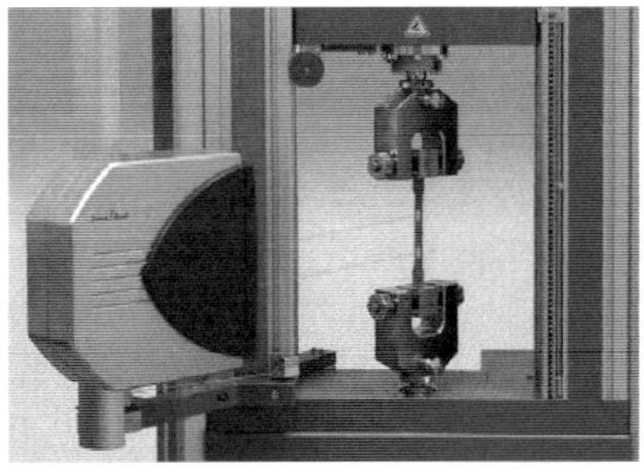

Figure 10. Testing machine and specimen.

Table 2. Tensile strength, Yield strength, and percentage elongation.

Test Parameters	LM25	AC4B
Yield strength Mpa	220	222
Ultimate strength Mpa	281	286
% Elongation in 4D GL	6	2.5

3.3. Hardness Test

The Brinell hardness test is furnished in Table 4. Brinell hardness test specimen in shown in figure 2.

Figure 10 shows the hardness value of three samples. It can be noted that sample 2 has the maximum hardness followed by sample 1 in all the trials.

Generally increasing the trend of hardness with C4B, 5% volume composite content of C4B performs minor hardness than sample of 2% volume. This may be emerged from existence of more porosity with greater content of C4B.

Table 3. Brinell hardness test.

Test parameters	LM25	AC4B
BHN (5mm ball/250 kg load)	93.9,92.8,92.3	89.7,90.2,89.2

3.4. Elongation

Percentage Elongation quantifies the ability of an element or compound to stretch up to its breaking point. In this research, the elongation of LM25 is comparatively low in ductile property and grain refinement. Nanocomposite C4B was tested and the test affirms the addition of nanocomposite C4B with LM25improves the elongation property. The % Elongation is plotted in a form of chart.

LM25 alloy, the bonding between the elements breaks up in the process called stir casting. In addition of C4B the electromagnetic bonding showed a rise in its elongation properties where C4B acts as an elastic component with Aluminium clearly manifestations the elongation properties.

Figure 11. LM25 is low in ductile property.

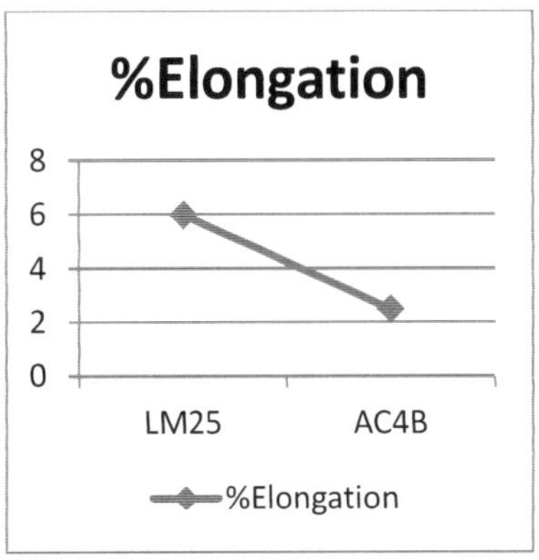

Figure 12. Elongation is plotted.

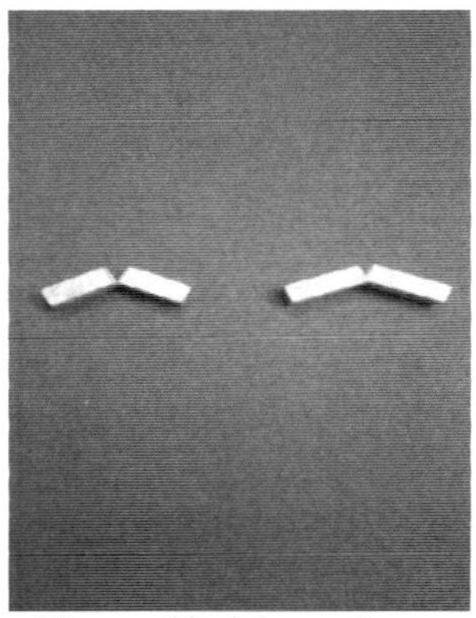

Figure 13. Stir casting.

3.5 Flextural test

Three point flextural testing machine is used for flexural test and as per ASTM: E290 Standard the specimen was cut. Test specimen is shown in figure 11-13 and flextural properties are shown in table 3.

4. Conclusion

The Aluminium alloy LM25 reinforced with AC4B nanocomposite is fabricated via AC4B using Gravity Die Casting method. The reinforced nanocomposite particles have shown significant results in terms of mechanical properties. However, the changes in the alloy is visualised and changes in tensile, hardness and yield strength tests are revealed. The other tests includes charpy impact test and brinell hardness test which proves that nanocomposite material shows good enhancement over mechanical properties of the materials.

References

[1] Wang Q G 2003 *Metallur. Mater. Trans. A*, **34(12)** 2887-99..

[2] Liu D, Atkinson H V, Kapranos P, Jirattiticharoean W, Jones H 2003 *Mater. Sci. Eng. A,* **361(1-2)** 213-24.

[3] Mrówka-Nowotnik G, Sieniawski J 2005 *J. Mater. Proc. Tech.* **162** 367-72.

[4] Hajjari E, Divandari M 2008 *Mater. Des.* **29(9)** 1685-9.

[5] Yar A A, Montazerian M, Abdizadeh H, Baharvandi H R 2009 *J. Alloys Comp.* **484(1-2)** 400-4.

[6] Reddy A C 2011 *Mater. Tech.* **26(5)** 257.

[7] Laser T, Nürnberg M R, Janz A, Hartig C, Letzig D, Schmid-Fetzer R, Bormann R 2006 *Acta Mater.* **54(11)** 3033-41.

IOP Conf. Series: Materials Science and Engineering **507** (2019) 012024 doi:10.1088/1757-899X/507/1/012024

Method for Evaluating the Degree of Metallization Corrosion Damage of Integrated Circuits under Accelerated Tests

Kolomiets V. I.[1]

[1]National Research Moscow State University of Civil Engineering, 129337, 26, Yaroslavskoye Shosse, Moscow, Russia

Abstract. This paper focuses on evaluation results of accelerated tests for integrated circuits metallization corrosion resistance. It has been proposed to use the average area of corrosion damage to bond pads and the average number of metallization damages as an indicator of the degree of metallization damage. The evaluation method is applicable for comparative tests of various design and technological options for integrated circuits manufacture.

1. Introduction

Currently, a significant part of industrial electronics and IT equipment hardware components used in industrial or civil construction facilities is digital integrated circuits (IC) such as microprocessors, microcontrollers, encoders/decoders, digital-to-analog converters, etc. It means that functional reliability of these facilities will be largely determined by ICs reliability.

Technology based on the use of complementary transistors with the structure of metal-oxide-semiconductor (CMOS IC) is now widely used in the manufacture of digital ICs. Its main advantages are low power consumption and a wide range of supply voltages. However, these advantages, on the other hand, also led to low corrosion resistance of IC metallization. As a result, special technological measures were required to increase its corrosion resistance, as well as methods to evaluate effectiveness of these measures.

There are two methods to improve ICs reliability:

- eliminate causes of operational failures by improving the technology and using more stringent process control;
- identify potentially unreliable ICs and remove them from the batch at the stage of accelerated screening tests (QA test) of finished ICs [1].

The first method is more efficient, however, the best result is achieved when it is combined with the second method.

To assess sealed CMOS IC resistance to metallization electrolytic corrosion (second method), generally, the median time to failure is used as resistance indicator [2]. The tests are carried out using heat and cold chambers where ICs run under normal electrical conditions for several hundred hours, which significantly increases the cost of testing.

To assess how ICs design and manufacturing process features affect metallization corrosion resistance (first method), it is reasonable to apply stress test methods, during which unsealed ICs are exposed to high temperature and humidity in combination with the supply voltage [3].

It is feasible to compare test results under these conditions based not on time between failures, but on the scale of metallization corrosion damage.

However, using mean time between failures as a corrosion resistance indicator for assessing results of such tests is not feasible. This leads to considerably higher cost of tests and lower integrity of results due to a larger number of parametric failures that are caused by a relatively thick water film forming on the surface of IC chip.

The method of metallization resistance assessment by bond pad surface discoloration as observed through an optical microscope has been described in several studies [4, 5]. However, it is not sufficiently accurate since evaluation of discoloration degree is subjective, and it is not suitable to assess damages observed in the case of electrolytic corrosion.

Due to the above-mentioned reasons, this study aims at developing a simple method for assessing the degree of corrosion damage to IC bond pad metallization specific for electrolytic corrosion.

2. Methods

The method of unsealed ICs accelerated testing in unsaturated water vapor at elevated temperatures with simultaneous electrical voltage used in this paper does not replace sealed ICs tests and is not intended to verify ICs reliability. Its purpose is a comparative assessment of aluminum metallization corrosion resistance for various design and technological options in IC manufacture.

The following conditions are selected as basic ones [6]:

- relative humidity of 85%, temperature of 85°*C* (implemented using a heat and humidity chamber);
- relative humidity of 85%, temperature of 120°*C* (implemented using autoclave).

If purpose-built equipment is not available, the conditions of 100% relative humidity, 85°*C* temperature are accepted with mandatory control of electrolytic corrosion process self-similarity.

Tests duration is determined by their purpose, laboratory equipment capabilities, IC design and technological features, and the selected conditions. The minimum test time is 72 hours for the heat and humidity chamber, 66 hours for the autoclave.

This paper proposes to use the value of corrosion damage area derived as the average of cathode $\Delta\overline{S}_{\kappa}$ and anode $\Delta\overline{S}_{a}$ bond pads area as an indicator that directly characterizes metallization resistance to electrolytic corrosion. This parameter may be used for an approximate quantitative assessment of corrosion resistance, as well as for comparative assessment of the rate of corrosion processes.

The bond pad area ΔS affected by corrosion is determined using:

$$\Delta S = \left[\left(S_o - S_н \right) / S_o \right] \times 100\% \tag{1}$$

where S_0 is the bond pad aluminum area observed in the optical microscope prior to start of test (not including the area closed from the observer by the thermo-compression bonding); $S_н$ is the undamaged aluminum bond pad area at the end of tests. An example of $\Delta\overline{S}_{\kappa}$ assessment is shown in figure 1, 2.

Further, the $\Delta\overline{S}$ average value of all tested ICs is calculated separately for the cathode $\Delta\overline{S}_{\kappa}$ and anode $\Delta\overline{S}_{a}$ bond pads:

$$\Delta\overline{S}_{\kappa} = \frac{1}{N_{\kappa}} \sum_{j=1}^{N_{\kappa}} \Delta\overline{S}_{j} = \frac{1}{nN_{\kappa}} \sum_{j=1}^{N_{\kappa}} \sum_{i=1}^{n} \Delta S_{ij} \tag{2}$$

$$\Delta\overline{S}_{\kappa} = \frac{1}{N_{\kappa}} \sum_{j=1}^{N_{\kappa}} \Delta\overline{S}_{j} = \frac{1}{nN_{\kappa}} \sum_{j=1}^{N_{\kappa}} \sum_{i=1}^{n} \Delta S_{ij} \tag{3}$$

where n is the number of tested ICs; N_{κ}, N_{a} is the number of cathode and anode bond pads in each IC.

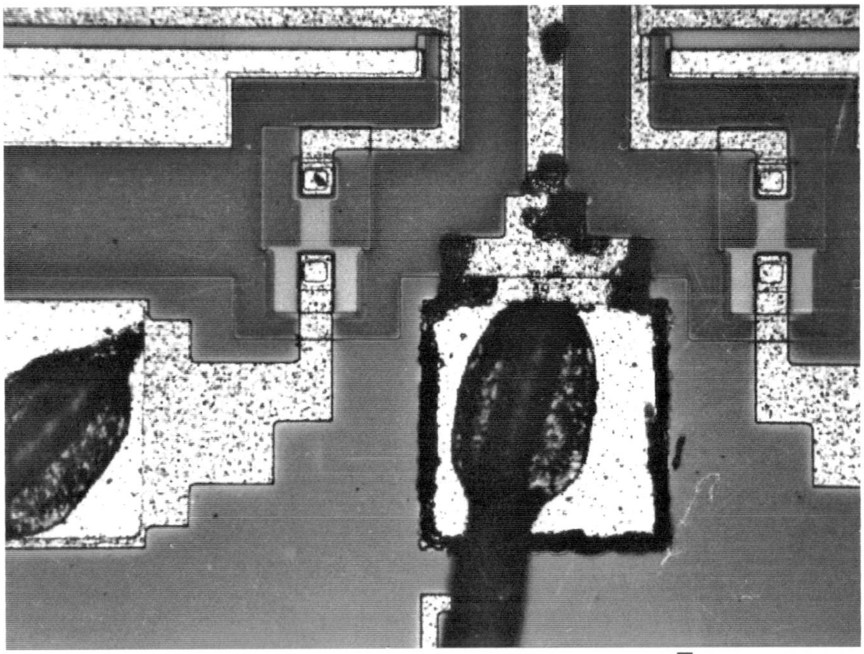

Figure 1. Corrosion damage of IC bond pads and interconnects ($\Delta\overline{S}_{\kappa}$ =20%, $\overline{M}$ =18).

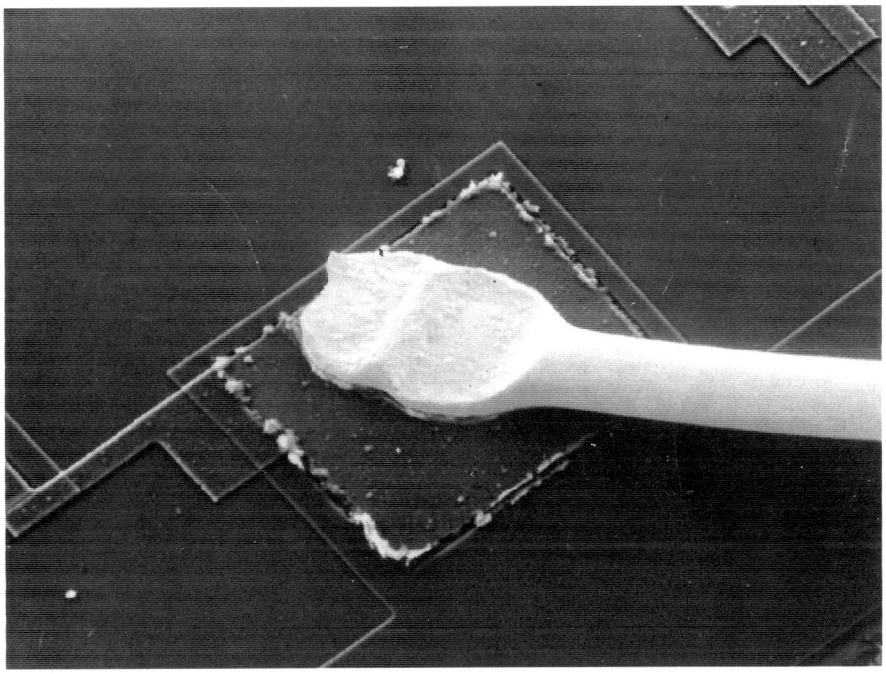

Figure 2. Corrosion damage of IC bond pads in SEM ($\Delta\overline{S}_{\kappa}$ =10%).

The average number of corrosion spots on the interconnects (excluding bond pads) of a single chip $\overline{M}$ is determined by:

$$\overline{M} = \frac{1}{n}\sum_{i=1}^{n} M_i \tag{4}$$

where n is the number of ICs from one batch undergoing corrosion resistance assessment; M_i is the number of spots on chip metallization of the ith IC.

When conducting tests for the purpose of quick response quality control of the IC manufacturing process, the level of metallization corrosion resistance is unsatisfactory if corrosion damage of aluminum is observed on bond pads and interconnects.

When conducting comparative tests of aluminum metallization corrosion resistance for various design and technological IC options, quantitative assessment of the degree of metallization corrosion damage is made separately for cathode $\Delta\overline{S}_\kappa$ and anode $\Delta\overline{S}_a$ contact pads, as well as $\overline{M}$ interconnects.

3. Discussion of the results

The proposed method for assessing the degree of metallization corrosion damage and indicators are sufficiently informative, reliable and reproducible.

In particular, it is possible to directly check the degree of metallization corrosion damage by visual inspection, as well as effectively use extensive statistical material. For example, when testing 20 ICs in 16-pin bodies, the $\Delta\overline{S}_\kappa$ and $\Delta\overline{S}_a$ values are calculated as the average of ⊔ 160 values each, which increases evaluation results integrity.

The method of assessing corrosion damage degree was used for corrosion resistance comparative assessment of 564 series IC *(Russian analogue CD4000A)* before and after the implementation of measures to increase IC corrosion resistance.

The table shows some results of corrosion resistance assessment of 564 series IC. The tests were carried out at a relative humidity close to 100%, which was due to the need to evaluate, first of all, metallization resistance to high humidity exposure.

Table 1. Results of metallization corrosion resistance evaluation of 564 series IC .

IC batch No.	Design and technological option	Test conditions			
		85°C/100%/15 V		121°C/100%/15 V	
		$\Delta\overline{S}_\kappa$, %	$\overline{M}$, pc.	$\Delta\overline{S}_\kappa$, %	$\overline{M}$, pc.
1	Before the implementation of measures	65	9.7	45	11
2		97	0.7	11	0.7
3		77	0	0.9	0.3
4				72	7
5	After the implementation of measures	0	0	0	0
6		0	0	0	0

Obviously, the proposed assessment method allows us to confirm the effectiveness of measures to increase the corrosion resistance of IC metallization (selection of metallization composition and application method , additional protective films, reducing the chip surface contamination level).

Also, the assessment method can be used to assess the test results of various additional surface protection systems for an IC chip. For example, such as silicone coatings, polymers and photoresists [7].

4. Conclusions

The proposed method for assessing corrosion damage degree based on average damaged area separately for cathode $\Delta\overline{S}_\kappa$ and anode $\Delta\overline{S}_a$ bond pads helps assess the degree of damages specific for electrolytic metallization corrosion. This method does not require big financial investments and complex diagnostic equipment and allows to effectively conduct the comparative assessment of aluminum metallization corrosion resistance for various design and technological options of IC manufacture.

The assessment method is easy to implement and has sufficient efficiency and statistical stability.

Reference

[1] Gorlov M, Stroganov A 2006 *Tech. Electr. Ind.* **1** 18-22.
[2] John W O 1996 *Semiconduct. Sci. Tech.,* **11(2)** 155–62.
[3] Luis A, Escobar and William Q. Meeker. 2006 *Statist. Sci.* **21(4)** 552–77.
[4] Mautz K 1998 *Proc. Symp. On Interconnect and Contact Metallization,* **97(31)** 240-251.
[5] Fauty J, Strouse S, Yoder J Int. 2001 *Journal of Microcircuit and El. Packaging,* **24(1)** 19-29.
[6] Kolomiets V 2018 *IOP Conf. Series: Materials Science and Engineering*, **317**.
[7] Zelyakova T I, Krutov L N, Krikunov N A 2006 *Proc. Int. Symp. "Reliability and Quality"* 52.

IOP Conf. Series: Materials Science and Engineering **507** (2019) 012025 doi:10.1088/1757-899X/507/1/012025

Challenges in Wi-Fi and Femto Offloading and Coexistence Issues

Kwame Sarfo[1, 3], **Song Lv**[2]
[1]School of Information and Communication. Wuhan University of Technology-China
[2]Hubei University, China
[3]Electrical/Electronic Eng. Kumasi Technical University, Ghana

Abstract. With reference to the increasing data transfer such as video, music, large files on mobile devices e.g. Smartphones, ipads and electronic tablets, there has been corresponding increment in data traffic in the cellular network as the number of users keeps escalating day by day. In the meantime, mobile internet is rapidly evolving towards embedded internet, expanding its reach from people to machines [1]. In fact, the wireless industry now expects 50 billion machine-type devices connected to the global network by 2020. This has brought a huge burden on cellular network operators. The most economical ways are rerouting the data through complementary technologies such as the use of Wi-Fi, Femto, Heterogeneous, opportunistic networks and so on. It is also noted that each of these networks has its pros and cons. These assist network operators to choose one or more systems over the other. This paper considers the challenges of these offloading networks in Wi-Fi and Femto networks and their coexistence issues with each other and other networks.

1. Introduction

The demand for higher data rate, lower latency and low consumption is directing mobile operators worldwide towards methods to handle the huge traffic and congestion that is often experienced by cellular networks. Many have suggested that offloading of huge traffic is the witty or advantageous way due to limited spectrum and cost of infrastructure. Offloading reduces the amount of data being carried on the cellular bands, freeing bandwidth for other users. It is also used in situations where local cell reception may be poor, allowing the user to connect via wired services with better connectivity. According to forecast global mobile data traffic will increase by 10-fold from 2014 to 2019 [2].

With respect to the initiator diversity of data offloading, we categorize data of loading techniques into four types viz data offloading through small cell networks, Wi-Fi networks, opportunistic mobile networks (HetNets) or Heterogeneous, respectively [3]. These are promising methods for now and the future. The evolution Wi-Fi and Femto networks are shown in figure 1. The cellular network functions in the Licensed spectrum and improving it involves the application of new and massive infrastructure which makes it economically unattractive. Most of these offloading technologies operate in the unlicensed spectrum and they require the installation of small user equipment (SUE) in vantage areas to offload the traffic.

2. Works done by other authors

Guillaume de la Roche and et.al talked about access method, comparison between methods, technical challenges and solution, and describe hybrid approaches [4]. While author's focus on access control strategy as crucial aspect for operator to give preferential access to femtocells for their subscribers [5]. In this article different mechanisms of access to femtocells were introduced and they did not consider the general types and coexistence issues. In [6], the authors proposed an integrated architecture exploiting the opportunistic networking paradigm to migrate data traffic from cellular networks to metropolitan Wi-Fi access points and narrowed it traffic control and without many other challenges. Again, Alexandros Kostopoulos and et. Studied the effect of Wi-Fi Offloading on Pricing Wireless Services and they considered maximizing users' rewards while meeting constraints on queue stability and energy consumption. But they did not focus on the detailed challenges and coexistence with other networks [7]. Yuan Wu and el studied the economics and incentive of traffic offloading and they established that it has a good potential benefits but could reduce the mobile unit data cost due to discontinuous coverage in Wi-Fi networks. They emphasized that it is vital to consider how much profit in terms of the data cost reduction [8]. Their reason was that offloading to Wi-Fi requires an arrangement between other operators who own various Wi-Fi infrastructure and this will be critical in terms of revenue sharing procedures.

3. Wireless Fidelity (Wi-Fi)

There are a number of reasons that attracts mobile network operators(MNO) to be interested in Wi-Fi offloading. First it helps operators to ease network congestion and improve the quality of service their customers receive in high-density areas where network congestion is high. Operators can also "expand" their network coverage, installing hotspots in areas with poor cellular coverage at a much cheaper cost. Again operators can form partnerships with other businesses/ISPs/MNOs in some areas to let their subscribers automatically offload onto the business/ISP/MNO Wi-Fi networks there by expanding their network without making any investment in huge infrastructure. This will help them generate revenue by giving other operators' customers roam onto their Wi-Fi network [9]. The performance of an 802.11/802.11b Wi-Fi system is affected by several external wireless-specific circumstances in its deployment. As with wired LAN systems, throughput and errors in Wi-Fi are product and setup dependent, which in turn depends on the Wi-Fi design and deployment considerations. These challenges include

Frequency management: First network operator use radio frequency (RF) devices in a given area and they must cooperate If they operate on the same frequencies at the same time. This makes their transmissions interfere with each other's receivers and the need for frequency management [10]. It is therefore helpful for these operators to identify any existing interference patterns and define the preferred locations of Wi-Fi APs. There are a number of factors that cause the batteries of devices such as mobile phones, laptop etc. to depletes or run down quickly. These include low signal strength, brightness the LCD back light, applications requiring regular updates, live wallpapers etc., the turning on of the Wi-Fi system of such devices causes a high drain of batteries especially when is turned on for an ample of time [11]. The serious aspect is that if the Wi-Fi is on, it continuously detects any Wi-Fi router of Wi-Fi zone. Also some mobile phones e.g. iPhone, Samsung Galaxy Note 8 etc. can be used as hotspots for many other Wi-Fi devices but the challenge is that hotspot causes the phone batteries to run down quickly and causes excessive heating in the device circuit [12]. Another challenge in this network is lack of mobility There are no automatic handovers on Wi-Fi devices, so users cannot roam around in the building without dropping the service from one AP to the other and as such users are compelled to reset their devices in order to use it in new areas and this is often time consuming and does not encourage users. For instance, if one is a hotel and using the Wi-Fi network, once he/she leaves the place or sometimes the floor, the individual has to go through a setup process to be connected to another Wi-Fi network. It becomes unbearable when one is in the market and wants to use cashless service to pay for the purchases, from one trader to the other a new set and reset must be done each time. Again one has to get a specific user's access code for every Wi-Fi provider in the place. IETF (Internet Engineering Task Force) came up with the specifications for MIP (Mobile Internet Protocol) and PMIP (Proxy Mobile Internet Protocol) to address the mobility issues in Wi-Fi [13] but as of now not much has been achieved. Again some Access Points (APs) have overlapping coverage to share the network, which would greatly increase the number of devices competing for access and collisions in overlapping areas could lead to deterioration of the whole network's throughput. Additionally, Wi-Fi service is Slow; the data rate on the Wi-Fi–air interface is high, in excess of 50 Mbps. However, the actual user data rate is not limited by the air interface but by the ADSL backhaul to the Wi-Fi access point. Typical user data rates on public-access Wi-Fi services in hotels and airport are often lower than 400 kbps, far below the 54 Mbps on the typical Wi-Fi air interface [14] and makes the system slow and could result in delay jitter which is variation in the delay encountered by similar packets following the same route through the network [15]. The Wi-Fi speeds may slow down if the device is too far from the router. The farther one is from the router, the more unreliable the connection and its throughput will become. Another cause for slowdowns is the lack of bandwidth. If everyone is home and using their computers, phones and televisions for data applications, speedy Internet is being spread thin and shared across these multiple devices and often slows down if many people are trying to stream videos, games etc. from, YouTube and Netflix, [16]. Another area where the Wi-Fi data offloading encounters a challenge is in transportation. For vehicular systems, there is a greater advantage of this system as passengers with mobile phones and other devices are able to offload data despite few challenges. However, in areas such as slow and high speed trains, there often exists the experience of loss of packets, extreme slow downloading and uploading and interruptions or breaks of the signal in certain terrains along the journey. The worst of it is that it could take much time to receive data and others services. The cause of such a problem is the lack of sufficient Wi-Fi infrastructure and repeater stations in such areas. Additionally, as passengers are on board in airplanes, they are warned to turn off some devices as mobile phones, laptops which make use of of Wi-Fi.

Another problem that the Wi-Fi systems encounters is congestion which is a network condition where traffic bottles up in queues to the point that it noticeably and negatively impacts the operation of the application when operators send huge loads [15]. The variable size of packets makes traffic prediction and guaranteed service provision difficult. So QoS is not always obtained. Congestion is usually caused by low memory, slow processors, bandwidth line and speed mismatch of devices [17]. Provision of QoS suffers as nodes operate on low frequency, Limited Battery Backups and where any node goes down, it may cause the frequent link breaks and end user suffers from poor quality of transmission [18]. As the packets go through the network, Packet loss could occur where they would be being dropped, get lost, or become corrupted while going through the network [15] and in such situations there is no Guarantee of Service. This is because Wi-Fi uses free open spectrum, so there is no guarantee that other Wi-Fi providers will not deploy Wi-Fi service in the same areas, using the same or adjacent frequencies and degrading the current Wi-Fi service. Therefore, there is no quality of service guarantee using Wi-Fi Hot-spot data areas in: Office buildings, Conference and meeting rooms. Furthermore, the solution to Wi-Fi offload is not straightforward [14]. Wi-Fi brings with it a new set of challenges such as how to automatically and seamlessly connect to cellular domains, how to integrate a Wi-Fi network with operator's policy and charging systems, and so on [19]. There are also security implementations: security features, such as choice of encryption, impacts on AP throughput. The performance drops when using encryption and depends on the vendor and AP model; it may be anywhere from 10% to 25% and may go as high as 50% [9/10] and as such the architecture must also provide mechanisms to defend against the possibility of theft, to prevent denial of service, and to anticipate equipment failure [15].

4. Femtocells

These are cellular network access points that connect standard mobile devices to a mobile operator's network using residential DSL, cable broadband connections, optical fibres or wireless last-mile technologies [30]. There are many reasons why customers and operators demand femtocells and the services, economies that can be provided by using them. Yet there are many challenges in achieving this potential. These are briefly highlighted here and examined in greater depth. There is the issue of co-layer Interference and this is describing as the unwanted signal received at a femtocell usually sent from other femtocells. This decreases the quality of its communication. Co-layer interference occurs mainly between immediate neighbours due to low isolation between houses and apartments. And the overall interference can be higher than any of the independent femtocell power level if they are many [20]. Again there is Cross-layer interference especially in CDMA co-channel deployed two-layer networks, due to the fact that both femtocells and macro cells use the same frequency band and due to power control, sudden high transmitting powers can cause the appearance of dead zones, reducing the feasibility of these networks [21]. The power transmission by Femto Access Points (FAPs) consist of the FAPs traffic power and pilot power which shows the data and cell coverage area. The effect of interference on collocated FAPs and MBS is dependent on these power levels. A higher pilot power results in a larger coverage area and has higher chances of causing interference [29].

Again Service availability is one key quality metrics customers use to admire and is a benchmark based on the percentage of time the network is fully capable of offering service. Denial of service often goes on in the network element and causes service disruption to users. This may be caused by attacks usually is in the form of overloading the network element with large amounts of protocol traffic to the point that the network device can no longer function properly. In addition, the authentication Header (AH) provides little for confidentiality, since it doesn't encrypt the packet in any way [22]. The figure 1 depicts a typical Femto set up.

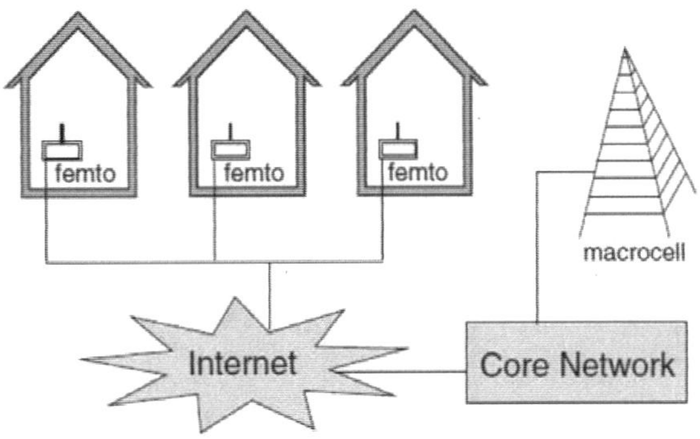

Figure 1. A simple Femtocell and macrocell [21].

5. Coexistence between Wi-Fi and small cell networks

Despite the individual challenges of these networks it is also of great concern that as these (the Wi-Fi and Femto exist together other challenges show up. First there is also the problem of interference management between femto and macrocells and mobility management and implementation issues (e.g. synchronization, signal processing and cost [23]. It is also known that if two identical signals are 100 % or 180 degrees out of phase they will completely cancel each another if combined. Since only Wi-Fi working in 2.4 GHz, it has the possibility to interfere with femto and ZigBee networks [24]. Again for users possessing dual-mode handsets endowed with both cellular and Wi-Fi capability (for example, the iPhone), it is hard to convince users to install femtocells since they already have Wi-Fi on these phones. On the contrary, significant numbers of cellular devices are not equipped with dual-mode capabilities and such users could be encouraging to go for femtocells for enhancing their indoor cellular coverage [25]. It is worth noting that technical issues related to the coexistence of Wi-Fi and heterogeneous cellular networks in unlicensed spectrum, such as efficient spectrum sharing and interference mitigation, have not been sufficiently addressed. Additionally, interference will range from macrocell to femtocell, femtocell to femtocell and femtocell to microcell [26]. Quality of service (QoS) based strategy is achieved by splitting the unlicensed spectrum between Wi-Fi and femtocell networks. But the split use of the spectrum between two systems prohibits a high cross-network throughput [26]. Another downside of the femtocell approach is that these are currently all operator specific, there is no simple choice for a building owner who wants to provide their customers with an operator wireless service can easily do that but Wi-Fi can easily provide that because it is compatible with all phones and devices, regardless of the home network operator [27]. Some Femtocells use standard cellular radio technologies, therefore any mobile device is capable of participating in the data offloading process, though some modification is needed to accommodate the different backhaul connection [28] On the contrary cellular radio technologies are founded on the ability to do network planning within licensed spectrum. Hence, it may turn out to be difficult, both technically and business wise, to mass deploy femtocell access points. Wi-Fi technology is different radio technology than cellular, but most Internet capable mobile devices now come with Wi-Fi capability.

6. Conclusion

In relation to the above studies it evident that there are numerous challenges that these offloading technologies encounter and these either results in the MNO not being able to maximize the returns with respect to huge capital investment or a lot of customers not being able to get complete satisfaction of their payments. Each technology has been analysed and the outcome is that both have challenges which mobile operator must tackle if their services will be attractive. The anticipated future 5G must be planned for and MNO must be poised to huge data offloading. From the elaborated problems, MNW operators can apply various means to achieve customer satisfaction by cooperating with other NW providers for use of infrastructure and also providing realistic incentives to customer groups.

7. Reference

[1] Ericsson 2011 More than 50 billion connected devices.
[2] Li J H, Dong X, Zhang K J, Liu J X, Effective capacity based Wi-Fi Delayed Offloading and Resource Re-Allocation in Heterogeneous Networks
[3] Huan Zhou, Hui Wang, Xiuhua Li and Victor C.M Leng, A Survey on Mobile Data Offloading Technologies ,Huan Zhou,Hui Wang, Xiuhua Li and Victor C.M Leng IEEE2018.
[4] K. D. Nguyen, H. N. Nguyen, and H. Morino, "Performance study of channel allocation schemes for beyond 4G cognitive femtocell-cellular mobile networks," in IEEE (ISADS'13), 1–6.
[5] Oh C Y, Chung M Y, Choo H, and Lee T J 2013 *Wireless Personal Commun.* **68(2)** 417– 32.
[6] S. Dimatteo, H. Pan, B. Han and V.O.K. Li, "Cellular traffic offloading through WiFi networks," in IEEE 8th International Conference on Mobile Ad Hoc and Sensor Systems (MASS), 192-201, Oct. 2011
[7] Alexandros Kostopoulos, Ioanna Papafili, Georgios Fortetsanakis, Maria Papadopouli1, The Effect of Wi-Fi Offloading on Pricing Wireless Services,
[8] Yuan Wu, Li Ping Qian, Jianwei Huang, Xuemin (Sherman) Shen, Energy-Aware Cooperative Traffic Offloading via Device-to-Device Cooperations: An Analytical Approach, **16**(1) 97 – 114, 2017
[9] https://www.alepo.com/wifi-calling-vs-wifi-offload.
[10] Richard Lau Ram Khare William Y. Chang,Service Assurance for Voice over WiFi and 3G Networks.
[11] https://www.lg.com/za/support/product-help/CT20230030-1434699598305-battery-life
[12] https://www.computerworld.com/article/2499772/mobile-wireless/how-to-use-a-smartphone-as-a-mobile-hotspot.html.
[13] Yan Li and Shaoyi Xu Traffic Offloading in Unlicensed Spectrum for 5G Cellular Network: A Two-Layer Game Approach * 2018
[14] Morten Tolstrup, Indoor Radio Planning A Practical Guide for GSM, DCS, UMTS, HSPA and LTE Second.
[15] Daniel Minoli, Hotspot Networks (Wifi for Public Access locations),
[16] https://www.cnet.com/how-to/common-wi-fi-problems-and-how-to-fix-them/
[17] Muhammad Fawad Dargai Malakand Khyber pukhtoonkhwa Pakistan, Fazli Rahim,Minimizing Delay in Heterogeneous Networks using Congestion Forecast.

[18] Paramjit Singh Waraich, Chandigarh Communication over Ad Hoc Networks under Quality of Services Constraints: A Review.
[19] https://hsc.com/Services/Product-Engineering-Services/Wireless-Engineering/WiFi
[20] Femtocells: Opportunities and Challenges for Business and Technology
[21] Jie zhang, guillaume de la roche, john wiley and sons,Femtocells: technologies and deployment 2010
[22] Joseph Boccuzzi, Michael Ruggiero Femtocells: Design & Application,
[23] Simon R. Saunders, Stuart Carlaw, Andrea Giustina, Ravi Raj Bhat, V. Srinivasa Rao and Rasa Siegberg,Femtocells: Opportunities and Challenges for Business and Technology© 2009 John Wiley & Sons Ltd. ISBN: 978-0-470-74816-9.
[24] Gaotao Shi Keqiu Li,Series editor Xuemin (Sherman) Shen,Signal Interference in WiFi and ZigBee Networks,Wireless Networks
[25] Vikram Chandrasekhar, Coexistence in Femtocell-aided Cellular Architectures,
[26] Haijun Zhang, Xiaoli Chu, Weisi Guo and Siyi Wang,Coexistence of Wi-Fi and Heterogeneous Small Cell Networks Sharing Unlicensed Spectrum
[27] https://www.thinksmallcell.com/Technology/offloading-traffic-to-wi-fi-and-femtocells-in-different-locations.html
[28] Att.com (Press release), Starbucks (NASDAQ:SBUX). AT&T Intellectual Property "More Than 12 Million AT&T, Starbucks Customers to Get Free Wi-Fi Access for a Rich In-Store Experience" Retrieved 2014-07-05.
[29] Masood Ur Rehman and Ghazanfar Ali Safdar,LTE Communications and Networks,Femtocells and Antenna Design Challenges 2018

IOP Conf. Series: Materials Science and Engineering **507** (2019) 012026 doi:10.1088/1757-899X/507/1/012026

Bond Behaviour of Steel Bars Embedded in Low Strength Concrete

Chichaya Boonmee[1] and Kittipoom Rodsin[2]

[1]Department of Teacher Training in Civil Engineering, Faculty of Technical Education, King Mongkut's University of Technology North Bangkok, Bangkok, Thailand
[2]Centre of Excellence in Structural Dynamics and Urban Management and Department of Civil and Environmental Engineering Technology, the College of Industrial Technology, King Mongkut's University of Technology North Bangkok, Bangkok, Thailand

Abstract. This paper aims at investigating bond behavior of steel bars embedded in low strength concrete compared to those embedded in normal strength concrete. The bond behavior of steel bars was studied using bar pull-out tests on 12 cubic specimens. The influence of bar diameter and concrete compressive strength on bond behaviour of embedded steel bars were investigated in this study. The load-slip behavior of all specimens was recorded during the test. The test result showed that the load-slip behavior of all specimens agreed well with the test report in literature. The splitting failure occurred in low strength concrete specimens when cracks in the surrounding concrete were caused by wedging action of the lugs of the steel bar. Failures occurred suddenly with the formation of the longitudinal cracks. For normal strength concrete specimens, the average peak axial load was significantly higher than those of low strength concrete specimens. The test results revealed that bond strength of the embedded steel bars was significantly dependent on compressive strength of concrete. It was noticed that when the strength of concrete is very low, the bond strength seems not to be proportional to square root of concrete strength (fc') as recommended by design provisions. The test results also showed that the size of steel bar has less influence on bond strength of steel bars compared to concrete compressive strength. In addition, the ACI code (ACI 308-08) was also used to calculate the bond strength of the test specimens. It was found that the experimental-to-design ratios were higher than 1.0 for all cases.

1. Introduction

The low strength concrete is prevalent in many old buildings and common in current detailing practice in regions of lower seismicity [1]. This type of concrete is believed to generally exist in many developing countries where the quality of design and construction material are very poor. The collapse of such low strength concrete under earthquake could be found in much country. For example, in 2008 Wenchuan earthquake, many buildings in Sichuan with poor quality concrete [2] were seriously damage and collapse under the earthquake. For example, in 2008 Wenchuan earthquake, many buildings in Sichuan with poor quality concrete [2] were seriously damage and collapse under the earthquake. After the Kobe Earthquake in 1995, seismic evaluations in Japan have found many existing RC buildings to have very low concrete strength (less than 13.5MPa) [3]. The poor quality concrete due to using very small size of poor quality aggregate was found as shown in figures 1(b). The problem may be aggravated when reinforcement detailing is also poor in beam and column. Under an earthquake these main structures may seriously damage as shown in figures 1(a).

IOP Conf. Series: Materials Science and Engineering **507** (2019) 012026 doi:10.1088/1757-899X/507/1/012026

(a) Poor quality concrete [5] in China (b) Poor graded sand in South East Asia[4] (c) Poor graded gravel in South East Asia[4]

Figure 1. The poor quality concrete in regions of lower seismicity.

Among many parameters influence reinforced concrete behaviour under an earthquake, bond behaviour of steel bars embedded in low-strength concrete is of most interest in this study. Therefore, the research presented herein is aimed at investigating bond behaviour of low strength concrete using a pull-off test compared to those embedded in normal strength concrete. The better understanding of bond behaviour of steel bar embedded in low strength concrete could be further used to improve the design and construction method to reduce the risk of bond failure in sub-standard concrete structure.

2. Experimental program

2.1. Test Specimen.

The experimental investigation was designed to study the influence of concrete strength and bar diameter on bond strength of embedded steel bars. A total of 12 pull-out specimens were casted using low strength concrete and normal strength concrete. The compressive strength of concrete in this study is 7.5 MPa and 26 MPa for low and normal strength concrete respectively. The deformed bars using in this study have diameter of 12 and 16 mm. Detailing of test specimens are shown in figure 2

Table 1. Test-Matrix for pull-out specimens and properties of materials used.

Mix Designation	Deformed bar	
	12	16
7.5 MPa	3	3
26 MPa	3	3
Total	6	6

From figure 2, the cross-section of the cube is 150 mm x 150 mm and the embedment length of the bar (l_e) is 140 mm. The steel rebar was embedded into the centre of concrete with 600 mm of the rebar length projecting out at one end and 10 mm of the rebar on the other end. This is to ascertain that the measurements for rebar slip at the other end could be correctly measured and the rebar can be tightly gripped for pull-out tests. Table 1 summarized the test matrix used in this study. All pull-out specimens were tested 28 days after casting.

IOP Conf. Series: Materials Science and Engineering **507** (2019) 012026 doi:10.1088/1757-899X/507/1/012026

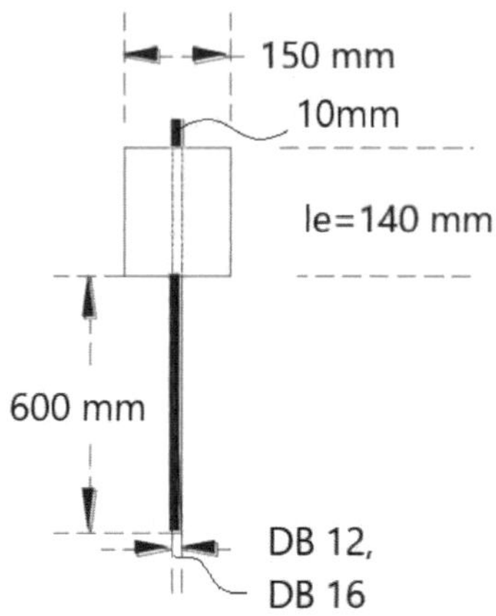

Figure 2. Detail of pull-out specimens

Figure 3. The universal testing machine.

2.2. Material properties and concrete mix proportions.

The materials used in this investigation include Type I cement, water, sand and aggregate for mix design of low strength concrete. In this study, dry ready mixed concrete was used for casting the normal strength concrete. Two different concrete mixes of 7.5 and 26 MPa compressive strength were designed and used. The concrete mix proportions and the basic properties of these mixes were determined following ASTM standards and presented in table 2.

Table 2. Summaries of mix properties for concrete.

Mix type	Design strength(MPa)	Cement(kg/m^3)	Water(kg/m^3)	Sand(kg/m^3)	Aggregate(kg/m^3)
7.5 MPa	7.5	160	180	1000	960
26MPa	26	220	175	834	1152

Table 3. Properties of Steel rebars.

Nominal Diameter (mm)	Elongation (%)			Yield Strength (MPa)			Ultimate Strength (MPa)		
DB12	30	30	30	538	542	537	683	688	689
DB16	32	32	30	563	547	555	654	654	654

The reinforcement properties of the test rebars were shown in table 3. The yield and ultimate strength of rebar were obtained from typical tensile strength test. At least 5 rebars from the same batch of those used in pull-out test were selected in this tensile strength test. Only average values were presented here in the table.

In this pull-out test, the universal testing machine as shown in figure 3 and 4 was used under displacement-controlled procedure. The concrete cubes were rested against steel plate and allowing the rebar penetrating through the hold as shown in figure 4. The bottom grip was used to hold the tip of the steel bar tight and gradually pull it down. The pulling force and bar slip at the end of the cube were monitored throughout the test.

3. Results and discussion

The test results of all specimens were presented in table 4. The peak load together with calculated bond strength and slip at peak load including mode of failure were also shown in table 4. The equation below was used to calculate bond strength.

$$\tau = \frac{F}{\pi d_b l_e} \tag{1}$$

where:

τ = experimental bond strength (MPa)

F = ultimate axial tension force (kN)

d_b = nominal rebar diameter (mm)

l_e = embedment length (mm).

According to various code provisions including ACI 318-08, the bond strength and the corresponding development length are related to the square root of the concrete compressive strength. This implies that the relationship between bond strength and the square root of compressive strength is linear. In this study the normalized bond strength was introduced in this study. The normalized bond strength was defined as the calculated experimental bond strength divided by $\sqrt{F_c'}$. The normalized bond strength (τ_{nz}) can be expressed as given in equation (2):

$$\tau_{nz} = \frac{\tau}{\sqrt{F_c'}} \qquad (2)$$

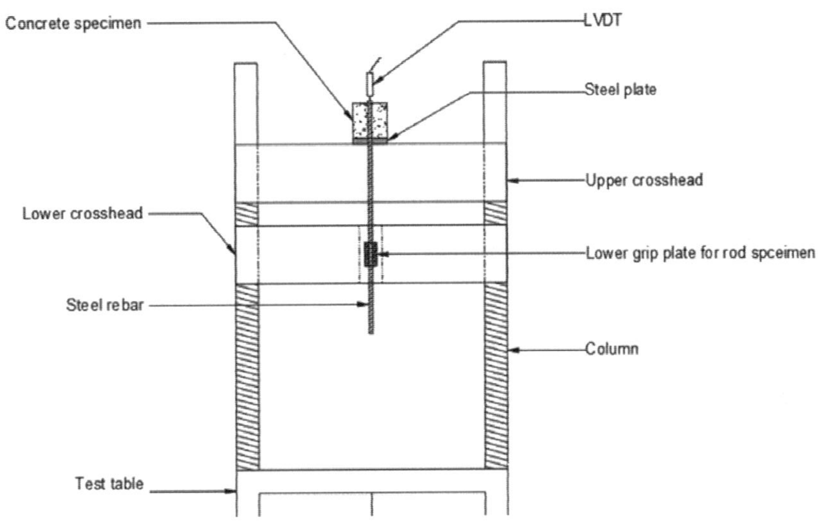

Figure 4. Schematic for details of Test set-up.

From the experimental results shown in Table 4, it can be observed that for the same embedment length, the bond strength of 7.5 MPa concrete ranged from 3.68 MPa to 3.87 MPa, whereas the bond strength for 26 MPa concrete ranged from 7.95 MPa to 9.88 MPa.

Table 4. Results of Pull-out tests for embedment length of 140 mm.

Specimen & Mix Designation	Compressive Strength (f_c') (MPa)	Nominal Bar Diameter (mm)	Peak Axial Load (kN)	Max. Bond Strength τ(MPa)	Normalized Bond Strength τ_{nz}(MPa)	Slip at Peak Load (mm)	Mode of Failure
LS7.5-DB12	7.5	12	20.43	3.87	1.41	1.07	Splitting
LS7.5-DB16		16	25.85	3.68	1.34	0.66	Splitting
NS26-DB12	26	12	52.10	9.88	1.94	1.23	Slipping
NS26-DB16		16	55.89	7.95	1.56	0.48	Splitting

IOP Conf. Series: Materials Science and Engineering **507** (2019) 012026 doi:10.1088/1757-899X/507/1/012026

From the crack pattern and damage observation, the bond strength was governed by a pull-out failure where the concrete between the steel ribs has be sheared off and the rebar slips in a frictional mode of failure. When the bar is embedded in low strength concrete, splitting through the entire concrete cover occurs. From failure mode in figure 5, it was observed that the specimens LS7.5-DB12, LS7.5-DB16 and NS26-DB16 were failed by splitting crack. Such cracks in the surrounding concrete are caused by the wedging action of the lugs of the steel bar. Failures occurred suddenly with the formation of longitudinal cracks. When cracking appears, the bond forces are directed outward from the bar surface and this causes anchorage failure resulting in splitting off of the confining concrete. However, only for the specimen NS26-DB12, the slipping failure of rebar occurs.

(a) Specimens for 12 mm rebar (Mix: 7.5)

(b) Specimens for 12 mm rebar (Mix: 26)

(c) Specimens for 16 mm rebar (Mix: 7.5)

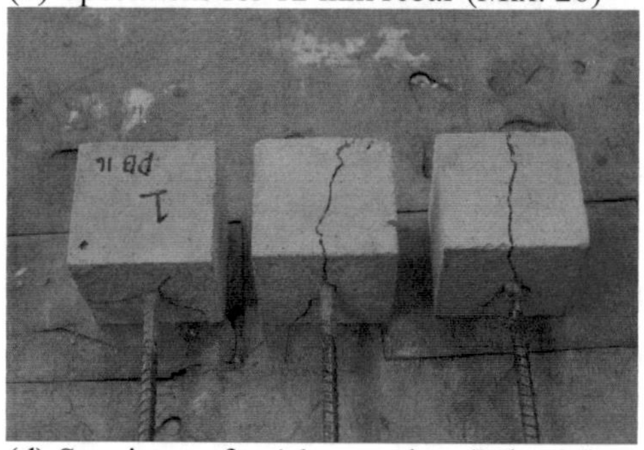

(d) Specimens for 16 mm rebar (Mix: 26)

Figure 5. Pull-out specimens after testing.

The load-slip relationships for all the specimens tested in this investigation are shown in figure 6 (a-d). It was found that, for all low strength concrete specimens, the average peak pull-off loads were 20.83 kN and 25.85 kN for specimen LS7.5-DB12 and LS7.5-DB16 respectively. For normal strength concrete specimens, the average peak pull-off loads were 52.10 kN and 55.89 kN for specimens NS26-DB12 and NS26-DB16 respectively. The average peak pull-off loads of normal strength concrete specimens were significantly higher that those of low strength concrete specimens.

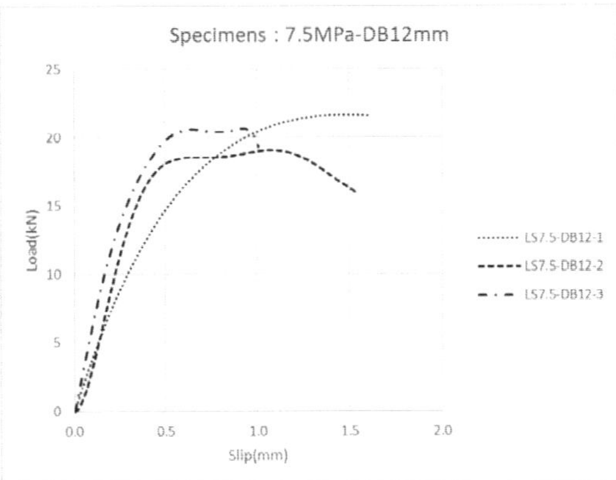

(a) Load-slip relationship for 12 mm rebar (7.5)

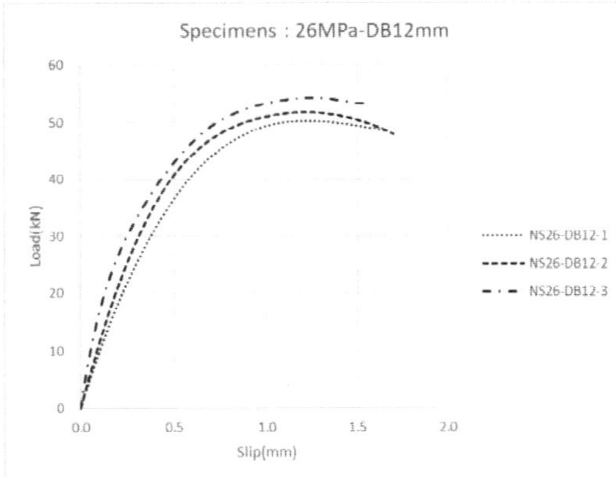

(b) Load-slip relationship for 12 mm rebar (26)

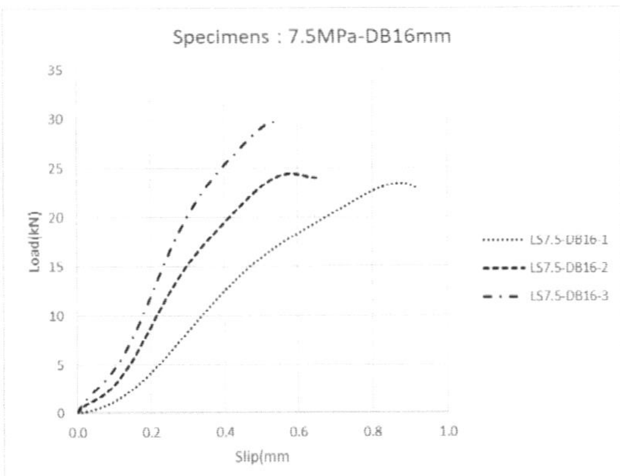

(c) Load-slip relationship for 16 mm rebar (7.5)

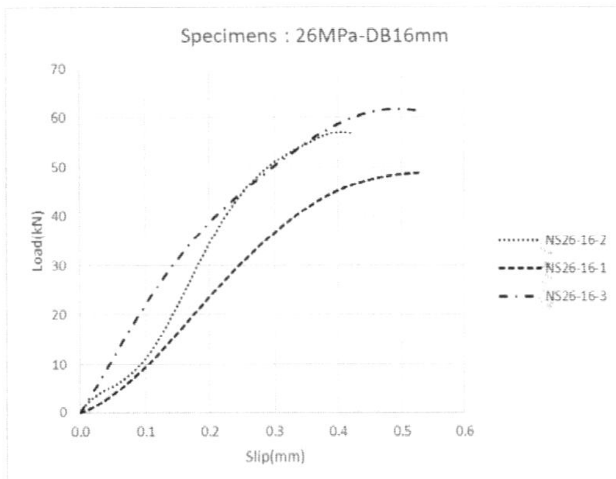

(d) Load-slip relationship for 16 mm rebar (26)

Figure 6. Load-slip relationship for steel bars.

4. Comparison with code recommendations and predicting equation

The bond strengths measured in this test were compared with the design bond strengths recommended by ACI 318-08. The expressions used in calculating the development length L_d are presented in Sections 12.2.2 in ACI 318-08 (equation 3) [6] for 19 bars and smaller:

$$L_d = \left(\frac{f_y \psi_t \psi_e}{1.4\lambda\sqrt{f_c'}} \right) d_b \tag{3}$$

Here, the bond strength F_b has been derived from these expressions for the ultimate state conditions from White and MacGregor in 2008[7]

$$F_b = \frac{1}{4}\frac{f_y d_b}{l_d} \tag{4}$$

where

f_y = the steel yield strength

ψ_t = reinforcement location factor (1.00)

ψ_e = coating factor (1.00)

λ = aggregate factor (1.00)

Equation 3 is intended for many practical construction cases which allows the designer to account for the actual values of the variables that control the development length. These expressions and their factors were determined in accordance with the conditions that pertain to the current tests.

The experimental-to-design bond strength ratios (EDR) was shown in figure 7 where the design strengths were calculated according to equation 4. The Figures show that the experimental-to-design ratios were higher than 1.0 for all cases when the bond strength was calculated using the ACI 12.2.2 equation of the ACI 308-08. It is also evident that the EDR ratios of low strength specimens is significantly lower than the EDR ratios of the normal strength specimens.

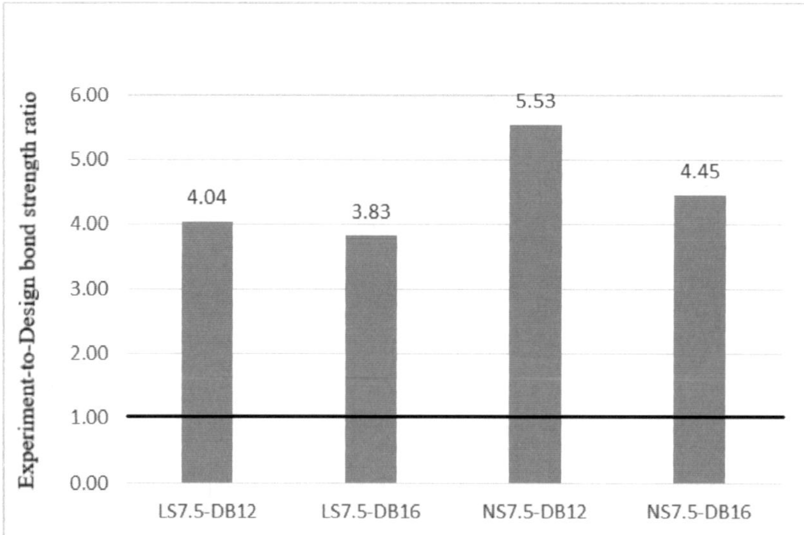

Figure 7. Experiment - to - Design bond strength ratio

5. Conclusions

This paper presents an experimental and numerical study on pull-out tests of steel bars embedded in low strength concrete (LSC) compared to those embedded in normal strength concrete. The research results are summarized as follows,

1. The load-slip behavior of all specimens was recorded and it was found that the load-slip behavior of all specimens agreed well with the test report in literature.

2. The splitting failure occurred in low strength concrete specimens when cracks in the surrounding concrete were caused by wedging action of the lugs of the steel bar. Failures occurred suddenly with the formation of the longitudinal cracks. For normal strength concrete specimens, the average peak axial load was significantly higher than those of low strength concrete specimens.

3. The test results revealed that bond strength of the embedded steel bars was significantly dependent on compressive strength of concrete. When the concrete strength is very low, the bond strength is also reduced. It was noticed that when the strength of concrete is very low, the bond strength seems not to be proportional to square root of concrete strength (fc') as recommended by design provisions.

4. The test results also showed that the size of steel bar has less influence on bond strength of steel bars compared to concrete compressive strength. In addition, the ACI code (ACI 308-08) was also used to calculate the bond strength of the test specimens. It was found that the experimental-to-design ratios were higher than 1.0 for all cases.

Acknowledgments

This research was supported by the research fund of King Mongkut's University of Technology North Bangkok.

6. References

[1] Wibowo A, Wilson J L, Gad E F and Lam N T K 2011 *Drift Capacity of Lightly Reinforced Soft Storey Structures*, Proceedings of the Ninth Pacific Conference on Earthquake Engineering. (Auckland: University of Auckland Press).

[2] Longwei C, Xiaoming Y, Zhenzhong C, Longqing H, Rui S, Lin D, Weiming W, Fanchao M and Hongjuan C 2009 *Earthq. Eng. Eng. Vibr.* **8** 219.

[3] Hong C and Araki H 2012 *Bond Characteristics between Low Strength Concrete and Plain Round Bar under Reversal Loading*, Proceedings of World Conference Earthquake Engineering, pp. 517.

[4] Willison R D 2008 *Handbook on Good Building Design and Construction In the Philippines* (Manila: German Technical Cooperation), pp. 54.

[5] EERI/GEER Reconnaissance Team Field Report 2008 *The Wenchuan, Sichuan Province, China, Earthquake of May 12 2008* (EERI Special Earthquake Report) pp. 6.

[6] ACI318-08 *Building Code Requirements for Structural Concrete and Commentary* (Detroit Michigan: American Concrete Institute)

[7] White J K, MacGregor J G 2008 *Reinforced Concrete: Mechanics and Design* (New Jersey: Prentice-Hall), pp. 1110.

IOP Conf. Series: Materials Science and Engineering **507** (2019) 012027 doi:10.1088/1757-899X/507/1/012027

Simulating Metamodel for Urban Bus Seats Design

C H Arroba[1], D F Núñez[1], F A Peña[1], E G Ruiz[1], S P Cabrera[1], M E Ureña[1]

[1]Ingeniería Mecánica, Facultad de Ingeniería Civil y Mecánica, Universidad Técnica de Ambato, Ecuador

Abstract. The finite element analysis is useful to predict the displacements and strength of the components of an urban seat bus. However, this analysis has a high computational cost when explicit dynamic technique is applied. This work uses Ansys LS Dyna software to carried out a complete 3D static test of an urban seat bus according with the Regulation No. 80 of the Economic Commission of the United Nations for Europe. To reduce the computational cost, we applied a bidimensional frame analysis (metamodel) using direct matrix method to obtain nodal displacements in 12 different models. The variables of this study were: thickness, sections and length of structural elements in order to optimize the strength of the seat. The results showed a range of variation about 0.1 % to 3.0% between finite elements analysis and metamodels Finally, it was proved that metamodels reduce the time analysis and can be used in a pre-design stage

1. Introduction

In the automotive industry, structural seat design is one of the most important aspect because it is necessary to ensure the users safety requirements. By other hand, the correct structural seat design could decrease the fuel consumption and environmental pollution due to reduction of the seat weight, especially in the large vehicles like the bus passengers [1, 2]. When the bus suffers a frontal crush, it can produce a "domino effect" on the seats, causing injuries in the passengers. It is due to a failure in the anchorage strength and breakage of the seat structure [3-5]. Bus seats must be able to resist the hit of head and knee of the back passenger; at the same time the structure will not be too rigid to absorb the energy during an accident, as well as the seat anchors must not be pulled off from the bus floor [6]. The optimization of the seat structure weight can be obtained substituting the conventional materials and applying modern design techniques. [1, 7, 8].

In order to guarantee the safety of the passenger seat, Regulation No. 80 of the United Nations Economic Commission for Europe establishes requirements for approval large vehicles seats, for passengers transportation concerning the resistance of the seats and their anchors. Its regulation stipulates: a) the seats must be installed facing forward in vehicles of categories M2 and M3, classes II, III and B; b) vehicles of categories M2 and M3, classes II, III and B, concerning to the anchorages of passenger seats and the seats installation [9].

The time used to structural design and optimization of the seat can be reduced applying metamodels, they have been used for particular applications, like in Abasolo's investigation [10] where, by a discretization of the component and using two freedom degrees elements and equivalent stiffness, obtained from the three-dimensional simulation of the component by finite element software, a metamodel was elaborated to simulate the sequences of screwed joints of wind turbines and the results obtained were

IOP Conf. Series: Materials Science and Engineering **507** (2019) 012027 doi:10.1088/1757-899X/507/1/012027

very similar comparing to the finite element method. Other research used a metamodel [11], to simulate dummy vibration modes in a car seat, obtaining a reduction of computational analysis time from 36000s to 36s with acceptable margin of error to compare with original model simulation.

The objective of this work is to develop a parameterized metamodel to estimate the structure displacements of the urban bus seat according to the static test described in the ECE R80 regulation, based on the results obtained by simulations using finite element analysis in the software Ansys Ls-Dyna.

2. Methodology

The model used is an individual urban plastic seat. It has two structural steel pillar to support the seat and the structure is attached to the bus floor by 10mm bolts (figure 1a, 1b)

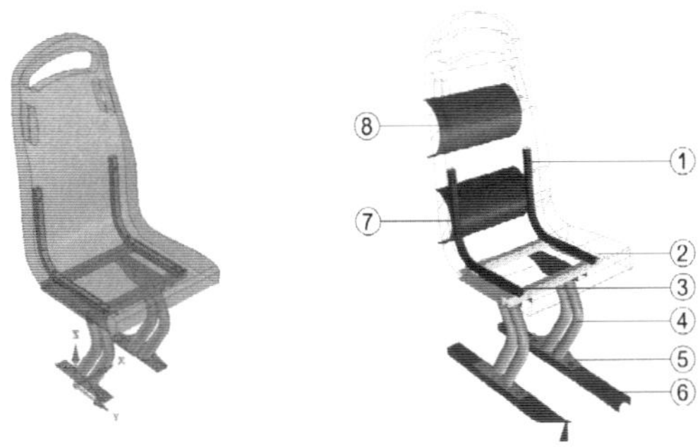

a) b)

Figure 1. Parts of the seat structure (1. Back support 2. Base support 3. Connection of the pillar base 4. Pillar 5. Connection between pillar and bus structure 6. Bus structure 7. Low impactor 8. Top impactor).

The 3D CAD geometry of the seat structure was modeled in Ansys SpaceClaim software. The conditions for the experiment are described in Appendix 5 about the requirements and procedure related to the static test in the ECE-R80 regulation [9]. An explicit dynamic analysis of the seat structure was carried out using shell elements. The maximum size of elements was 5mm, the shear correction factor was 0.83333, the number of integration points 3, the thickness of elements is considered uniform. The thickness of back support, base support, connection of base to pillar, pillar, connection between pillar and bus structure were 2 mm and materials according to table 1.

The bus structure (part 6, figure 1b) are restricted in all degrees of freedom. The seat elements and impactors had automatic contact (* CONTACT AUTOMATIC SINGLE SURFACEID); In addition, penetration between parts was restricted. The impactors were assigned as totally rigid bodies [14].

The properties of materials used in the seat components were defined by modeling materials MAT ELASTIC and MAT PIECEWISE LINEAR PLASTICITY, the last one allows define the behavior of the elastic and plastic zone [15]. These properties are shown in Table 1. The applied load in the upper impactor was $F1 = 1400N$ and in the lower impactor $F2 = 4000N$ those in the horizontal direction and constant during 0.2 seconds.

Table 1. Features of material models used in Ansys-LsDyna.

Properties of MAT_ELASTIC for seat belt							
Density [kg/m^3]	Young module [MPa]	Poisson coefficient					
1390	110000	0.3					
Properties of MAT_ELASTIC for the joints							
Density [kg/m^3]	Young module [MPa]	Poisson coefficient					
7890	200000	0.3					
Properties of MAT_PIECEWISE_LINEAR_PLASTICITY for ASTM A500 steel							
Density [kg/m^3]	Young module [MPa]	Poisson coefficient	Plastic deformation of break	Yield strength [MPa]	Plastic deformation of failure	C Deformation velocity	P Deformation velocity
7890.00	200000.00	0.30	0.283	320.00	0.370	40.00	5.00
Properties of MAT_PIECEWISE_LINEAR_PLASTICITY for ASTM A500 no liner behavior							
0.0000	0.01092	0.03959	0.04388	0.04817	0.1431	0.2545	0.4000
320.0	320.00	347.16	351.22	355.28	382.07	341.91	0.0

The simulation results of the complete seat let to generate a metamodel with the quality of reproducing the displacements of the structure elements after applying the loads for static test, considering the structural element where the Force 1 of the impactor take place will be displaced horizontally, also to rotating it around the z-axis, and where Force 2 of the lower impactor is located, there will be a horizontal displacement and a rotation around z too. Consequently, the metamodel that is needed for the proposed case must have elements with three freedom degrees in the plane, we concluded, the displacements that influence in greatest way on the overall seat behavior of the during the test are: along the axis X, Y axis as well as around the Z axis.

The frame type element was used for the schematization of the metamodel, it counts with three freedom degrees, equivalent section and equivalent stiffness correspond to the seat structure parts determined from the Ls-Dyna simulation. The elements were divided according to the points which is necessary to apply loads or know their displacement, taking as a reference the complete seat simulation result. One node was placed in each point to determine the vertical displacements Y, horizontal X and rotational around the Z axis, obtaining a model that is shown in figure 3.

From the obtained metamodel, parts thicknesses, profiles sections, pillar angle, elements length, magnitude and application of loads were parameterized, facilitating the structure optimization and verifying its resistance. The identified variations in the parametrized metamodel were verified by three-dimensional Ls-Dyna simulation.

3. Simulated metamodel

3.1. Metamodel obtaining

By seat structure simulating in Ls-Dyna were identified the displacements in the points where the loads were applied with values around 113 mm in the upper impactor and 83 mm in the lower impactor as shown in figure 3a). From the three-dimensional model, nodes 3667, 2964 and 389 were labeled, which provide us the maximum displacements required for verifying compliance with the ECE R80 regulation. The other nodes were determined based on variation of section and stiffness of the parts, this supported to determine the more approximated metamodel to the three-dimensional

IOP Conf. Series: Materials Science and Engineering **507** (2019) 012027 doi:10.1088/1757-899X/507/1/012027

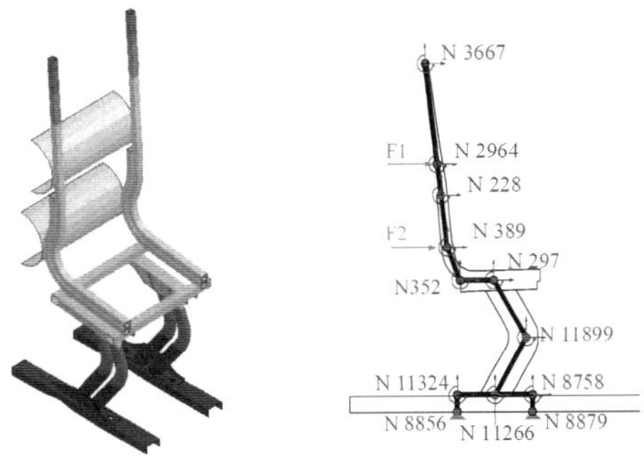

a) b)

Figure 2. a) Deformed structure b) Obtaining metamodel nodes.

The metamodel of 12 nodes and 11 elements shown in figure 3. The element L1 with its nodes 1 and 2 represents a bolt, node 1 is placed as fixed support restricting in all degrees of freedom, while node 2 will have the six degrees of freedom simulating the interaction between the pillar connection with the bus structure and bolt. Element 2 and its nodes 3 and 4 represent the other bolt, node 3 is considered similar to node 1 and node 4 as node 2, representing the interaction between the connection on the other side.

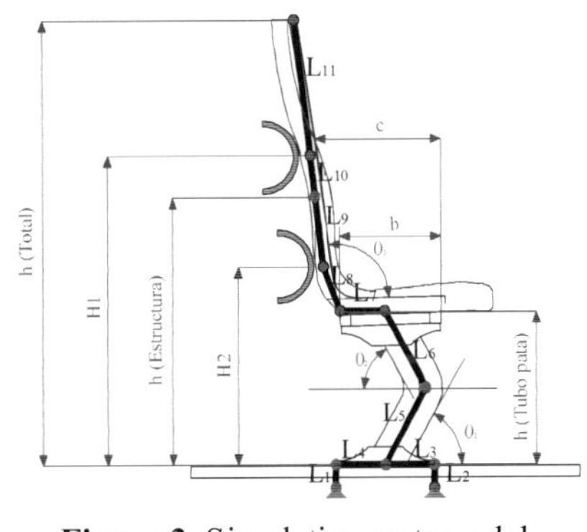

Figure 3. Simulation metamodel.

The element L3 and L4 with the nodes 2, 4 and 5 represents the connection between pillar and structure of the bus where the pillar component joins. The pillar is represented in two sections by elements L5 and L6 with nodes 5, 6 and 7; the last one represents the connection of the base with pillar, the element L7 with its respective nodes 7 and 8 represents the base support, the elements L7, L8, L9, L10, and L11 with the nodes 7, 8, 9, 10, 11 and 12 represents the back support. This discretization was made considering the points of equivalent section change, equivalent stiffness change, contact of the impactor with the back support, the maximum seat height and results of three-dimensional finite elements analysis.

3.2 Metamodel acceptance

The equation that solves the metamodel, considering the discretization of figure 3, is necessary to determine the displacement on the X axis of node i, δ_{xi};; displacement on the Y axis of node i, δ_{yi}; θ_i rotational displacement around the z axis of node i, obtaining a 12x3 freedom degrees system.

The forces produced by the upper and lower impactor on the back support are called F_{x11}, F_{x9} respectively. The constitutive matrix equation of each element corresponds to (1) and (2):

$$F_i = [k_i] \times [\delta] \tag{1}$$

$$[k_i] = \frac{E_i}{L_i}\begin{bmatrix} A_iC_i^2 + \frac{12I_i}{L_i^2}S_i^2 & \left(A_i - \frac{12I_i}{L_i^2}\right)C_1S_1 & -\frac{6I_i}{L_i}S_i & -\left(A_iC_i^2 + \frac{12I_i}{L_i^2}S_i^2\right) & -\left(A_i - \frac{12I_i}{L_i^2}\right)C_iS_i & -\frac{6I_i}{L_i}S_i \\ \left(A_i - \frac{12I_i}{L_i^2}\right)C_iS_i & A_iS_i^2 + \frac{12I_i}{L_i^2}C_i^2 & \frac{6I_i}{L_i}C_i & -\left(A_i - \frac{12I_i}{L_i^2}\right)C_iS_i & -\left(A_iS_i^2 + \frac{12I_i}{L_i^2}C_i^2\right) & \frac{6I_i}{L_i}C_i \\ -\frac{6I_i}{L_i}S_i & \frac{6I_i}{L_i}C_i & 4I_i & \frac{6I_i}{L_i}S_i & -\frac{6I_i}{L_i}C_i & 2I_i \\ -\left(A_iC_i^2 + \frac{12I_i}{L_i^2}S_i^2\right) & -\left(A_i - \frac{12I_i}{L_i^2}\right)C_iS_i & \frac{6I_i}{L_i}S_i & A_iC_i^2 + \frac{12I_i}{L_i^2}S_i^2 & \left(A_i - \frac{12I_i}{L_i^2}\right)C_iS_i & \frac{6I_i}{L_i}S_i \\ -\left(A_i - \frac{12I_i}{L_i^2}\right)C_iS_i & -\left(A_iS_i^2 + \frac{12I_i}{L_i^2}C_i^2\right) & -\frac{6I_i}{L_i}C_i & \left(A_i - \frac{12I_i}{L_i^2}\right)C_iS_i & A_iS_i^2 + \frac{12I_i}{L_i^2}C_i^2 & -\frac{6I_i}{L_i}C_i \\ -\frac{6I_i}{L_i}S_i & \frac{6I_i}{L_i}C_i & 2I_i & \frac{6I_i}{L_i}S_i & -\frac{6I_i}{L_i}C_i & 4I_i \end{bmatrix} \tag{2}$$

Where:

Ci = Cosine of the angle at which the element is oriented
K_i= Element equivalent stiffness
A_i= Area or equivalent section of the element
L_i= Element length
I_i= Inertia of the element
E_i= Elastic module of each element
S_i= Seno of the angle to which the element is oriented
C_i= Cosine of the angle at which the element is oriented

The assembled global equation of the metamodel includes the elements from 1 to 11 corresponds to a 36x36 matrix. From the solution of the mathematical model, considering the boundary conditions of each node, obtaining the displacements in the X, Y, and rotation around the Z axis that compared to the results of the Ls-Dyna simulation a maximum error of 2.98% was recorded, indicated in table 2.

Table 2. Obtained displacements in Ls-Dyna comparing with metamodel.

Node		Ls-Dyna Simulation		Metamodel		% error	
Metamodel	Ls- Dyna	Horizontal	Vertical	Horizontal	Vertical	Horizontal	Vertical
1	8856	-0,26	3,01	0,00	0,00	1,00	1,00
2	11324	0,41	11,17	-0,13	0,02	1,30	1,00
3	8879	0,02	-3,41	0,00	0,00	1,00	1,00
4	8758	-0,11	-3,96	0,21	-0,02	2,89	0,99
5	11266	-0,07	3,23	0,06	-0,11	1,95	1,03
6	11899	20,92	-9,34	23,90	-13,87	0,14	0,48
7	297	43,82	0,60	48,52	0,34	0,11	0,43
8	352	44,52	13,24	48,52	17,77	0,09	0,34
9	389	61,13	25,77	67,97	37,20	0,11	0,44
10	228	88,07	24,28	85,06	38,69	0,03	0,59
11	2964	103,30	22,29	112,42	41,09	0,09	0,84
12	3667	167,10	29,50	159,21	45,18	0,05	0,53

The experiment was repeated modifying the thickness of the back support, base support and pillar with values of 1, 2, 3mm, obtaining a maximum variation of 2% between the metamodel and LS-Dyna results shown in table 3. The greatest variation occurs at lower thicknesses and this could be due to effect of plastic deformation in the elements.

Table 3. Results variation of displacements analyzed by metamodel and FEM.

Part	Thickness (mm)	Variation
Pillar	1	2%
	2	1%
	3	1%
Base support	1	1%
	2	1%
	3	2%
Back support	1	2%
	2	1%
	3	1%

4. Conclusion

A metamodel and parametric model has been created for an urban bus seat, it allows knowing the displacements of the seat structure after applying the loads established in the ECE R80 regulation and verifying the compliance with its resistance, with variations less than 2% compared with results of the finite element analysis. In addition, a new methodology has been developed, by using of metamodel and parameters variation, allows be able to optimize thicknesses, sections, lengths of the elements, improving the model structure design and reducing computational costs and time in the design stage.

References

[1] Oliveira J, Teixeira P, Lobo G, Duarte J and Reis A, 2013 *Key Eng. Mater.* **554** 681-688
[2] Hu C Q 2014 *Appl. Mech. Mater.* **477** 99-104.
[3] Minakshi Das A E, Seshagiri Rao G V R 2014 *Int. J. Eng. Sci. Res. Tech.* 208-212.
[4] Walber M, Meira Jr A D d, Goedel F, Fazio E A, Izquierdo F A and Iturrioz I 2014 *Latin American J. Sol. Struct.* **11(9)** 1463-89.
[5] Xu Z W, Sheng Z Q, Zhang H H and Liu Y X 2009 *Appl. Mech. Mater.* **16** 178-82.
[6] Veeranjaneyulu M and Ravikanth D 2015 *Int. J. Res.* **2(10)** 1084-8.
[7] Kim H S, Lee Y S, Yang S M, and Kang H Y 2016 *Int. J. Prec. Eng. Manuf. Green Tech.* **3(1)** 75-9.
[8] Patil P M and Kumar M S 2016 *J. Adv. Eng. Res.* **3(1)** 72-7.
[9] ECE, 2013, "Regulation N° 80: Uniform provisions concerning the approval of seats of large passenger vehicles and of these vehicles with regard to the strength of the seats and their anchorages" Ed: United Nations.
[10] Abasolo M 2011 "*Metamodel for the simulation and optimization of the sequences of screwed joints in wind turbines towers*," PhD Thesis, University of the Basque Country.
[11] Kang S J and Chun B K 2000 *Thin-Walled Struct.* **37(2)** 113-25.

IOP Conf. Series: Materials Science and Engineering **507** (2019) 012028 doi:10.1088/1757-899X/507/1/012028

A Study of Exhaust Waste Heat Recovery in Internal Combustion Engines

Quang Khong Vu[1], Dien Vu Minh[2], Tien Nguyen Duy[1] and Tuan Pham Minh[1], Luong Nguyen The[1]

[1]Department of Internal Combustion Engine, School of Transportation Engineering, Hanoi University of Science and Technology, 1 Dai Co Viet Road, Hanoi, Viet Nam
[2]Department of Automobile Engineering, Hanoi University of Industry, Km 13, 32 Road Minh Khai - Tu Liem – Hanoi, Viet Nam

Abstract. This research presents an investigation of an energy recovery solution from exhaust gases in internal combustion based on the heat exchange to distil fresh water from the sea water. Consequently, an optimization of the flow field design of the heat exchanger was performed using the commercial computational fluid dynamics (CFD) software. The result of this research showed that the energy recovery performance of the exchanger strictly depends on the engine load, which can reach a value of approximately 33% at the full load condition of the research engine. These results might forecast a future application of the optimized exchanger's flow-field design in automotive waste heat recovery in order to improve the internal combustion engine efficiency.

1. Introduction

Recently, internal combustion engine has been considered as the major power sources for worldwide applications such as industry, transportation in over the world. However, the energy crisis and environmental pollution are highlighted to be one of the major global problems in the 21st century. As a result, finding the solution to enhance the internal combustion engine efficiency is the great challenge of scientists for sustainable development [1]. Many previous studies have been conducted to apply modern technologies and harvested some certain successes as mentioned in [2-4]. The use of cooling water and exhaust heat to improve the thermal efficiency of internal combustion engines is a potential solution to solve the above problems [5-9]. Especially, water distillation from the seawater using the exhaust gases energy has been performed in worldwide because it not only improves the internal combustion engine efficiency but also solve the problem pertaining to the freshwater scarcity in over the world [10].

However, due to the limitation of the technology, this solution has been only applied in the high-power ships. In the small ships, the efficiency of heat utilization is low because of the low energy from the exhaust gases, although the freshwater demand is still very high. This research aims to optimize the flow-field design of the exchanger applied in small ship engines in order to improve its efficiency. These results might forecast a future application of the optimized exchanger's flow-field design in automotive waste heat recovery in order to improve the internal combustion engine efficiency.

IOP Conf. Series: Materials Science and Engineering **507** (2019) 012028 doi:10.1088/1757-899X/507/1/012028

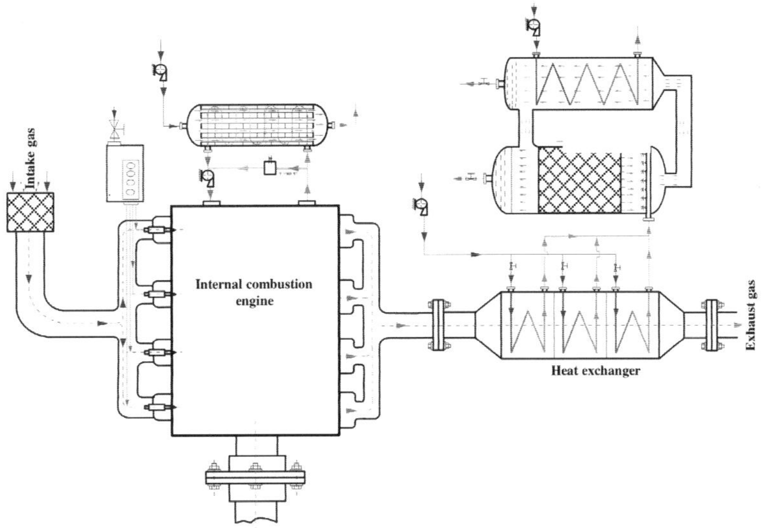

Figure 1. Equipment utilizing exhaust heat

2. The fundamental theory of the simulation software

The heat exchange process of the heat exchanger can be clarified into three stages as follows: The heat transfer process between the exhaust gas and heat exchanger wing, heat conduction in the wing and the heat transfer from the wing to the heating solvent. Meanwhile, it is so difficult to determine the heat transfer coefficients due to their variation according to their shapes, dimensions and gas ambulatory ability of the seawater inside the heat exchanger. However, ANSYS Fluent can solve these problems based on the Navier-Stokes to describe the energy exchange. The governing equations being used in this research are as follows: the continuity, momentum and energy equation.

3. Modelling of the simulation model

3.1. Proposed Model for Utilizing Exhaust Heat

The design of the heat exchanger is described in Fig. 2a. It includes 3 separate compartments. The exhaust gas normally comes into the center of tube and exchange heat with the wings inside the heat exchanger. Based on the parameters of the heat exchanger design, the 3D model was built in a CAD software and transferred to ANSYS Fluent for the numerical analysis

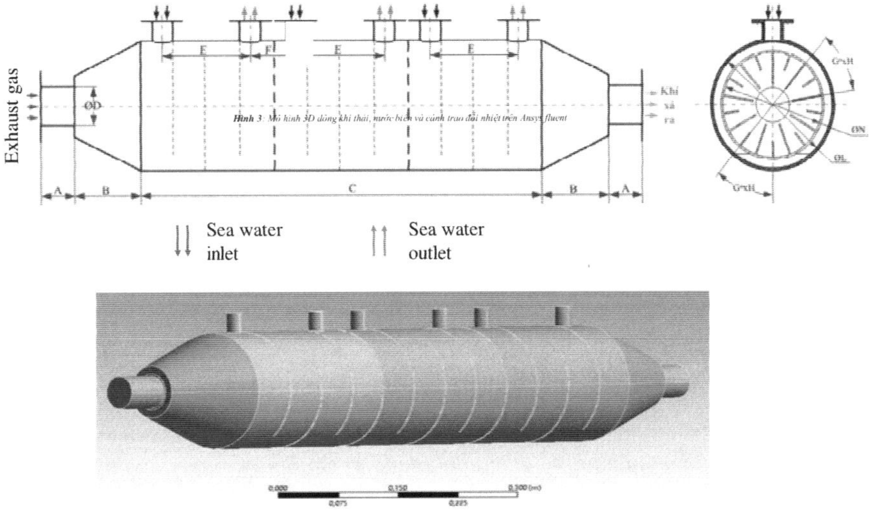

Figure 2. Design of heat exchanger tube

3.2. Boundary conditions for the simulation

In this research, the D243 engine was selected for the exhaust gas energy calculation. Based on the parameters of this engine, the engine model was built in AVL-Boost to simulate its operation. Consequently, the exhaust gas flow was calculated and considered as the inlet boundary conditions for the heat exchanger as shown in Table 1.

Table 1. Exhaust gas parameters according to the load

% load	Temperature (K)	Speed (m/s)	Flow (kg/s)
25	492	49	61.01
50	602	55	60.05
75	673	60	59.30
100	733	70	58.18

4. Result and discussion

4.1. The heat exchange analysis

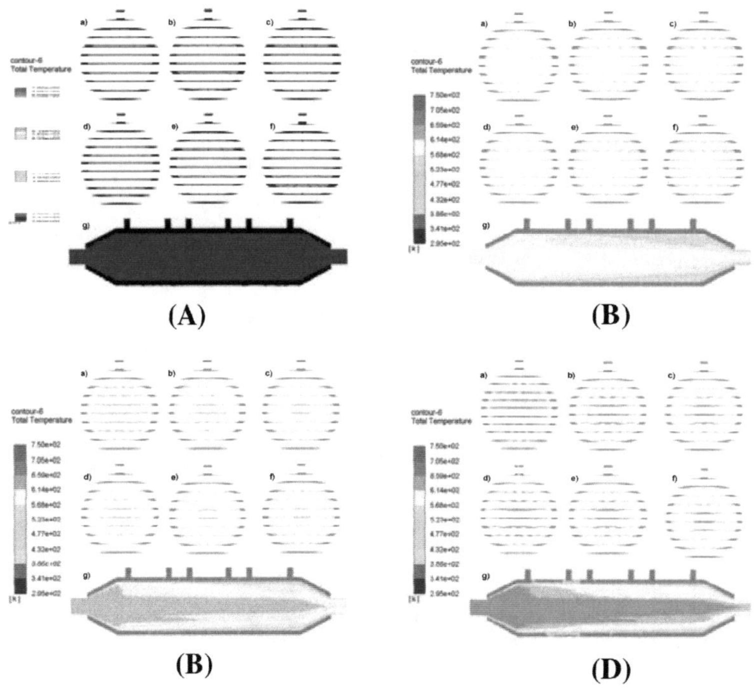

Figure 3. Temperature distributions inside the heat exchanger at different reference locations (a, b, c, d, e, f), (A) 25% load, (B) 50% load, (C) 75% load, and (D) 100% load.

Fig. 3 shows the temperature distributions of the exchanger corresponding to a variety of the engine loads and reference locations. Consequently, at a reference location, the emission temperature is distributed symmetrically. However, it tends to decrease from the center of the tube to the wing. This trend can be observed more clearly at the rear cross sections or when increasing the engine load as shown in Fig. 3. Furthermore, the engine load increase also results in the increase of temperature; as a result, the peak temperature in case of 100% load reaches to 7500 C compared to 5000 C in case of 50% load.

4.2. Effect of engine load on the heat exchanger efficiency

Fig. 4 shows the temperature decrease of the exhaust gas and the heat of the sea water received from the heat exchange process with the exhaust gas as described in Fig. 4.

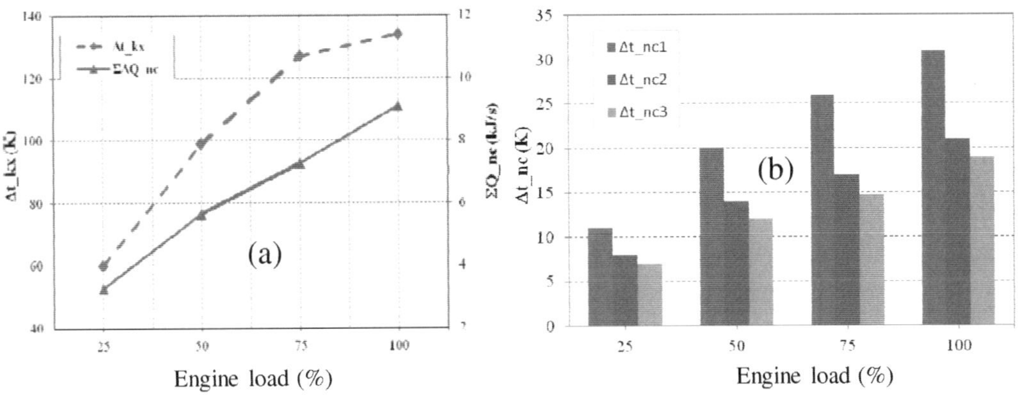

Figure 4. The decrease of exhaust gas temperature and heat exchange of the sea water according to the engine load

It is clearly observed that the decrease temperature level of the exhaust gas tends to reduce when increasing the engine load. Meanwhile, the received heat of the seawater is increased. These phenomena are caused by the temperature increase of the exhaust gas; as a result, the temperature difference between the exhaust gas and seawater is increased. In addition, the exhaust gas velocity is also increased when increasing the engine load, and thus the coefficient of heat transfer is also increased. Fig. 4b shows the effect of load on the engine load on the exhaust gas heat utilization. It is increased when increasing the engine load and can reach to a value of 33% at the full load condition of the engine. At the engine load of 50% to 75%, it is approximately 29% to 31%.

4. Conclusion

This research presents an investigation of an energy recovery solution from exhaust gases in internal combustion based on the heat exchange to distil fresh water from the sea water. The result clearly illustrates that The ability to utilize the exhaust gas heat strictly depends on the engine load and the exchanger design, which can reach a value of approximately 33% at the full load condition. These results are the foundation for applying the system design for Vietnam fishing boats

References

[1] Orr B, Akbarzadeh A, Mochizuki M and Singh R 2016 A review of car waste heat recovery systems utilising thermoelectric generators and heat pipes *Appl. Therm. Eng.* **101** 490–5

[2] Chintala V and Subramanian K A 2017 A comprehensive review on utilization of hydrogen in a compression ignition engine under dual fuel mode *Renew. Sustain. Energy Rev.* **70** 472–91

[3] Nguyen Duc K, Nguyen Duy V, Hoang-Dinh L, Nguyen Viet T and Le-Anh T 2019 Performance and emission characteristics of a port fuel injected, spark ignition engine fueled by compressed natural gas *Sustain. Energy Technol. Assessments* **31** 383–9
[4] Duc K N and Duy V N 2018 Study on performance enhancement and emission reduction of used fuel-injected motorcycles using bi-fuel gasoline-LPG *Energy Sustain. Dev.* **43** 60–7
[5] Duc K N, Tien H N and Duy V N 2018 Performance enhancement and emission reduction of used motorcycles using flexible fuel technology *J. Energy Inst.* **91** 145–52
[6] Anh T, Duy V, Thi H and Xa H 2018 Experimental Investigation on Establishing the HCCI Process Fueled by N-Heptane in a Direct Injection Diesel Engine at Different Compression Ratios *Sustainability* **10** 3878
[7] Wang E H, Zhang H G, Fan B Y, Ouyang M G, Zhao Y and Mu Q H 2011 Study of working fluid selection of organic Rankine cycle (ORC) for engine waste heat recovery *Energy* **36** 3406–18
[8] Ghazikhani M, Hatami M, Ganji D D, Gorji-Bandpy M, Behravan A and Shahi G 2014 Exergy recovery from the exhaust cooling in a DI diesel engine for BSFC reduction purposes *Energy* **65** 44–51

IOP Conf. Series: Materials Science and Engineering **507** (2019) 012029 doi:10.1088/1757-899X/507/1/012029

Modelling Fuel System Injector Hydro-Mechanical Processes

O. Vrublevskyi[1]

[1] Department of Technical of Science, University of Warmia and Mazury, Olsztyn, Poland

Abstract. This study analysed the characteristics of the fuel system of a self-ignition engine with electronic control, taking into account the characteristics of the processes in the injector and the injection control system. The results of simulations of hydro-mechanical processes in an injector with an electromagnetic controller are presented. The proposed model takes into account the propagation processes of the pressure impulse in the fuel and in the parts of the injector subjected to deformation during injection. To describe the elastically-deformable state of the injector parts, an equivalent model of the injector valve stem was used. The results of calculations are proposed for a common simulation of hydrodynamic and electro-dynamic processes occurring during fuel injection into the Common Rail (CR) system.

1. Introduction

Currently, fuel systems with electronic control are almost exclusively used in combustion engines. At the same time, research is ongoing to improve existing solutions, increase the reliability of components of this system, particularly of the CR-type electromagnetic injector [1].

This paper presents the mathematical model of the injector, which differs from other models in a number of original solutions. In hydro-mechanical fuel systems with direct injection [1], the time interval during which the elementary injection process is carried out is 0.05 - 0.1 ms, and for systems with electronic control, transient processes that are carried out in the injector volume, take a few microseconds. It follows that it is impossible to use known calculation methods to model CR injection systems [1, 2], without their proper adaptation. For a description of hydro-mechanical processes in the feed system with electronic control, dynamic models should be applied, which take into account the value of electromagnet force, fuel pressure in the injector volume, as well as the speed of displacement of the needle, control valve and anchor.

Research into injection processes with electronic control is also presented in the paper [3]. Modelling of the displacement of injector parts takes place with regard to elasticity and damping parameters. Differential equations were solved using the implicit scheme used for modelling rigid systems. The method of calculating the fuel flow in the pipeline using the Lax-Friedrichs method is also presented [4].

Other studies have examined modelling the fuel injection process using the AMESim program [5, 6]. Higher modelling accuracy can be achieved considering the degree of deformation of the injector piston and the phenomenon of cavitations [7, 8]. The results of the theoretical and experimental research on pressure conversion in the injector input cross-section occurring during the fuel injection seems to be valuable in this work.

In the AVL HYDSIM program [9], the calculation of the hydrodynamic process was proposed using the Godunov method. Using this method allows to map most of the factors that occur in the real hydrodynamic process. This method may also be used for the modelling and calculation of fuel equipment with electronic control [10].

2. Objective and scope of the study

Please follow these instructions as carefully as possible so all articles within a conference have the same style to the title page. This paragraph follows a section title so it should not be indented.

The objective of the study is to develop a mathematical model of hydrodynamic processes that occur in an injector with electronic control. The modelling takes into account factors that affect the transformation of the fuel pressure in the injector's volumes, as well as the displacement speed of the needle, control valve and anchor (figure 1).

To develop a method for calculating the parameters of a fuel system, a mathematical model is necessary, and its basis is a hydro-mechanical scheme (figure 1). This diagram takes into account changes in fuel pressure in the control chamber P_s and, in the storage tank P_z, the displacement values of the anchor h_k, the anchor silencer ho, valve h_z, bucket increaser h_t and needle h_i.

Determining the displacement characteristics of the injector valve requires knowledge of the Fm force value of the magnetic field of the electromagnet, which changes during the injection, depending on the current value, the distance between the anchor and the core and the speed of the anchor displacement.

3. Mathematical model of the injector

3.1. Fuel flow

The fuel flow parameters in the high pressure duct are determined by the following equation [4]:

$$\frac{\partial^2 C(x,t)}{\partial t^2} - a_c^2 \frac{\partial^2 C(x,t)}{\partial x^2} + K_c \frac{\partial C(x,t)}{\partial t} \tag{1}$$

where: $C(x, t)$ - fuel flow speed, t - time, x - coordinate of the distance in the high - pressure duct, K - hydrodynamic drag coefficient, α - velocity of pressure wave propagation in the fuel.

In the solution of equation (1), the method of characteristics is used [11]. The assumptions made in the diagram (figure 1) require the modernization of this method. In the fuel system [1, 2], for which the method of characteristics has been developed, the pressure impulse spreads from the pump to the injector (in one direction). This determines the order according to which the volume of the system is calculated.

Analyzing the hydro-mechanical scheme of the CR injector (figure 1), it should be noted that the pressure impulse moves from control chamber **B** to volume **C** via two paths:

– an injector piston/valve spindle and needle in the form of an elastic wave at a speed of about 5 m/ms;

– in the high pressure duct at a speed of about 1.4 m/ms.

From that, it follows that pressure waves reach the atomizer at different times. The second important feature of the scheme is that the injection process is controlled not in the fuel reservoir A but in the chamber **B**.

In the modernized method of characteristics (figure 2), two initial points corresponding to the control volume **B** and the atomizer **C** are determined. The channel between these volumes is divided into two sections of equal length. At the point where these segments connect, there is pressure of P_{tr}. At this point, the condition of pressure continuity and fuel flow velocity is assumed.

In the first section of the high-pressure channel there is a tee, which is the beginning of additional characteristics that lead to the volume **A** (up to the pressure P_z). The actual waveforms of the characteristics will be disturbed by the locally variable value of the speed of sound. The solution of the problem regarding the flow of fuel in the high pressure duct requires consideration of the initial and final conditions. The following initial conditions were adopted (ignoring the fuel return): the fuel flow velocity is zero, the fuel pressure in the **A**, **B**, and **C** volume are equal to the pressure of P_z. The final conditions are determined by differential equations of the effected fuel flow rate (effected fuel portions) balance in the injector volumes and the equations of movement of the needle, the injector piston/valve spindle and the valve taking into account their deformation. In the volumes of high pressure **A**, **C**, **B** the compressibility of the fuel is also considered. In the low pressure **E**, **K**, **L** volumes, the fuel inertia is taken into consideration when the valve, anchor and silencer displace. This is related to the existing adhesion forces between the fuel and the surfaces of the injector parts. It is therefore necessary to take into account the mass of adhering fuel to these parts during their movement.

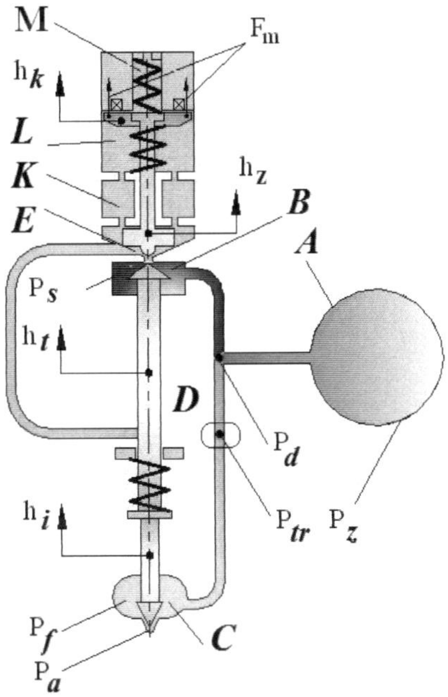

Figure 1. Hydro-mechanical scheme of the CR injector.

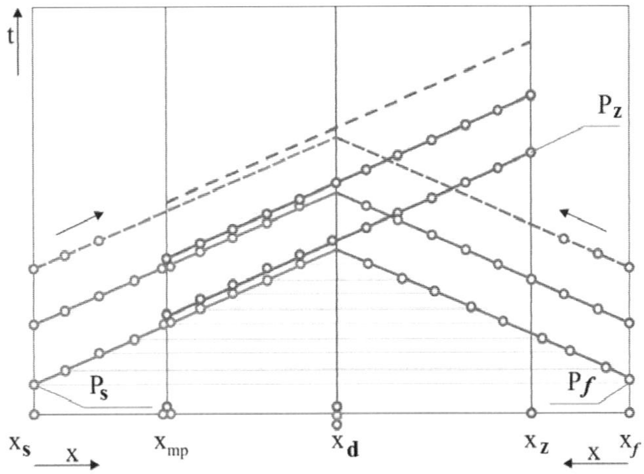

Figure 2. Network of characteristics used for equation (2) calculations.

3.2. *Modelling of elastic elements*

The spring stress (needle, check valve and silencer) is determined taking into account the variability of their turns, to which the equivalent model of the valve stem is used [12] and its analytical equations. Analogous models are used to describe the elastically-deformable condition of the needle, the injector piston/valve spindle and the control valve stem.

Equations for calculating the response q in springs and injector pistons/valve spindles with fluctuation records take the following form:

$$q\ (t) = z_c^M \cdot \begin{pmatrix} h_s + 0{,}5\,\Delta h_1(t) + \sum_{j=1}^{t/T_s} df^{\,j} \cdot \Delta h_1(t - jT_s) - 0{,}5 \cdot (1 + df) \times \\ \times \sum_{j=0}^{t/T_s - 0{,}5} df^{\,j} \cdot \Delta h_2(t - (j + 0{,}5) \cdot T_s) \end{pmatrix}$$

$$\Delta h_{1,2}(t) = \left(T_s \cdot \dot{h}_{1,2}(t) + \frac{2 \cdot (1 - df)}{1 + df} \cdot h_{1,2}(t)\right) \qquad (2)$$

where: m - mass; z - coefficient of elasticity; T_s - basic period of variation; df = 0.9 ... 0.97 - damping decrement; h_s - initial deformation; $h_{1,2}(t)$ - displacement of end cross-sections.

4. Application of modelling

Based on the obtained calculation results (figure 3) it can be concluded that the deformation of the injector piston Δh (0.17 - 0.2 mm) corresponds to the needle displacement (0.25 mm).

In the work [5] it was indicated that the deformation of the injector piston is 36% when displacing the needle caused by pressure in the fuel reservoir of 140 MPa. The omission of the causes of this deformation leads to an error in determining the initial moment of displacement of the needle. In works [3, 5, 13], the causes of failure sections (or small steepness of the graph, as shown in figure 4) were identified in the fuel flow rate characteristics. The authors of the indicated works see the reasons for this effect in changing the dynamics of the valve's movement at the turning point, in wave processes in the pipeline and in the presence or absence of the needle reaching its seat.

The obtained results also show that unstable injection may be the result of deformation of the 'injector piston' and 'valve stem'.

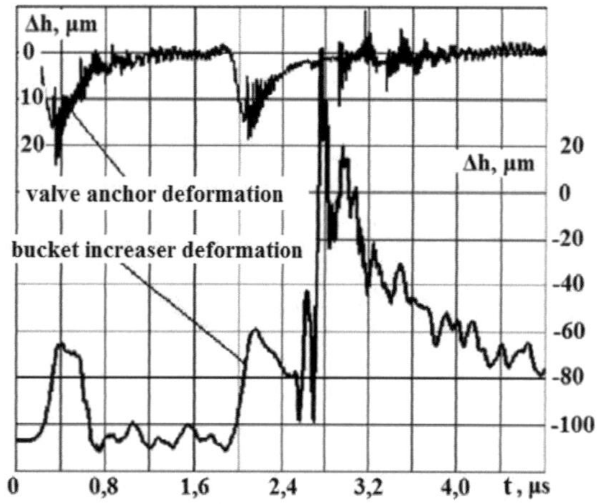

Figure 3. Deformation of injector parts.

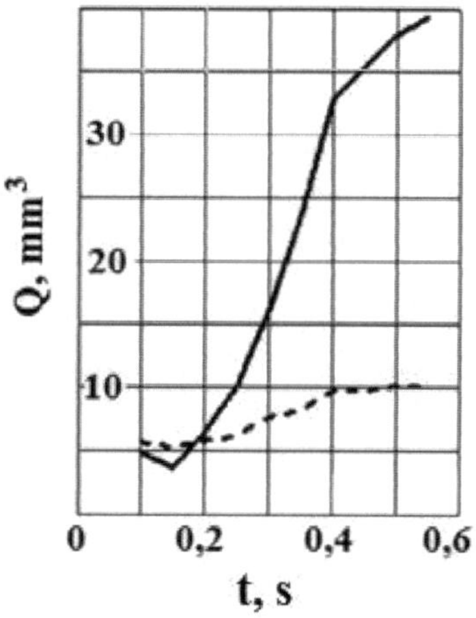

Figure 4. Dependence of the affected fuel portion on the duration of the control pulse.

These factors were taken into account for the development of a mathematical model, which was used in the analysis of hydrodynamic processes occurring in the Bosch injector [1]. Selected results of process modelling, at a fuel reservoir pressure of 143 MPa, are shown in Figure 6. Based on this model, a method and program for dynamic calculations have been developed.

The application of this method allowed the main criteria for the selection of CR injector construction parameters to be determined, as well as to identify the occurring damages, which contributes to increasing the reliability of the injector design.

The conducted research shows that the asymmetry of the magnetic chain reduces the impact of inaccuracy in the manufacture of the injector elements on the valve's motion characteristics.

In order to reduce the impact of eddy currents, the shape of the anchor can be modified by introducing cuts (grooves in its foot). In order to maintain the stability of the injection process, it is necessary to limit the speed of the valve movement during the impact of the cone lock on the bottom of the sleeve. It should also be noted that the anchor seat design used in multi-phase injection is of a low affectivity.

Effective electronic control of the injector is possible only when the control chain, solenoid and hydrodynamic characteristics are taken jointly into account. At the same time, as noted in [6], the criteria of efficiency cannot only take into account whether the required injection intensity is achieved, but it must also consider the stability of the main indicators and reliability of the fuel system.

The above problem can be solved by calculating the change of magnetic chain properties [14] and the control algorithm. While controlling the short fuel injection pulse, the control valve must move at maximum speed, and after the valve has hit against its seat, the magnetic field attraction force must be reduced to zero.

When controlling by the main pulse, the valve should set at a low speed, thus being protected against damage. In addition, the supporting current shall be corrected by taking into account the fuel pressure in the reservoir and the magnetic chain shall be designed by taking into account the necessary reserve of the magnetic force.

IOP Conf. Series: Materials Science and Engineering **507** (2019) 012029 doi:10.1088/1757-899X/507/1/012029

5. Conclusion

A method and program for calculating hydrodynamic phenomena occurring during the operation of an electronically controlled injector were developed.

To determine the flow of fuel in the pipe, a modernized method of characteristics was used, which takes into account the hydraulic properties of the injector.

The stem equivalent model was used for determination of the elastically-deformable state of the needle, bucket increaser and the valve stem as well as for calculation of the stress of springs.

The main principles for the selection of the injector construction parameters were also established. In order to increase the injection pressure, it was proposed to use wave processes that exist in the tubing connecting the fuel reservoir with the injector.

References

[1] Mollenhauer K, Tschöke H. 2010 *Handbook of Diesel Engines*. Springer-Verlag. Berlin. Heidelberg.
[2] Heywood J. 1988 *Internal combustion engine fundamentals*. McGrow-Hill. New York, USA.
[3] Coppo M, Dongiovanni C, Negri C 2004 *J. Eng. Gas Turbines Power*. **126(4)** 874–85.
[4] Roache P J 1982 *Computational Fluid Dynamics*. Hermosa. Albuquerque, USA.
[5] Seykens X L J, Somers, L M T, Baert R S G 2005 *MECCA*. **3(2)** 3.
[6] Bingqi T, Liyun F, Xiuzhen M, Hayat Q, Yun B, Yang L 2014 *China Int. J. Smart Sens. Intell. Syst.* **7(2)** 820–36.
[7] Salvado F J, et al. 2013 *Energy Conv. Manag.* **66** 246–56.
[8] Brusiani F, Falfari S, Pelloni P 2014 *Energy Proc.* **45** 749–58.
[9] AVL Boost – Hydsim. 2013 *User's Guide*. Graz. Austria.
[10] Borchsenius F, Stegemann D, Gebhardt X 2010 *MTZ*. **6** 42–5.
[11] Baumann J 2006 *Injection quantity correction in common rail systems using magnetoelastic pressure sensors*. Dissertation.
[12] Grunauer A, Tartakovskyi I, Grigoriev O 1985 Theory Mech. Mach. **39** 7–22.
[13] Jiping L, Shuiyuan T, Yong Z, Jing H, Zhonghua J 2011 *Int. J. Comput. Intell. Syst.* **4(6)** 1282-9.
[14] Bianchini G M, Pelloni P, Filicori F, Vannini G 2000 *SAE paper No.* 2000-01-2042.

IOP Conf. Series: Materials Science and Engineering **507** (2019) 012030 doi:10.1088/1757-899X/507/1/012030

Wear analysis of welded layers containing node carbide, in the abrasive soil mass

J Napiórkowski[1], K Ligier[1], M Lemecha[1]

[1]Faculty of Technical Sciences, University of Warmia and Mazury in Olsztyn ul. M. Oczapowskiego 11, 10-719 Olsztyn, Poland.

Abstract. The paper presents the research of the wear intensity of welded layers in natural soil abrasive mass. The samples made of 38GSA steel with the welded surface layers contains carbide forming elements. In case of the test abrasive mass, XHD 6715 material showed the wear 2.5 times lower than El-Hard 61 and twice smaller than El-Hard 65. The analysis of the wear indicates that the welding layer containing vanadium and chromium carbides, molybdenum and niobium carbides has proven to be the most resistant to wear in the test abrasive mass.

1. Introduction

The issue of the wear of parts operating within abrasive soil mass is very complex due to the multiplicity of factors affecting the wear process [1-3]. For compact abrasive soil masses, the abrasive particles may have an influence as fixed grains resulting in losses in the material due to processes typical for abrasive wear (scratching, microcutting, and ridging). In the soil masses with low compactness, the wear is caused by the impact of loose abrasive particles on the surface of the material being subjected to abrasion, which results in the wear through decohesion (more frequently) and fatigue.

The properties of the surface layer have a decisive influence on the service life of the parts used in machinery and tools. The outer layer, properly selected considering operating conditions, should be characterized by appropriate tribological features arising from the physicochemical properties. In order to ensure that an operating part has appropriate features contributing to its resistance to the wear processes, various methods of padding are applied involving the uniform application of an alloy onto the material surface [4, 5].

One of the main components determining the tribological properties of a padding layer is the chemical composition of an alloy being applied [6]. The formation of the outer layer microstructure is largely determined by the carbide-forming elements including inter alia vanadium, niobium, tungsten, molybdenum, and chromium [7-10].

The carried out studies on the effect of carbides forming elements on the abrasive wear resistance were mainly related to tool steels. In [11] it was shown that the increase of vanadium contents in a high-speed steel increases its resistance to the abrasive wear tested by the "pin on disc" method. The paper [12] indicates the influence of the matrix properties on abrasion resistance. The best wear resistance was obtained for vanadium carbides embedded in a martensitic matrix.

Taking into account the carried out analysis, it can be assumed that the content of elements favorable for the formation of carbides in the weld should improve the resistance of the surface layer of the tool to the wear in natural abrasive soil masses.

The aim of the work is to analyze the wear of surface layers with varied content of niobium tungsten and vanadium in a soil abrasive mass.

2. Research materials

Hardfacing materials were applied to hardened 38GSA steel. The substrate material was characterized by the structure of martensite with bainite and troostite. The hardness of the substrate material amounted to 414 HV10. The chemical composition of the steel was as follows: C – 0.38%, Mn – 1.07%, Si – 1.17%, P – 0.028%, Cr – 0.18%, Cu – 0.16%, Al – 0.02%.

The additional materials containing C + Cr + Mo + W + V + Nb were applied on the steel surface by arc welding with El-Hard 61, El-Hard 65, XHD 6715 electrodes (table 1).

Table 1. The chemical composition of welded layers.

material	Chemical composition %					
	C	Cr	Mo	W	V	Nb
El-Hard 61	5,2	29,0	0,71	-	-	7,0
El-Hard 65	4,5	34,0	6,0	2,0	1,0	6,0
XHD 6715	5,0	21,0	8,50	6,0	1,9	7,0

3. Research methodology

The wear intensity was examined in a laboratory by the "rotating bowl" method [2] (figure 1). 30 x 25 x 10 mm cuboidal shape samples were tested. The testing machine was filled with natural soil abrasive mass equivalent to dry soil with the granulometric composition as per PN- EN ISO 14668-2(2004):

- light soil: silt: 1.69%; dust: 20,83%; sand: 77.48%.

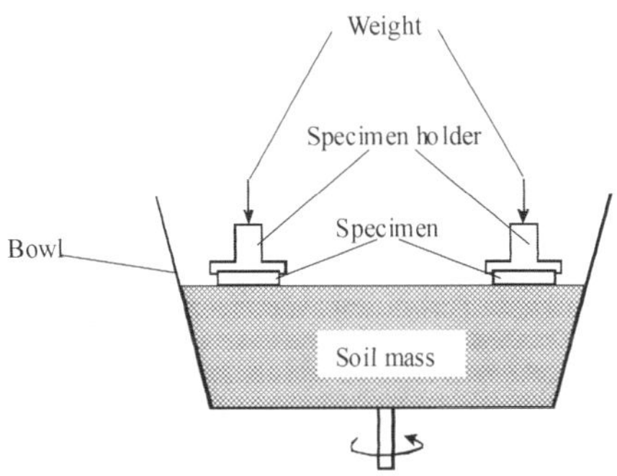

Figure 1. The diagram of a "rotating bowl" experimental test-stand.

The following friction parameters were taken into account: velocity 1.4 m/s, friction distance 10,000 m, weight 49 N (5 kg weight mass). The test-stand was equipped with elements to mix and compact the abrasive mass. Test runs were repeated six times for each tested material. The soil moisture was maintained at 10%. The pH of the abrasive soil mass was in the range from 6.3 to 6.9.

The measurements of the wear were taken every 2000 m. The unit wear was calculated from the following formula:

$$Z_j = \frac{Z_w}{s * P} \; [g/km * cm^2], \qquad (1)$$

where: Z_w – mass wear [g], s – friction path [km], P – the surface area of the examined sample [cm^2]

The hardness of the materials was measured by means of a type HV-10D Vickers hardness tester, in accordance with the PN-EN ISO 6507-1:1999; quality standard, an indenter load of 98 N was used, lasting for 10 s. The light microscopy examinations were conducted using a Neophot 52 microscope coupled with a Visitron Systems digital camera.

The scanning electron microscopy examinations and the microanalysis of the chemical composition were conducted using a JEOL JSM – 5800 LV scanning microscope, coupled with an Oxford LINK ISIS – 300 X-ray microanalyzer.

4. Test results

On the basis of the conducted research, the following structures of the examined layers were identified:

- 38GSA steel – martensite, bainite and small contents of troostyt (figure 2).
- El–Hard 61 –alloy ferrite, which passes in the structure of the layer consisting of ferrite and carbides (ledeburitic structure) (figure 3). In the microstructure of the padded layer, the separation of chromium and niobium carbides was identified.
- El-Hard 65 – in the alloy ferrite there are visible of large primary chromium carbides (type M7C3), and other fine carbides (figure 4) and niobium carbides.
- XHD 6715 – in the alloy ferrite matrix there are large secretions of primary chromium carbides (Figure 5) and other types of carbides (Mo, Nb, W). Small areas of ledeburite are also visible.

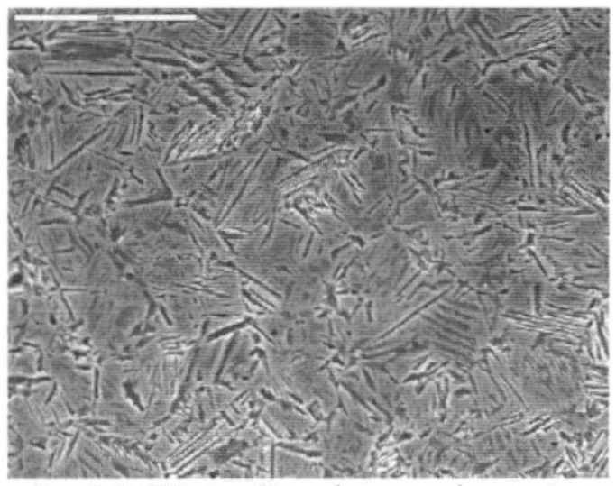

Figure 2. 38GSA Steel – microstructure of martensite, bainite and a small content of troostyt. SEM microscopy.

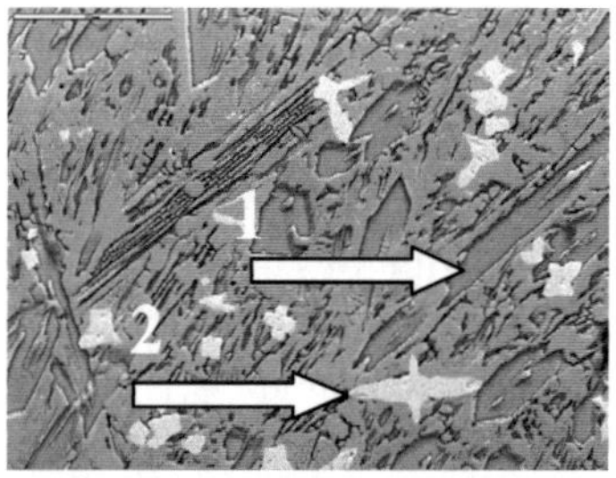

Figure 3. The microstructure of El-Hard 61 welded layer, chromium carbide (1-grey) and niobium chromium (2-white) exuded. SEM microscopy.

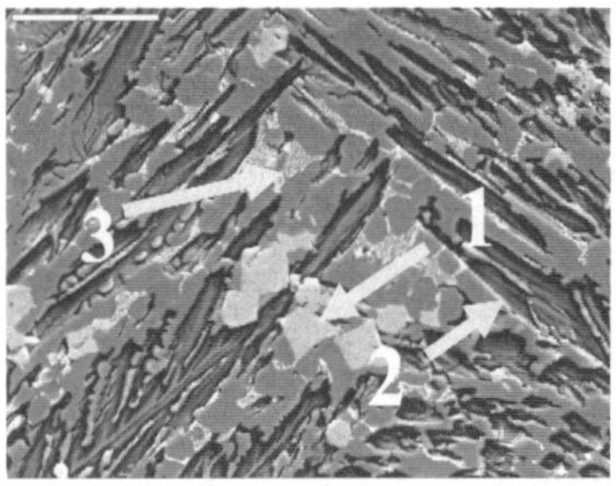

Figure 4. XHD 6715 welded layer microstructure, large primary chromium carbides (2), in ferrite groundwork and niobium carbides (1). Visible small areas of ledeburite (3), SEM microscopy.

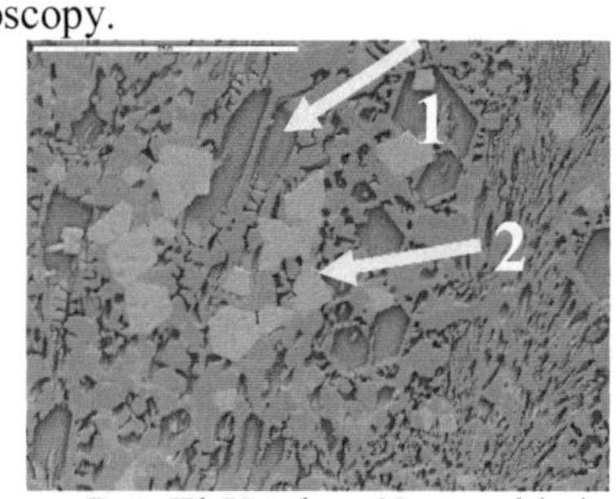

Figure 5. El-Hard 65 welded layer microstructure, mixture of ferrite and M7C3 – [Fe,Cr7C3] carbides (1) and niobium carbides (2), SEM microscopy.

The highest hardness of the surface layer was found respectively for XHD 6715 (820 HV10), then EL-Hard 65 (682 HV10) and EL-Hard 61 (632 HV10).

The results of the study of the value of the weight loss of the tested materials as a function of the friction distance are presented in figure 6.

IOP Conf. Series: Materials Science and Engineering **507** (2019) 012030 doi:10.1088/1757-899X/507/1/012030

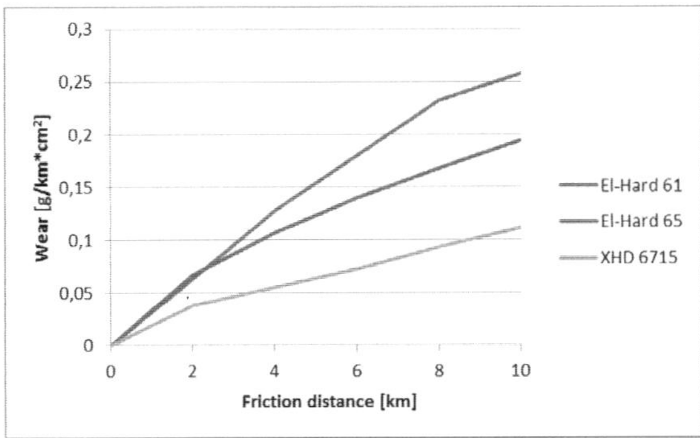

Figure 6. Change of wear value in light soil.

The highest wear values were observed for the El-Hard 61 padding (figure 6). The light soil abrasive mass is characterized by a high content of hard, irregular shaped abrasive grains. In this kind of soil furrowing and drawing are dominating in by mechanical wear processes (figure 7). For this type of abrasive mass, XHD 6715 material showed wear 2.5 times lower than El-Hard 61 and twice smaller than El-Hard 65. The El-Hard 61 welded layer contains only niobium. The other two layers, in addition to niobium, also contain vanadium and tungsten, while the XHD6715 material contains the largest amount of the mentioned elements.

The scratches created on the surface of the tested materials are the largest in the case of the material El-Hard 61 (figure 7a). On this surface, there are also visible traces of the removed carbides.

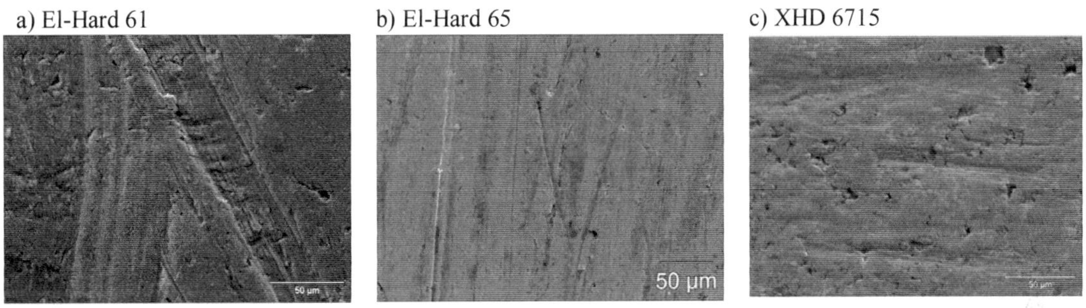

Figure 7. Welded layers surface following wear test in light soil

5. Summary

Given that the process of thepad welding was carried out at a relatively low temperature (high carbon and chromium contents), no structures of the metallic matrix comprising austenite were obtained. The metallic matrix of the obtained layers comprises alloy ferrite and, for the XHD6715 layer, ledeburite as well. As for the padding weld with significant tungsten (6%) and vanadium (1.9%) contents, the increased amount of carbides was obtained, with a significantly smaller size as compared to other layers, within the metallic matrix of alloy ferrite and small-sized ledeburite. This was due to the decrease of the carbon content of the metallic matrix, at the expense of the formation of a greater amount of carbides. Such a structure ensures the greatest hardness of the outer layer and resistance to wear within the soil mass.

The use of niobium alone in the outer layers of Fe-Cr-C does not ensure high resistance to the wear due to the chipping of hard niobium carbides from the relatively soft metallic matrix. In order to increase this layer resistance to the wear, cooling the obtained padding weld with water in order to obtain an austenitic matrix should be considered.

6. Reference

[1]. Napiórkowski, J. 2010 *Tribologia*, **41**(5) 53-62.
[2]. Napiórkowski, Jerzy (ed.). *Badania i modelowanie procesów zużywania ściernego i zmęczeniowego (Research and modeling of abrasion and fatigue wear processes), in Polish.* Wydawnictwo Uniwersytetu Warmińsko-Mazurskiego, 2014.
[3]. Natsis A, Petropoulos G, Pandazaras C 2008 *Tribology International*, **41(3)** 151-157.
[4]. Buchely M F, et al. 2005 *Wear*, **259(1-6)** 52-61.
[5]. Hutchings I M, *Friction and wear of Engineering Materials*, Cambridge, 1992, 133–171.
[6]. Napiórkowski J, Ligier K, Lemecha M, 2015 *Mechanik* **4** 92-97.
[7]. Filipovic Mirjana, et al. 2013 *Mater. Des.* **47** 41-48.
[8]. Zhiguo Zhang, et al. 2014 *China Foundry*, **11(3)** 179–184.
[9]. MikołAjczak P, Napiórkowski J, 2016 *Eksploatacji i Niezawodność - Maintenace and Reliability*, **18(4)** 544–551.
[10]. Krawczyk J et al. 2016 *Tribologia*, **6** 69-81.
[11]. Wei Shizhong et al. 2006 *Tribology International*, **39(7)** 641-648.
[12]. Ji Ying Ping et al. 2012 *Wear* **294** 239-245.

IOP Conf. Series: Materials Science and Engineering **507** (2019) 012031 doi:10.1088/1757-899X/507/1/012031

Management of the ecological safety of urban zones in terms of sources of power supply for buses

J Napiórkowski[1] and P Szczyglak[1] and A Langer[1]

[1] University of Warmia and Mazury in Olsztyn, Faculty of Technical Sciences, Poland
E-mail: jerzy.napiorkowski@uwm.edu.pl

Abstract. All articles *must* contain an abstract. The abstract text should be formatted using 10 point Times or Times New Roman and indented 25 mm from the left margin. Leave 10 mm space after the abstract before you begin the main text of your article, starting on the same page as the abstract. The abstract should give readers concise information about the content of the article and indicate the main results obtained and conclusions drawn. The abstract is not part of the text and should be complete in itself; no table numbers, figure numbers, references or displayed mathematical expressions should be included. It should be suitable for direct inclusion in abstracting services and should not normally exceed 200 words in a single paragraph. Since contemporary information-retrieval systems rely heavily on the content of titles and abstracts to identify relevant articles in literature searches, great care should be taken in constructing both.

1. Introduction

A side effect of the widespread use of means of transport is air pollution. One method for its reduction in urban environments is the widespread of public transport. A bus emitting little more exhaust fumes than a passenger car transports as many people as 70 cars at the same time. Exhaust gases constitute a load of more than 15 thousand chemical compounds. Means of road transport are responsible for the emission of many hazardous substances, including nitrogen oxides and dioxides, carbon dioxide, and particulate matter suspended in the air [1, 2, 6]. All liquid, solid and gaseous substances whose content in the air exceeds the average content in clean air are called pollutants. Their harmfulness for the environment results from many factors, such as state of matter, extent of dispersion, chemical and toxic properties, concentration in the air, climatic and meteorological conditions, time of interaction, etc. [3, 7].

Vehicle exhaust fumes are more harmful to people than pollution which originates from industry, as they spread at high concentrations at low heights in the immediate vicinity of people. Basic pollutants include:

- Nitrogen oxides NO_x;
- Carbon oxide CO ;
- PM – 10.

In the case of long-term exposure all emitted substances cause the development of cancer cells. High concentrations of exhaust fumes lead to the formation of smog (London, photochemical) and to the greenhouse effect. In tables 1 acceptable levels of substances in the air in terms of health protection in accordance with applicable legislation are presented. Urban buses should be distinguished by low noise emissions, engines equipped with filters, and catalysts limiting the content of hazardous substances in exhaust fumes [4, 5, 8].

Table 1. Substance emission values for new vehicles with compression-ignition engines

Standard	Effective from	CO [g/km]	HC + NO_x [g/km]	PM [g/km]
Euro I	07/92	3.16	1.13	0.14
Euro II	01/96	1.00	0.70	0.08
Euro III	01/00	0.64	0.56	0.05
Euro IV	01/05	0.50	0.30	0.009
Euro V	09/09	0.50	0.23	0.005
Euro VI	08/14	0.50	0.17	0.005

In the paper an analysis of the content of hazardous substances in public transport buses, depending on the type of energy source in terms of the ecological safety of the population, is presented.

2. Materials and Methods

The subject of the study comprised buses of public transport (table 2) belonging to the same class (midi) and supplied with diverse sources of energy (table 3). The study was conducted under normal operating conditions in two cities with similar populations, using an active-passive experiment. During the study the actual consumption of energy means by the full tank method, and travelled distance as indicated by odometer, were recorded. The studies were performed in May and September.

The average annual mileage for all buses was 7,700,000 km. Annually, each bus covers an average route of 46,107 km, which is 126 km daily.

The study included the following buses and their characteristics and characteristics of fuels are presented in Tables 4 and 5:

a) MAN NL202 – a bus supplied with diesel fuel;
b) Scania OmniLink – a bus supplied with ethyl alcohol (ethanol) C_2H_5OH;
c) Scania OnmiCity – a bus supplied with compressed gas CNG (Compressed Natural Gas);
d) Solbus Solcity – a bus supplied with liquefied gas LNG (Liquefied Natural Gas);
e) Solaris Urbino 12 – a bus supplied with fuel oil.

Table 2. Characteristics of the study vehicles

Buses Parameters	MAN NL 202	Scania OmniLink ethanol	Scania OmniCity CNG	Solbus Solcity	Solaris Urbino 12
Engine	D0826LUH03	DC 9.01	DC 9.01	Cummins ISB6.7EV	Cummins ISB6.7
Engine displacement [l]	6.9	8.9	8.9	8.9	8.9
Number of cylinders	6	8	5	6	6
Engine power [kW]	156	169	169	184	209
EURO standard	2	5	5	6	6
Number of standing	66	70	80	82	82

places					
Number of seats	30	29	30	37	37
Year of production	1990	2005	2006	2008	2014

Table 3. Characteristics of fuels

Characteristic	Diesel fuel	Ethanol	CNG	LNG
Chemical formula	$C_{15}H_{32}$, $C_{18}H_{38}$	C_2H_5OH	CH_4	CH_4
Calorific value	43 [MJ/kg]	26.8 [MJ/kg]	45-50 [MJ/kg]	45-50 [MJ/kg]
Flame temperature [K]	328 <	2185	630 - 640	600
Viscosity at 296K [cSt]	3	1.51		$10.22x10^6$
Octane rating [*OR*]	72	108	130	130
Density	830 [kg/m^3]	789.3 [kg/m^3]	0.7175 [kg/Nm^3]	450 [kg/m^3]
Boiling point	175	78.37	-161.49	-161.49
Melting point	Not determined	- 114	-182.6	-182.6

3. Results

Vehicles, depending on the applied energy source, are characterised by a varied amount of consumed energy carriers (table 4).

Table 4. Consumption of energy carriers

Vehicle	Type of fuel	Average consumption of fuel		Energy consumed by vehicle [kWh/100km]
Scania OmniLink	Ethanol	73.89 [dm^3/100km]	58.392 [kg/100km]	1564.905
Scania OmniCity	CNG	58.60 [nm^3/100km]	42.045 [kg/100km]	1892.025
MAN NL 202	ON	39.64 [dm^3/100km]	32.901 [kg/100km]	1414.743
Solbus Solcity	LNG	51.66 [nm^3/100km]	37.06 [kg/100km]	1667.7
Solaris Urbino 12	ON	36.30 [dm^3/100km]	30.129 [kg/100km]	1295.547

The largest amount of fuel was used by the bus supplied with ethyl alcohol, followed by CNG and LNG, while the lowest consumption was recorded in the case of vehicles supplied with diesel fuel. The high consumption of ethanol is caused by its low calorific value compared to other energy carriers. After the determination of the energy value stored in particular energy carriers consumed during the test at a distance of 100 km, it was found that the amount of energy stored in the used ethanol was not the highest, and equalled 1564.905 kWh. The highest amount of consumed energy was recorded in the case of the bus supplied with CNG (1892.025 kWh). Slightly less energy was used by the vehicle supplied with (1667.7 kWh), while the lowest amount was used by the modern compression-ignition engine (1295.547 kWh).

In tables 5–7 characteristics of the generation of hazardous substances during the test are presented.

IOP Conf. Series: Materials Science and Engineering **507** (2019) 012031 doi:10.1088/1757-899X/507/1/012031

Table 5. Characteristics of generated hazardous substances over a distance of 100 km [g/km]

Daily amount of generated substances (type of fuel)	MAN NL 202 (ON)	Scania OmniLink (Ethanol)	Scania OmniCity (CNG)	Solaris Urbino (ON)	Solbus Solcity (LNG)
PM	35.32	1.36	0.85	1.38	0.48
HC+NOx	285.07	-	29.01	46.83	20.43
CO	797.17	135.34	85.32	137.74	42.5

Table 6. Characteristics of hazardous substances generated during one day [g/km]

Daily amount of generated substances	MAN NL 202 (ON)	Scania OmniLink (Ethanol)	Scania OmniCity (CNG)	Solaris Urbino (ON)	Solbus Solcity (LNG)
PM	44.5	1.70	1.075	1.74	0.6
HC+NOx	359.18	-	36.55	59.01	25.73
CO	1004.43	170.52	107.51	173.55	53.55

Table 7. Characteristics of hazardous substances generated during one year [g/km]

Daily amount of generated substances	MAN NL 202 (ON)	Scania OmniLink (Ethanol)	Scania OmniCity (CNG)	Solaris Urbino (ON)	Solbus Solcity (LNG)
PM	16242.5	620.50	392.36	635.1	219
HC+NOx	131100.7	-	13340.75	21538.65	9391.45
CO	366616.95	62239.80	39241.15	63345.75	19545.75

The most ecological bus was Solbus Solcity, which for 100 kg emits particulate matter at an amount lower by 41% compared to a bus supplied with ethyl alcohol, by 77% compared to a bus supplied with compressed natural gas CNG, and nearly three times less than a bus supplied with diesel fuel. The bus which was supplied with diesel fuel emitted the largest amount of hazardous exhaust fumes. The level of HC+NO_x was over 2.29 times higher, while for CO it was 3.24 times higher in comparison to the bus supplied with LNG

A bus supplied with diesel fuel produces 0.42 kg of particulate matter, 12.15 kg of HC+NO_x and 54 kg of CO yearly. In turn, these values for the supplied bus in the case of Olsztyn for 10 buses are 4.2 kg, 120 kg of HC+NO_x, 540 kg of CO.

4. Conclusions

Management of the safety of big city populations begins as early as at the time of planning the purchase of public transport means. The selection of the appropriate fleet will help to minimise the negative impact of the urban communication system on the environment. Public transport, particularly buses, will continue to be the primary means of population movements. The purchase of buses supplied with conventional fuels, in heavy traffic, contributes to increased emissions of nitrogen oxides, particulate matter, carbon dioxide and noise. The use of vehicles supplied with ethanol is ecologically justified, but the high operating costs of this type of vehicle means that public transport companies give up this solution, an example of which is the decommissioning of buses supplied with ethanol after the completion of the test programme. Therefore, the protection of the environment of humans, and the issues of the diversification of energy sources should constitute the basic reasons for the purchase of buses supplied with gas fuel, particularly LNG.

References

[1] Bonnel P, Weiss M, Provenza A 2011 *In – use Emissions Requirements in the New and Future European Motor Vehicle Emissions Regulations: State of Play* 8th Annual SUN Conference, Ann Arbor.
[2] Bougher T, Khalek I A, Trevitz S, Akard M 2011 *Verification of a Gaseous Portable Emissions Measurement System with a Laboratory System Using the Code of Federal Regualtions* Part 1065 SAE Technical Paper Series 2010–01–1069.
[3] Kozak M, Merkisz J, Bielaczyc P, Szczotka A J 2008 *Environmental Performance of Diesel-Fuel Synthetic Oxygenates: Maleates and Carbonates* FISITA World Automotive Congress Monachium Deutschland 14-19.09.2008.
[4] Massod M 2007 *Hydr. Energy* **32** 2539–2547.
[5] Nylund N O, Koponen K 2012 *Fuel and Technology Alternatives for Buses* VTT Technology 46 P.294+app. p. 94 Espoo Kopijvä Oy Kuopio.
[6] Pronk A, Coble J, Stewart P A 2009 *J. Exp. Sci. Envir. Epidem* **19** 443–457.
[7] Pütz R, Schwürzinger J 2012 *Options for the fully electrified operation of ruban bus lines (EBSF Study)* UITP Brussels.
[8] Stieb D, Evans G, Sabaliauskas K, Chen L, Campbell M, Wheeler A, Brook J, Guay M 2007 *J. Expo Sci. Environ. Epidemiol.* **18** 495–505.

IOP Conf. Series: Materials Science and Engineering **507** (2019) 012032 doi:10.1088/1757-899X/507/1/012032

The influence of type of abrasive soil mass on the wear intensity of multi-phase welded coating

J Napiórkowski[1], K Ligier[1], M Lemecha[1]

[1]Faculty of Technical Sciences, University of Warmia and Mazury in Olsztynul. M. Oczapowskiego 11, 10-719 Olsztyn, Poland.

Abstract. One of the major problems when machining soil mass is wear of working tools. The process consists of many factors related to, amongst other things, the work item, the properties of the soil and the working process parameters. Processes which occur in the surface layer material are associated with changes in its mechanical properties, physical and chemical properties. The paper analyzes the process of wear steel 38GSA and weld electrode EstelMn60, ENSW2Mo-B, EStelMoNb60. Studies were conducted in laboratory conditions by the "spinning bowl" method. During the tests carried outwear of samples depending on the road friction and soil mass. The results found that the highest wear values were observed for gravel soil for the ENSW2Mo-B welded layer and it is 2.5 times higher than in the heavy soil. The lowest wear value was characteristic of the EStelMn60 welded layer, which in the chemical composition contains carbide elements. In heavy soil, the highest wear was observed for ENSW2Mo-B the smaller for EStelMoNb60.

1. Introduction

One phenomenon accompanying the operation of working elements in the soil is their intense wear, which is a process of physicochemical qualitative and quantitative changes occurring on the friction surface [1, 2].

It is generally accepted that the wear of operating parts processing an abrasive soil mass results from abrasive wear. The basic forms of the wear include scratching, ridging, microcutting, and local fatigue. These processes are associated with the mechanical impact of abrasive material grains on the surface layer of the material. Depending on the type of an operating part, including the impact of the pressure exerted by the abrasive material, chemical processes associated with the impact of soil environment can occur as well [3-5]. The soil is a complex, animate creation of nature, in which continuous processes of decomposition and synthesis occur for both mineral and organic compounds, along with their displacement [6].

Due to the multiplicity of factors affecting the course of wear for working elements in soil mass, so far there has been no formalized theory of mutual interactions between the soil and the working element. Most papers indicate that the wearing properties of soil result from its grain sizes, cohesiveness, and humidity [7].

The course and intensity of wear in the soil mass depend on properties of the soil processed (moisture content, compactness, pH and grain size distribution), type of the impact of the working element on the soil, design and technological solutions concerning the working element.

Users of working elements expect the manufacturers to provide design solutions that offer high durability. Therefore, the materials of the working elements of tools for treating soil abrasive mass should demonstrate high abrasive wear, as well as impact performance.

The process of the rational selection of the construction and technological forms of working elements, and of planning the course of operation, is possible when their wear process is known. The identification and description of wear in soil come down to a comprehensive description of the process of friction involving: the soil - the working element - the parameters of work [8].

One of the commonly applied ways to form the chemical composition as well as to change hardness and microstructure of the outer layer of the parts interacting within an abrasive environment is the formation of abrasion resistant layers using pad welding [9, 10]. This solution offers one of the methods to increase the durability of machine parts exposed to the effect of the abrasive environment. Abrasive wear resistance can be increased by the application of welding methods, making it possible to modify the chemical composition of the base material by introducing additional components increasing the resistance to abrasive wear as well as the change in hardness and microstructure of the material.

When selecting the material for operating parts to be used within an abrasive soil mass, it is necessary to carry out an analysis of the wear processes occurring at the interface between the soil and the operating part and to determine the environmental forces, including the properties of the abrasive mass. Wear resistance is determined inter alia by the element content, resilience, the micro-structure character and the matrix hardness [11].

2. Research materials

The study was subjected to 38 GSA steel. The tests involved samples 30 x 20 x 10 mm. This is martensitic steel which contains bainite and troostite. The chemical composition was as follows: C - 0.38%, Mn - 1.07%, Si - 1.17%, P - 0.028%, S - 0.02%, Cr - 0.18%, Cu - 0.16%, Al - 0.022%. Additional materials containing carbides were applied to the steel surface by arc welding with electrodes EStelMn60, ENSW2Mo-B, EStelMoNb60 (table 1).

Table 1. The chemical composition of the padding welds.

Ingredients content. %	Materials		
	EStelMn60	ENSW2Mo-B	EStelMoNb60
C	3.40	0.94	5.31
Mn	4.32	1.63	1.28
Si	1.68	1.06	1.46
Cr	31.60	5.41	21.90
Mo	0.04	6.83	7.66
Ni	0.30	0.17	0.06
Al.	0.03	0.01	0.05
Cu	0.09	0.12	0.09
P	0.05	0.01	0.02
S	0.01	0.05	0.02
B	0.03	-	-
W	0.05	-	2.00
V	-	-	1.70
Ti	-	-	0.03
Nb	-	-	7.00

3. Research methodology

The study was conducted under laboratory conditions using a "rotating bowl" method wear testing machine (figure 1).

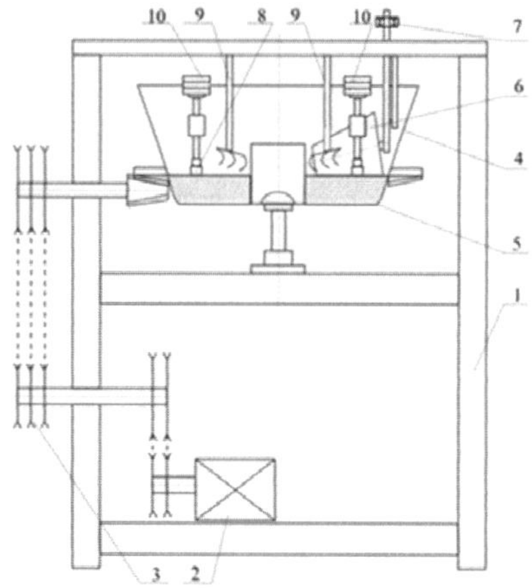

Figure 1. A stand for examination of wear in abrasive soil pulp: 1 - frame, 2 - motor, 3 – belt transmission, 4 - bowl, 5 - soil, 6 – pressing rollers, 7 – roller pressing, 8 – grip with the sample, 9- scarifying fork, 10- sample weight.

Research samples are mounted on two independent sections fixed on a detached wishbone suspension. The stem of the rocker is equipped with a special stem enabling the use of weights to add extra load to the sample.

Friction tests were conducted at a load of 53 kPa, a cutting speed of 2 ms^{-1}. Measuring sections of 2, 4, 6, 8 and 10 km were used.

The tests were carried out in two types of abrasive masses. A moisture content of the soil was determined to be equal to 5%, and the soil pH to be equal to 7.

The grain size distribution, determined with Mastersizer 2000 the laser particle size analyser, was as follows (PTG 2008 classification):

- Gravel soil: gravel: 56.6%; sand: 25.4%; dust: 9.4%; silt: 8.6%
- Clay soil: sand: 9.0%; dust: 23.6%; silt: 67.4%

Mass wear was determined based on the following formula:

$$Z_m = m_p - m_k\ [g], \qquad (1)$$

where: Zm – mass wear [g], mp – sample mass before testing [g], mk – sample mass after testing [g]

The hardness of the materials was measured by means of a type HV-10D Vickers hardness tester, in accordance with the quality standard PN-EN ISO 6507-1:1999; an indenter load of 98 N was used, lasting 10 s. Microscopic examinations by means of light microscopy were conducted using a Neophot 52 microscope coupled with a Visitron Systems digital camera.

Examinations by means of scanning electron microscopy and the microanalysis of the chemical composition were conducted using a JEOL JSM – 5800 LV scanning microscope, coupled with an Oxford LINK ISIS – 300 X-ray microanalyser.

4. Analysis of the obtained results

On the basis of the conducted research, the following structures were identified:

- 38 GSA stell - martensite, bainite and a small content of troostyt;

- EStelMn 60 – Varied structure on the thickness and length. Fusion zone: alloy ferrite with numerous carbides (Fe, Cr)3 C. Padding weld: primary carbides eutectic mixture [α + (Fe, Cr)7 C3], mainly with plate-like structure + residual austenite. Hardness 620÷726 HV10;

- ENSW2Mo-B – Varied dendritic structure. Fusion zone: alloy ferrite with carbides located on grain borders. Padding weld: alloy ferrite with small molybdenum carbides on grain borders and inside them. Hardness 720÷762 HV10;

- EStelmoNb 60 – Dendritic structure. Fusion zone: alloy ferrite. Padding weld: alloy ferrite with numerous carbides of the Mo2C, VC, TiC type. And with eutectics [α + (Fe, Cr)7 C3].

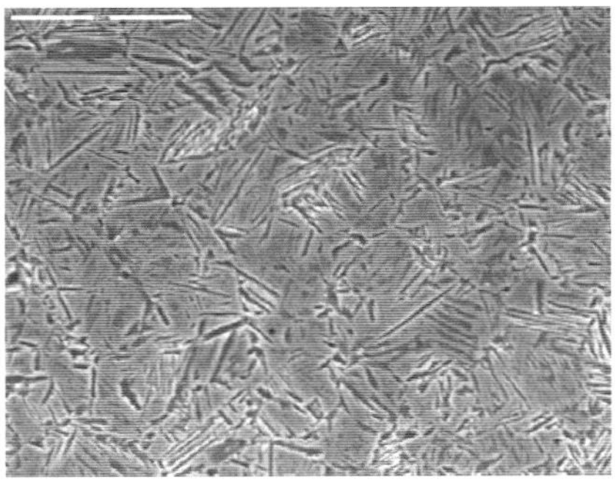

Figure 2. The microstructure of 38 GSA Steel

Figure 3. The microstructure of EStelMn60 welded layer

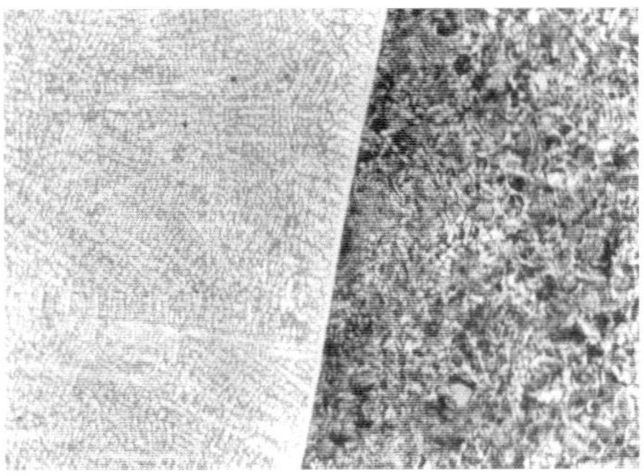

Figure 4. The microstructure of ENSW2Mo-B welded layer

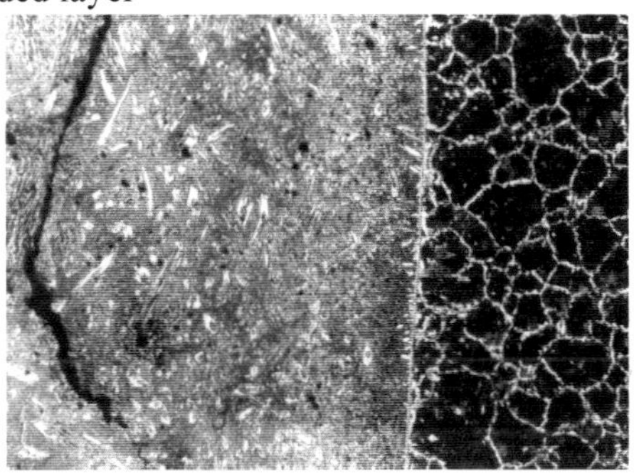

Figure 5. The microstructure of EStelMoNb60 welded layer

The results of the study of the value of the weight loss of the tested materials as a function of the friction distance are presented in figures 6 and 7.

As a result of the tests, a significant influence of the type of abrasive mass on the intensity of wear of the tested materials was found.

The highest wear values were observed for gravel soil (figure 6) for the ENSW2Mo-B welded layer and it is 2.5 times higher than in the heavy soil. The lowest wear value was characteristic of the EStelMn60 welded layer, which in the chemical composition contains carbide elements.

This is mainly the result of the interaction of high hard content and at the same time an irregular shape of the sand fraction. The soil is dominated by mechanical wear processes through microplowing and microfacture.

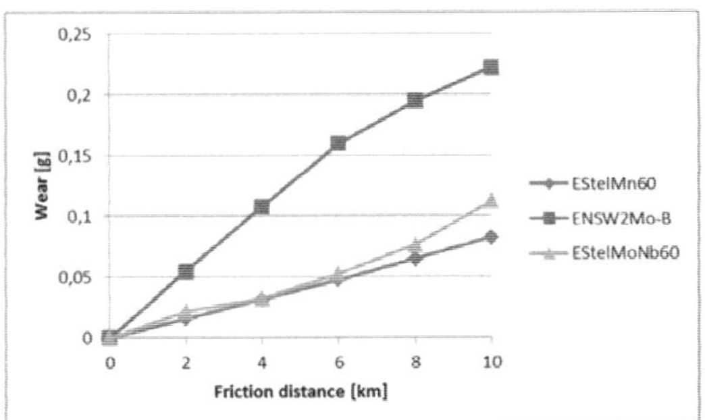

Figure 6. Change of wear value in gravel soil

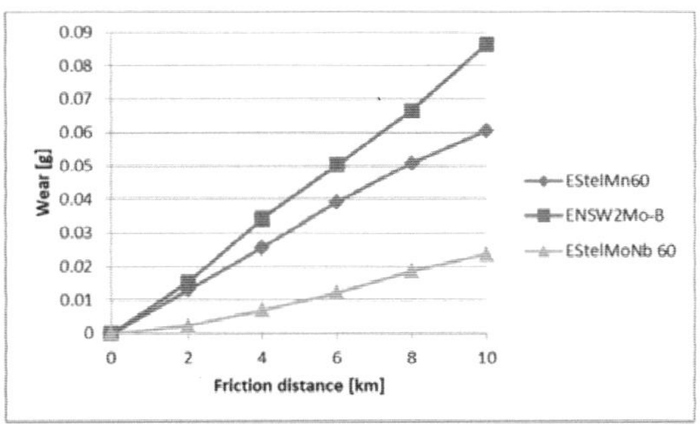

Figure 7. Change of wear value in clay soil

In heavy soil (figure 7), the highest wear was observed for ENSW2Mo-B and it was 0.086 g, the smaller for EStelMoNb60, which was over 3.5 times lower.

5. Summary

The description of the use of weld layers in soil mass is very difficult. It is caused by a multitude of factors affecting the process of material destruction. Many of them result from soil properties (grain size, compactness, humidity), other from material properties. The conducted research has shown over three times greater the wear of weld layers in diversified soil conditions: gravel and clay soil.

It can be concluded that the intensity of wear of weld layers in soil depends primarily on the share and distribution of primary carbides and eutectics on their cross-section.

Reference

[1]. Napiórkowski Jerzy 2005 Polish Soc. Agricult. Eng. **9(12)** 3-171.
[2]. Napiórkowski Jerzy (ed.). *Research and modeling of abrasion and fatigue wear processes*. publishing house University of Warmia and Mazury in Olsztyn, 2014.
[3]. Kovaříková Ingrid et al. 2009 Mater. Sci. Tech. **1** 1-8.
[4]. MIKOŁAjczak Paweł, Napiórkowski J, *Eksploatacja i Niezawodność,* **18(4)** 539-543.
[5]. Nalbant, Muammer; Palali, A. Tufan 2011 Turkish J. Agricult. Forestry, **35(3)** 215-223.
[6]. Bhakat A K, Mishra A K, Mishra N S 2007 Wear **263(1-6)** 228-233.
[7]. Lemecha M, Napiórkowski J, Konat, Ł. 2017 *Tribologia* **3** 101-109.
[8]. Stachowiak G W, Batchelor A W. *Engineering tribology*. 3rd ed. Amsterdam, Boston: Elsevier Butterworth-Heinemann 2005.
[9]. Napiórkowski J, Drożyner P, KołAkowski K, Mikolajczak P. 2012 Int. Virt. J. Sci. Tech. Innovat. Ind. **6(11)** 27–29.
[10]. Napiórkowski J. Ligier, K. Tribologia, 2018.
[11]. Goel G, Cherukuri Harish P, Toro A. 2012 A numerical study of abrasive wear in tillage tools due to soil-tool interaction. In: Simulia, Rhode Island Convention.

Study on Effects of Water Drop for High Concrete Face Rockfill Dams in Narrow Valley

Yucong Gao[1], Rong Chen[1], Dongxue Hao[1], Myoungsoo Won[2]
[1]Northeast Electric Power University, School of Civil Engineering and Architecture, Jilin, China
[2]Kunsan National University, Department of Civil Engineering, North Jeolla Province Gunsan, South Korea

Abstract. In order to analyze the influence of sudden drawdown of dam water level of high concrete face rockfill dam (CFRD), the paper took a narrow valley high CFRD project for example and conducted a comparative study of stress and deformation of panel and dam between different drop amplitude of dam water level. An elastoplastic hardening soil (HS) constitutive model was used to predict the behavior of CFRD. Then, through analyzing the calculation result of six calculation Cases, the paper summarized the effect of dam water level drawdown speed and amplitude on deformation and stress respectively of panel and rockfill zones. The results showed that owing to the arching effect of the CFRD in narrow valley, total deformation of concrete face slab occurring near the dam crest. However, the pressure and tensile stress of concrete face slab along the dam axial direction have a significant influence on the separation of the slab pieces, and potential seepage at the joints in the slab. The study indicates that the hardening soil model could better to analysis of the behavior of high CFRDs and to predict the performances of high CFRDs.

1. Introduction

Concrete faced rockfill dams, which covered by concrete slabs at the upstream face as part of an impermeable barrier for the water, have been broadly constructed in many parts of the world due to their various advantages, such as good adaptability to topography, inherent flexibility and lowest-cost [1-3]. Compare with other types of dams, CFRD has own advantage, however, some high CFRDs, such as Campos Novos (Brazil, 202m height), Barra Grande (Brazil, 185m height) and Tianshengqiao (178m, China), have significant structural failure because of concrete-slab fractures which are clearly related to stresses on the concrete facing [4]. Because of the particle size, material type and purpose of the rockfill, the dam body is usually divided into zones, shown in figure 1 [5]. Since the 2 and 3A zones have little effect on deformation of concrete face, therefore those two zones are not taken into account in models.

2. Numerical modeling

The finite element program PLAXIS 3D is employed in this paper and describes the effect of different drop amplitude of dam water level on high CFRD in narrow valley. The earliest research was conducted CFRDs in 2D plane strain conditions, however, 2D plane strain conditions are not representative of reality because of the cross-valley arching effect which plays an important role in high CFRDs, therefore, 3D conditions are more appropriate than 2D conditions for modeling such a dam.

This CFRD was analyzed with the height of 180.0m and crest length of 540m is similar to NN2 CFRD located in the Lao People's Democratic Republic. This dam has an upstream slope of 1V:1.4H (vertical: horizontal) and a downstream slope of 1V: 1.4H. The dam is constructed on a narrow valley with 40m wide and abutment slopes are 35.8° for both left and right abutments. The CFRD simulation has been conducted under 3D modeling with 75233 4-node tetrahedral elements and 105237 nodes. Because of the foundation and abutment are not main point of this research, they were assumed to be linearly elastic, with a Poisson's ratio of 0.2 and a Young's modulus of 5.0GPa. The concrete face slab was built by 3D solid element with thickness varied from 0.3m at the dam crest to 0.8m at the toe, and assumed to be linearly elastic, with Poisson's ratio of 0.2 and a Young's modulus of 20.0GPa. The 3D CFRD mesh for this study generated by PLAXIS is illustrated in figure 2. To simplify the dam simulations, the construction stage was divided into 37steps, there is one steps for the first impoundment stage and the dam water level reach 170m. The drop amplitude of dam water level in the same velocity as 5m/day are 25m, 50m, 75m, 100m, 125m and 150m, respectively.

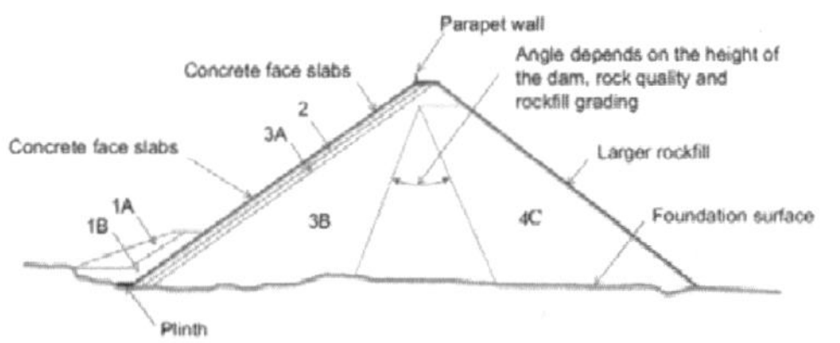

Figure 1. Typical CFRD configuration.

Figure 2. 3D finite element mesh.

An elastoplastic hardening soil (HS) constitutive model was used to predict the behavior of CFRD. In theory, the dilatancy behaviour has an influence on the volumetric strain for real behavior of dense sand. Therefore, the HS model is more suitable in representing the rockfill behaviour than the hyperbolic elastic model, especially for the analysis of high rockfill dam [6]. Such parameters for the CFRD simulation in this study are shown in table 1. The unit weights of the rockfill are 21.0kN/m^3 for zone 4C and 21.5kN/m^3 for the zone 3B and others.

Table 1. Parameters for the CFRD.

	φ	ψ	E_{50}^{ref} (MPa)	E_{oed}^{ref} (MPa)	m	R_f	Other
3B	47.09	3.2	80	55	0.29	0.78	E_{ur}^{ref}= 3E_{50}^{ref}, P_{ref}=100kP$_a$,c=1kPa
4C	42	-5	20	17	0.68	0.68	OCR=1, K_0^{nc}=1-sinφ, ν=0.3

3. Concrete face slab deformation and stress

The overall pattern of deformations and stress indicate that, as example by 75m water level drawdown in figure 3, owing to the load of impounding, the concrete face slab both settles and moves towards the downstream, with the maximum values of pressure stress perpendicular to the dam axial direction occurring approximately in the mid-section and total deformation occurring near the dam crest, since the arching effect of the CFRD in narrow valley. However, the stress along the dam axial direction shows that pressure stress occurring in the mid-section of the slab and tensile stress occurring near the abutment. They have a significant influence on the separation of the slab pieces, and potential seepage at the joints in the slab.

a) Total deformation b) Stress perpendicular to the dam axis c) Stress along the dam axis

Figure 3. Total deformation and stress of the concrete face slab due to impounding

Figure 4 shows the maximum values of deformation and stress of the slab. The maximum values of total deformation decrease with the drop amplitude of water level decreased. However, the maximum values in the water level of 145m which after drop amplitude of 25m has discrepancy. The figure 4 (b) indicates that both pressure and tensile stress, the increasing amplitude of maximum values of stresses along the dam axis decreased as lower dam water level after drawdown. The figure 4 (c) shows that the water level after drawdown has little influence on maximum stress of perpendicular to dam axis.

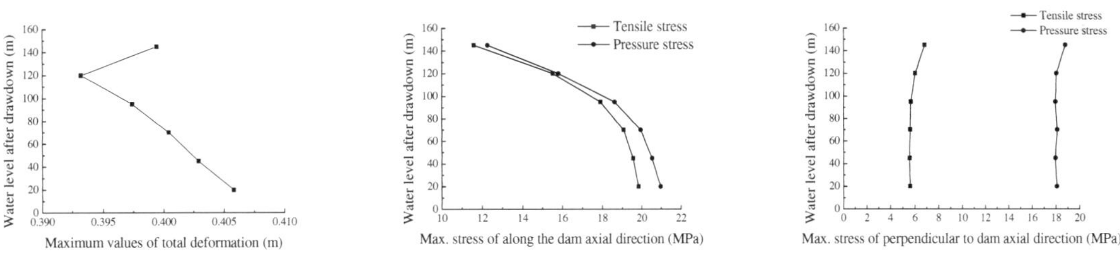

a) Maximum values of total deformation b) Maximum values of stress c) Maximum values of stress

Figure 4. Maximum values of deformation and stress after different amplitude of water level drawdown.

4. Settlement behavior of CFRD

Examples by 75m drop amplitude of dam water level, the figure 5 (a) and (b) show the settlement of CFRD after drawdown for transection of perpendicular (ToP) and transection of along (ToA) the dam axis, respectively. The maximum settlement of rockfill for ToP simulated from both models occurs in zone 4C at downstream side since the stiffness of rockfill in zone 4C is lower than those in zones 3B. The maximum settlement of rockfill for ToA occurs in the mid-section of rockfill and mirrored at the valley axis. The figure 5 (c) shows that maximum amplitude values of settlement for ToP and ToA decreased with the lower of drop amplitude of dam water level.

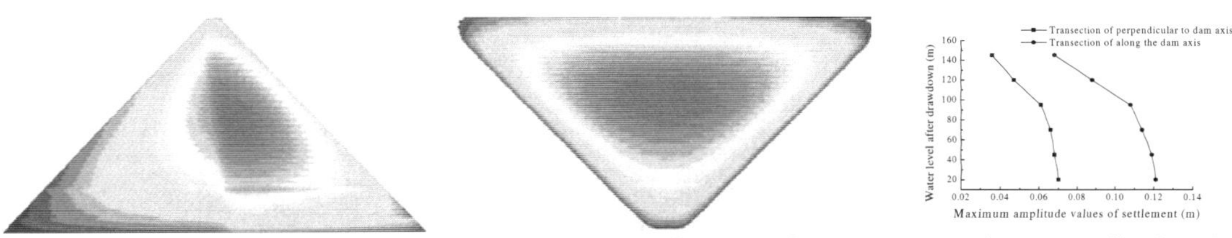

a) Settlement for ToP after drawdown b) Settlement for ToA after drawdown c) Maximum amplitude values of settlement

Figure 5. Settlement behavior of CFRD after drawdown.

5. Summary

In this study, 3D finite element analysis of high CFRD is carried out and computed the effect of water drop on stress and deformation of high CFRD. The rockfill material is represented by the hardening soil model. The results show that since the narrow valley, the total deformation of concrete face slab occurring near the dam crest; the stress along the dam axial direction shows that pressure stress occurring in the mid-section of the slab and tensile stress occurring near the abutment. They have a significant influence on the separation of the slab pieces, and potential seepage at the joints in the slab; and the maximum amplitude values of settlement for transection of perpendicular and along the dam axis decreased with the lower of drop amplitude of dam water level. The results obtained from 3D non-linear finite element analysis using hardening soil model are found to be in relatively good agreement in predict the performances of high CFRDs and satisfactorily reflect the deformation characteristics of rockfill masses.

References

[1] Hunter G, Glastonbury J, Ang D. and Fell R 2003 *The performance of concrete face rockfill dams*. The University of New South Wales, Sydney, Australlia.
[2] Pinto N L S, Marques F P 1998 *Int. J. Hydropower Dams*, **5(6)** 28-31.
[3] Swaisgood J R 1995 *Estimating deformation of embankment dams caused by earthquakes*. Association of State Dam Safety Officials, Western Regional Conference, Red Lodge, MT, USA, pp. 1-7.
[4] Ma H, and Cao K 2007 *Sci. Chi. Ser. E: Tech. Sci.* **50(Suppl 1)** 20-33.
[5] Cooke J B, and Sherard J L 1987 *J. Geotech. Engrg,* **113(10)** 1113-33.
[6] Pramthawee P, Jongpradist P, Kongkitkul W 2011 *Songklanakarin J. Sci. Technol.* **33(3)** 325-34.

IOP Conf. Series: Materials Science and Engineering **507** (2019) 012034 doi:10.1088/1757-899X/507/1/012034

Preparation of Cu/Co/M (M=Al, Fe, Cr, NiFe) mixed metal oxides and their catalytic activity on selective catalytic reduction of NOx by hydrocarbon

H Liu[1], Z Huang[1, 2]

[1]School of Chemical Engineering and Materials Science, Beijing Institute of Technology, Zhuhai, Zhuhai, 519088, China
[2]Zhuhai Jinggong detection technology limited company, Zhuhai, 519088, China

Abstract. Cu/Co/M (M=Al, Fe, Cr, NiFe) mixed metal oxides (Cu/Co/M-MMOs) were prepared by calcination of Cu/Co/M layered double hydroxide (Cu/Co/M-LDHs) precursors. Their composition, thermal stability, structure and morphology were characterized by ICP, TGA, XRD, IR, BET and SEM, respectively. The results indicated that all precursor samples show good hydrotalcite structure with carbonate inside the layer, and the carbonate is removed by calcination at 773 K. While the mixed metal oxides have spinel structure with specific areas of 20-110 m^2 g^{-1}. Cu/Co/M-MMOs have hexagon structure with particle size in the range of 50-100 nm. The order of catalytic activities of Cu/Co/M-MMOs on selective catalytic reduction of NOx is: Cu/Co-MMOs>Cu/Co/Ni/Fe-MMOs>Cu/Co/Fe-MMOs>Cu/Co/Al-MMOs≈Cu/Co/Cr-MMOs.

1. Introduction

With the rapid development of economy and increasing concern of environment, more and more attention has been paid to air pollution. One of the principal environmental concerns in air pollution is the minimization of NOx emissions. Research in the field of NOx abatement has grown significantly in the last two decades [1]. Several methods have been used to reduce NOx emission. One of the main methods for the removal of NOx is selective catalytic reduction of NOx with NH_3 (ammonia-SCR) [2, 3]. SCR of NOx is a highly efficient way to reduce NOx emission in the oxygen-rich exhausts [4]. Mixed metal oxides based catalysts have been widely studied in the area of catalytic removal of NOx due to their high hydrothermal stability and moderate catalytic activity [5]. But ammonia-SCR is not environment friendly. Hydrocarbon is a kind of reductant for SCR of NOx (HC-SCR) attracts more and more interest due to its low price and easy obtained as well as its formation is similar to that found in the exhaust [3, 6]. Layered double hydroxides derived mixed metal oxides catalysts have attracted much attention as potential candidates for the catalytic removal of NOx from lean-burn and diesel exhaust gases under excess oxygen conditions [7]. They have showed good catalytic activities on SCR of NOx at low temperature and in the presence of SO_2 and H_2O [8, 9]. Zhang Z. reported behavior of soot combustion with NOx/O_2 on potassium-supported Mg-Al hydrotalcite mixed oxides using in situ FTIR, the results showed that the presence of NOx in O_2 favors the soot combustion at lower temperatures (<573 K) [10]. Huang reported that $MnFeO_x@TiO_2$ catalyst with a core-shell structure shows excellent K tolerance on SCR of NOx by NH_3 as reductant [11]. Gao have done research on activity and SO_2 tolerance of Co- or Ni-doped MnO_x-CeO_2 catalysts for SCR of NOx with NH_3 at low temperature, the results showed that Co/Ni-modified MnOx-CeO_2 catalysts showed higher SCR performance and SO_2 tolerance [12]. However only few researches on HC-SCR of NOx by using hydrotalcites derived mixed metal oxides catalysts. Therefore, series of Cu/Co/M-MMOs (M=Al/Fe/Cr/NiFe) were prepared in this paper and their catalytic activity of SCR of NOx using hydrocarbon as reductant was studied.

2. Experimental

2.1. Preparation of Cu/Co/M-LDHs precursor and Cu/Co/M-MMOs

Cu/Co/M-LDHs were synthesized by co-precipitation method. Metal nitrate of Cu, Co, Al, Fe, Cr, Ni and NaOH as well as Na_2CO_3 were used as raw materials. Two solutions, namely, solution (A) containing desired amount of metal nitrates were added slowly into solution (B) containing the precipitating agents (NaOH as well as Na_2CO_3). The pH value of the mixed solution was maintained constant around 9 under vigorous mechanical stirring. After aging for 24h, the obtained slurry was filtered and washed thoroughly with distilled water before being dried overnight in an oven under 373 K. The obtained Cu/Co-LDHs, Cu/Co/Al-LDHs, Cu/Co/Fe-LDHs, Cu/Co/Cr-LDHs and Cu/Co/Ni/Fe-LDHs were grinded into powder. Cu/Co/M-MMOs were prepared by calcination of Cu/Co/M-LDHs precursors in a temperature controlled muffle furnace at 773 K for 4 h. Finally, the catalysts (signed as Cu/Co-MMOs, Cu/Co/Al-MMOs, Cu/Co/Fe-MMOs, Cu/Co/Cr-MMOs and Cu/Co/Ni/Fe-MMOs, respectively) were cooled down to ambient temperature and ready for activity test.

2.2. Characterization of Cu/Co/M-LDHs precursor and Cu/Co/M-MMOs

Inductive Coupled Plasma (ICP) was used to determine the elemental composition of as prepared solid precursor sample (JY, model ULTIMA). Thermo gravimetric analysis (TGA) of the as-synthesized sample was carried out on a HCT-2 thermo balance at a heating rate of 10 $Kmin^{-1}$ from room temperature to 873 K in a flowing nitrogen atmosphere. X-ray diffraction (XRD) patterns were determined using a PANalytical X'Pert PRO MPD diffractometer equipped with Cu Kα radiation (λ=1.5406 Å) operating at 40 kV and 40 mA. Fourier Transform Infrared spectra (FT-IR) were recorded on a Brucker VECTOR22 spectrometer in 4000-450 cm^{-1} using KBr pellet technique. Scanning electronic microscopy (SEM) images were taken with a Hitachi S-4800 field-emission electron microscope. The BET surface area was evaluated by N_2 adsorption in a constant volume adsorption apparatus (Micrometer ASAP2010).

2.3. Catalytic test of Cu/Co/M-MMOs

Cu/Co/M-MMOs catalysts were grinded to a particle size of <250 μm. The catalytic test of the sample was took place in a fixed bed reactor, which was made of quartz. 0.24~0.26g of catalysts was placed into the reactor. Prior to the tests, the catalyst was pre-treated by heating in 5% O_2/Ar with a total flow rate of 50 cm^3 min^{-1} up to 573K for 1 h and then cooled down to 323K in flowing Ar. The pre-treated catalyst was introduced to the NOx reduction feed consisting of 600 ppm NO, 600 ppm C_3H_6, 12.0% O_2, 7.0% H_2O, 6.0 % CO_2, and He balance. The total flow rate of the NOx reduction was 200 cm^3 min^{-1}. The reaction temperature was increased with step of 50K from 323 to 823 K holding every temperature for 0.5 h. The concentrations of NO and NO_2 in the reactor effluent was continuously measured by a heated vacuum Chemiluminescent NOx analyzer (4000VM, SINGAL). Figure 1 show the scheme of the catalytic test of the Cu/Co/M-MMOs catalysts.

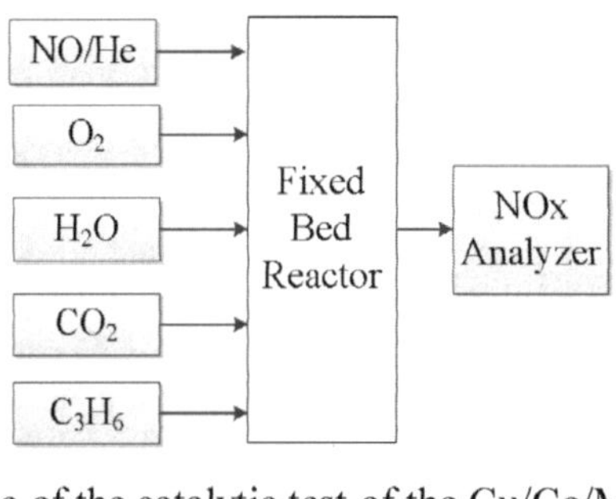

Figure 1. Scheme of the catalytic test of the Cu/Co/M-MMOs catalysts.

3. Results and discussion

3.1. Elemental composition of Cu/Co/M-LDHs

Elemental composition of the prepared Cu/Co/M-LDHs was determined by ICP. ICP results of elemental composition of Cu/Co/M-LDHs are listed in table 1. It can be observed that all the metal contents in prepared sample solids are similar with those in solution. Cu/Co-LDHs sample has higher Cu/Co molar ratios in solid than that in solution. The fact that Cu is the only M^{2+} metal leads to the hardness of LDHs formation.

Table 1. ICP results of elemental composition of Cu/Co/M-LDHs.

sample	Molar ratios of n (Cu) : n (Co) : n (M)	
	In solution	In solid
Cu/Co-LDHs	4:1	4.88:1.00
Cu/Co/Al-LDHs	1:2:1	1.00:1.88:0.95
Cu/Co/Fe-LDHs	1:3:2	1.00:2.92:1.99
Cu/Co/Cr-LDHs	1:3:2	1.00:2.80:1.83
Cu/Co/Ni/Fe-LDHs	3:9:4:8	3.24:9.04:4.00:8.32

3.2. Thermal stabilities of Cu/Co/M-LDHs

Thermal stabilities of Cu/Co/M-LDHs were investigated by TGA. TGA curves of Cu/Co/M-LDHs precursors are shown in figure 2. It can be observed that different samples have different weight loss stages. Only one weight loss stage is observed for Cu/Co-LDHs. Two weight loss stages are observed for Cu/Co/Fe-LDHs and Cu/Co/Ni/Fe-LDHs samples. While three weight loss stages are observed for Cu/Co/Fe–LDHs and Cu/Co/Cr-LDHs samples. It can be concluded that the thermal decomposition of the LDHs precursors proceeded in one to three steps. The low temperature weight loss at (< 473 k) is due to the departure of weakly bonded water molecules and surface water. The second one at high temperature (473–673 K) corresponds to the dehydroxylation of the brucite–like layer as well as part of the interlayer carbonate anions. The high temperature weight loss at (> 673 k) is further dehydroxylation of the layer. The total weight loss is around 30% (by wt.). Thus, Cu/Co/M-LDHs precursors were converted into Cu/Co/M–MMOs after calcinations at 773K.

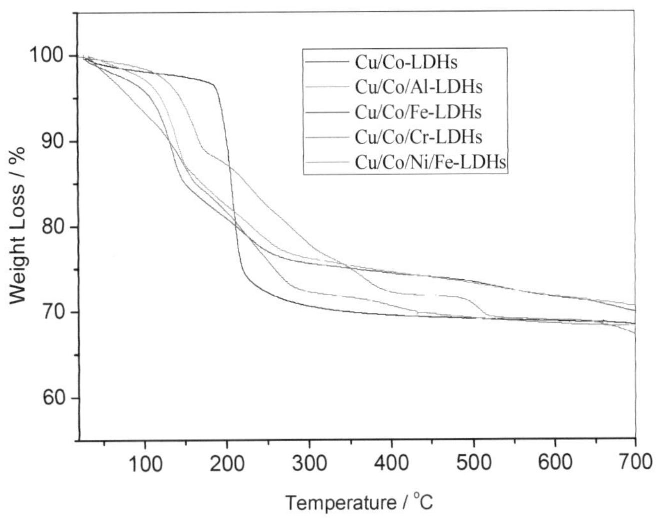

Figure 2. TGA curves of Cu/Co/M-LDHs precursors.

3.3. Structure of Cu/Co/M-LDHs and Cu/Co/M-MMOs

XRD was used to characterize the crystalline structure of the sample. XRD patterns for Cu/Co/M-LDHs precursors are shown in figure 3. It can be seen that all the patterns in the figure show typical LDHs structure with peaks at 2θ = 11.8 °, 23.6 °, 34.7 °. It is indicated that the samples are well crystallized with high rhombohedral symmetry layer. The first peak of Cu/Co-LDHs shifts to higher angle due to the Jahn-Teller effect of Cu. Furthermore, CuO phase was found in Cu/Co-LDHs sample. It can be also explained by the Jahn-Teller effect of Cu which leads to the hardness of Cu^{2+} to introduce into the LDHs layer. Thus, the Cu^{2+} formed into CuO phase and existed in the sample.

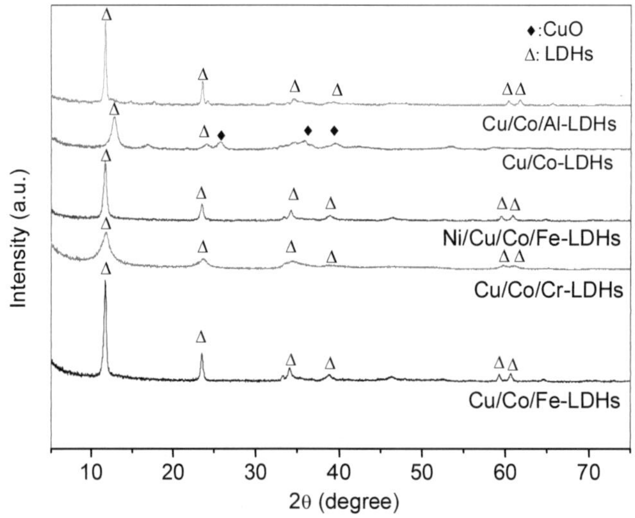

Figure 3. XRD patterns for Cu/Co/M-LDHs precursors.

XRD patterns for Cu/Co/M-MMOs are shown in figure 4. It can be observed that all typical peaks of LDHs are disappeared. Broad peaks at 2θ = 32 °, 37 °, 45 °, 59 ° and 66 ° were observed for all samples which was assigned to spinel phase. This means that Cu/Co/M-LDHs precursors are formed into spinel phase after calcination. NiO phase was seen for Cu/Co/Ni/Fe-MMOs sample. Sharp peaks of CuO phase were found for Cu/Co/Cr-MMOs sample. This drastic change in the diffract patterns reveals that thermal decomposition of the LDHs takes place *via* a highly disordered structure with small particle sizes.

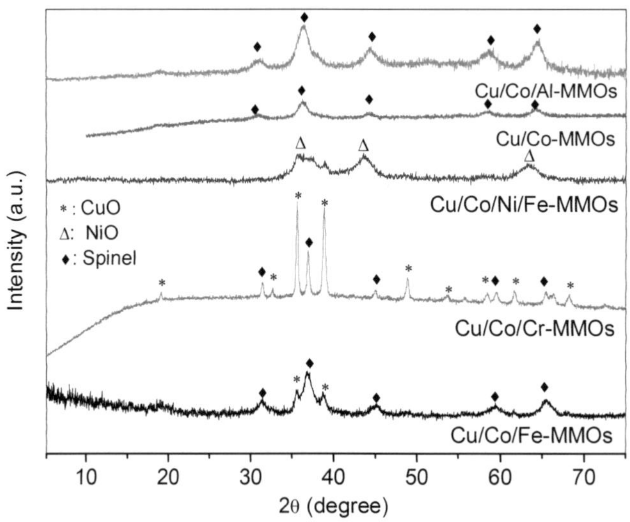

Figure 4. XRD patterns for Cu/Co/M-MMOs.

IR spectra were also taken in order to analyze the anions between layers. IR spectra are shown in figure 5. The broad peaks with low intensity at 3400, 1350 and 1480 cm^{-1} are assigned to –OH in the layer, as well as H_2O and CO_3^{2-} between the layers. For Cu/Co-LDHs and Cu/Co/Al-LDHs samples, absorption peak around 1400 cm^{-1} is appeared, which is assigned to NO_3^-. It is indicated that both carbonate and nitrate are existing in the layers.

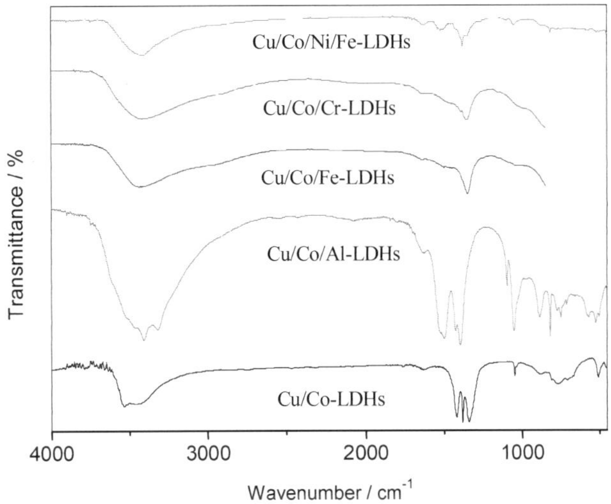

Figure 5. IR spectra for Cu/Co/M-MMOs.

BET results for specific surface areas and pore diameters of Cu/Co/M-MMOs are listed in table 2. It is showed that Cu/Co-MMOs and Cu/Co/Al-MMOs have lower specific surface areas of 20~40 m^2g^{-1}. Cu/Co/Fe-MMOs, Cu/Co/Cr-MMOs and Cu/Co/Ni/Fe-MMOs have higher specific surface area of 80~110 m^2g^{-1}. The pore diameters are around 5~10 nm.

Table 2. BET results for specific surface areas and pore diameters of Cu/Co/M-MMOs.

Sample	BET results	
	BET surface area(m^2g^{-1})	Pore diameters(nm)
Cu/Co-MMOs	20.9	5.73
Cu/Co/Al-MMOs	37.5	5.44
Cu/Co/Fe-MMOs	85.9	7.30
Cu/Co/Cr-MMOs	107.8	10.24
Cu/Co/Ni/Fe-MMOs	94.4	7.65

3.4. Morphology of Cu/Co/M-MMOs

Morphology of Cu/Co/M-MMOs was also investigated by SEM. SEM photos are shown in figure 6. It can be observed that Cu/Co/Fe-MMOs, Cu/Co/Al-MMOs, and Cu/Co/Ni/Fe-MMOs show hexagon flakes with particle size in the range of 50~100 nm. Cu/Cu/Cr-MMOs show homogeneous regular particle with smaller size of 20~30 nm. Cu/Co-MMOs show both particles and bigger rod shape in the photo. It is not that well dispersed like the other samples. This is why Cu/Co-MMOs have lower specific surface areas than the other samples, which is in correspondence with BET results.

3.5. Catalytic activities of Cu/Co/M-MMOs

The catalytic activities of Cu/Co/M-MMOs were studied under NOx analyzer. NOx conversion rate curves catalyzed by Cu/Co/M-MMOs are shown in figure 7. It can be observed that different Cu/Co/M–MMOs samples show different NOx conversion rate. Cu/Co-MMOs sample exhibits the highest NOx conversion of 23%. It shows better NOx conversion rate than the other samples though it has the lowest specific surface areas. It also shows higher NOx conversion rate at lower temperature range. Thus it can be concluded that higher specific surface area is not necessarily corresponding to higher catalytic activity for Cu/Co/M-MMOs catalysts. Cu/CoFe-MMOs, Cu/Co/Al-MMOs and Cu/Co/Cr-MMOs have similar NOx conversion curves. SCR of NOx by using hydrocarbon as reductant show lower catalytic activities than by using NH_3. The catalytic activity order is: Cu/Co-MMOs>Cu/Co/Ni/Fe-MMOs>Cu/Co/Fe-MMOs>Cu/Co/Al-MMOs≈Cu/Co/Cr-MMOs.

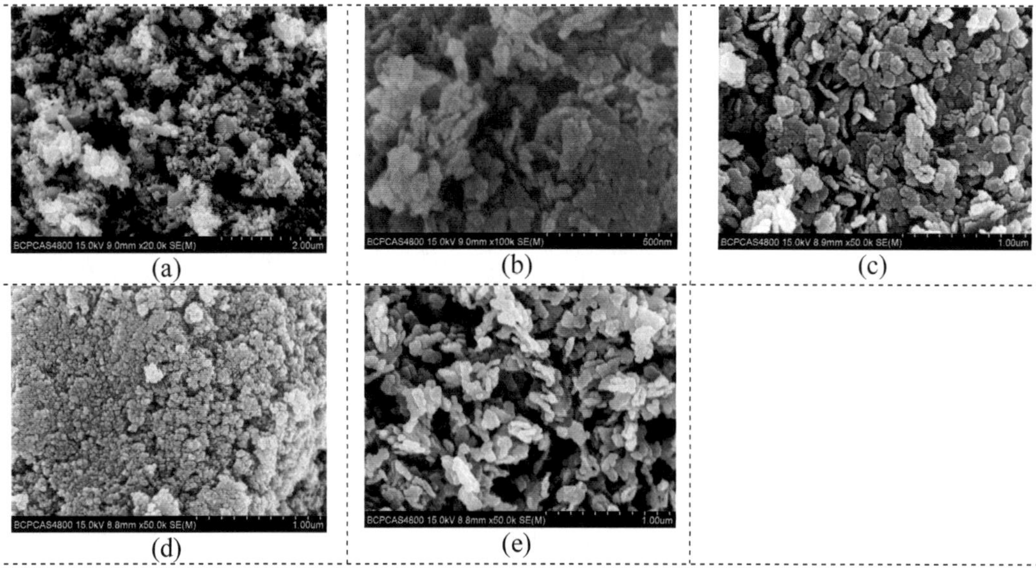

Figure 6. SEM photos for Cu/Co/M-MMOs, (a) Cu/Co, (b) Cu/Co/Al, (c) Cu/Co/Fe, (d) Cu/Co/Cr and (e) Cu/Co/Ni/Fe-MMOs.

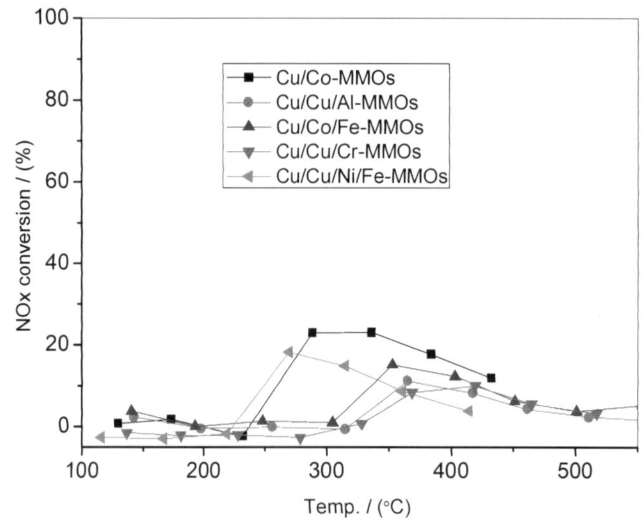

Figure 7. NOx conversion curve catalyzed by Cu/Co/M-MMOs.

4. Conclusions

TGA, XRD and IR results show that Cu/Co/M-LDHs precursors and Cu/Co/M-MMOs catalysts were successfully prepared. Cu/Co/M-MMOs catalysts have spinel phase. The element compositions of catalysts are similar with those in solution. The catalysts have homogeneous distribution of hexagon flakes and particles size in the range of 20~100nm. Their specific surface areas are in the range of 20~110 m^2g^{-1}. Cu/Co/M–MMOs catalysts show some activities on SCR of NOx by hydrocarbon. Cu/Co/M–MMOs prepared with different metal show different catalytic activities. Cu/Co–MMOs sample exhibits the highest NOx conversion of 23%. The order of catalytic activity of Cu/Co/M is: Cu/Co-MMOs>Cu/Co/Ni/Fe-MMOs>Cu/Co/Fe-MMOs>Cu/Co/Al-MMOs≈Cu/Co/Cr-MMOs. SCR of NOx by using hydrocarbon as reductant show lower catalytic activities than using NH_3.

Acknowledgments

This work was financially supported by the program of excellent young teachers in higher education institution in Guangdong province (2015). English language improvement by Dr Shirong Wang from Perkin Elmer, Australia is also appreciated.

References

[1] Roy S, Hegde M S, Madras G 2009 *Appl. Energy* **86** 2283.
[2] Gómez-García M A, Pitchon V, Kiennemann A 2005 *Environ. Int.* **31** 445.
[3] Mrada R, Aissat A, Cousin R 2015 *Appl. Catal.* A 542.
[4] Xie P, Deng L, Zhong X 2017 *Nat. Conf. on Envir.Catal. Envir. Mater. (Changsha)* pp. 346.
[5] Zhou H, Su Y, Deng W 2016 *Envir. Sci. Tech.* **39** 93.
[6 Liu F, Yu Y, He H 2014 *Chem Commun.* **50** 8445.
[7] Liu H, Huang Z, Li L 2014 *Adv. Mater. Res.* **1033-1034** 1058.
[8] Fornasari G, TrifiròF F, Vaccari A 2002 *Catal. Today* **75** 421.
[9] Liu Y, Yang X, Zhang Z 1999 *Chinese J. Catal.* **20** 450.
[10] Zhang Z, Zhang Y, Su Q 2010 *Envir. Sci. Tech.* **44** 8254.
[11] Huang C, Guo R, Pan W 2018 *J. Ener. Ins.* Online DOI:10.1016/j.joei.2018.09.005.
[12] Gao F, Tang X, Yi H 2017 *Chem. Engin. J.* **317** 20.

IOP Conf. Series: Materials Science and Engineering **507** (2019) 012035 doi:10.1088/1757-899X/507/1/012035

Prototype and Evaluation of Cutting Force Testing Machine for Hand Work Tools

Takanori Yamazaki[1], Takahito Sato[2],

[1]School of Science and Engineering, Tokyo Denki University, JAPAN
[2]Graduate School of Science and Engineering, Tokyo Denki University, JAPAN

Abstract. The goal of our study is to find a cutting edge angle with high cutting ability and durability for hand work tool. For this purpose, we developed a device to measure the cutting force of the hand work tool and evaluate its performance. At first, we examined the fixed method to correctly measure cutting force. Next, we perform many cutting experiments and evaluated the performance of the prototype force testing machine. After the validity of the testing machine has been proved, the relation between the cutting edge angle and the cutting force was examined for the three manufacturer.

1. Introduction

Pliers and Nipper are widely used as the hand work tool. In recent years even if various processes are automated, the blade sharpening process is manual operation. In case of the blade sharpening process, it focuses that the no gaps when aligning blades. That is, there is only an approximate standard for the cutting edge angle. It seems that this process causes variations in the product quality.

In this study, our goal is to find a cutting edge angle with high cutting ability and durability. For this purpose, we need a device to evaluate the cutting force, but there is no the dedicated device for now. In this study we manufacture a prototype of cutting force testing machine for the hand tool and evaluate the cutting force when cutting the copper wire.

2. Prototype cutting force testing machine

In order to clarify the relationship between the cutting edge angle and cutting force, an apparatus for evaluating cutting force is required. There is no device to measure the cutting force of a work tool such as Nipper, it seems to measure it by attaching a special jig to the tensile testing machine. Therefore, we designed and manufactured a cutting force test machine at first.

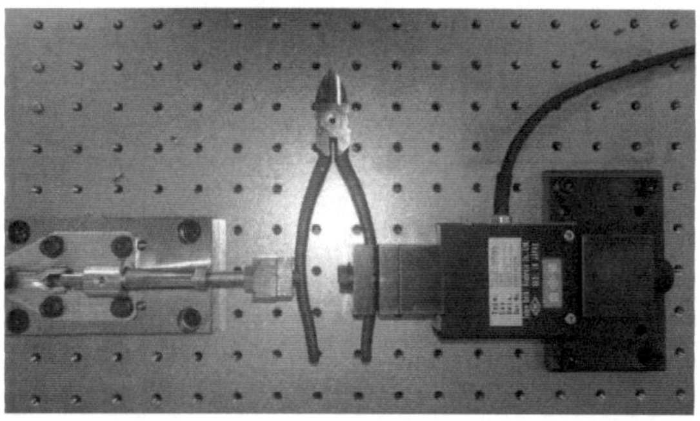

Figure 1. Prototype cutting force testing machine (Photo).

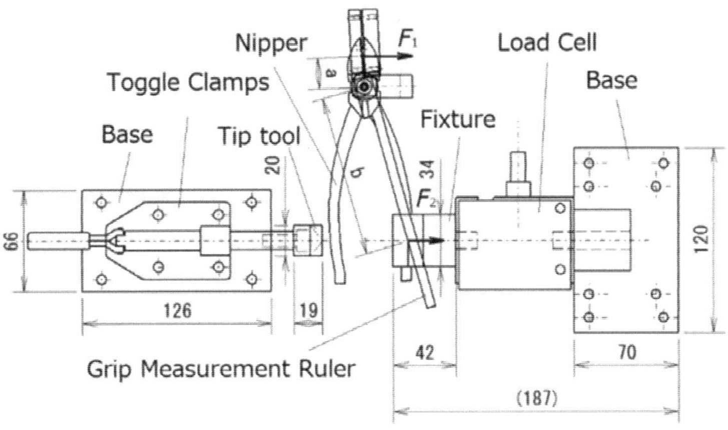

Figure 2. Prototype cutting force testing machine (Drawing).

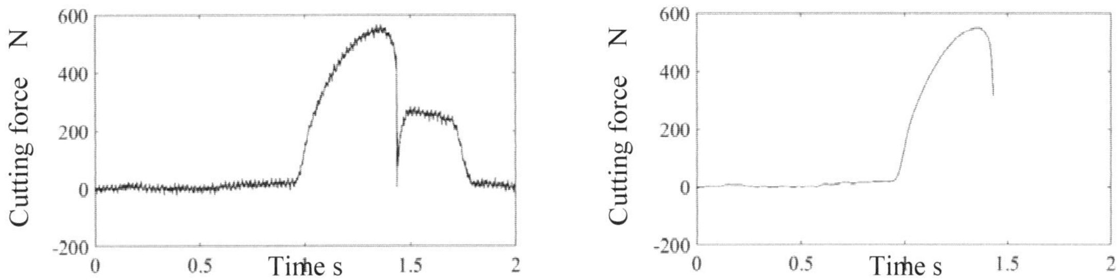

(a) Without signal processing (b) With signal processing

Figure 3. Force applied to blade.

2.1. Device configuration.

The cutting force test machine manufactured is shown in figure 1. The right side of the figure is the fixed side, it is comprised of base, load cell and fixture. The load cell (TCLZ-10KNA, rated capacity: 10 kN, rated loutput: 4000×10^{-6}) and the dynamic strain amplifier (DC-97A, rated output: 1~10 V, sensitivity: 50~10000 μS) made by TML. Fixtures are replaceable and can be selected from the surfaces, line and point fixation. The left side of the figure is the load imparting side, it is comprised of base, toggle clamp (MC07-3, stroke: 30 mm) made by MISUMI and tip tool.

The relationship between the force applied rod and the force acting on edge is shown in the equation (1). Here, F_1 is the force acting on edge, F_2 is the force applied rod, a is the distance from the pivot point to a wire, and b is the distance from the pivot point to a rod.

$$F_1=F_2\,(b/a) \quad (1)$$

Figure 3(a) shows an example of the cutting force measured when the copper wire is cut in actual, where the experimental condition is a = 20 mm, b = 100 mm. The part where measurement value goes up and down largely, it is considered that the copper wire was cut and caused by hitting the left and right blades. Therefore, we removed this part because it has nothing to do with the sharpness of the blade. Because the measured value contains noise, we smooth it using a moving average filter. The waveform through this filter are shown in figure 3(b). Assuming that the maximum value after these numerical processes is F_{max}, the force actually acting on the rod can be obtained by the equation (2). Here, V_I is the input voltage, R_o is the rated output voltage.

$$F_2=V_I\,(F_{max}/R_o) \quad (2)$$

IOP Conf. Series: Materials Science and Engineering **507** (2019) 012035 doi:10.1088/1757-899X/507/1/012035

Table 1. Shape of fixture for hand work tool.

Surface	Line	Point
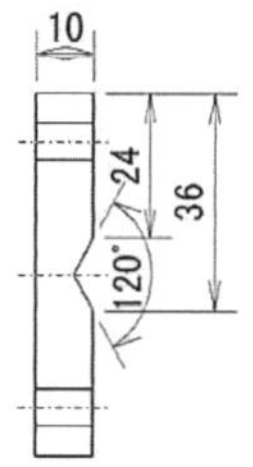	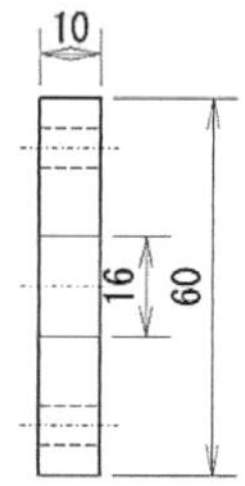	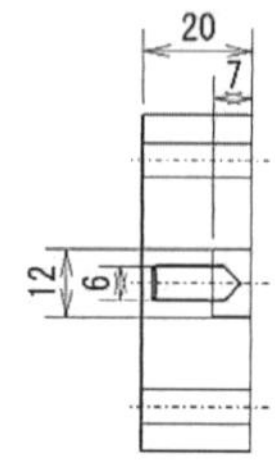

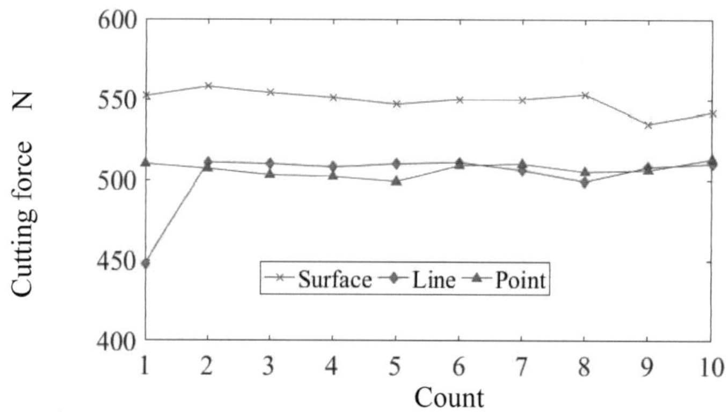

Figure 4. Cutting force with different fixed method.

Table 2. Evaluation of cutting force with different fixed method.

Fixture	Surface	Line	Point
Average force [kN]	0.549	0.502	0.506
Standard deviation	0.00621	0.0184	0.00418

2.2. Comparison of Fixed Method.

To decide fixed method of the work tool when measuring cutting force, three kinds of fixture were designed and compared cutting forces. In this experiment, using the plastic nipper (classified in slant edge cutting nippers), the cutting force was measured for three kinds of fixture. The shape of fixture for hand work tool are summarized in table 1.

From the results shown in figure 4 and table 2 showing the cutting force by different fixed methods and its evaluation, it was found that the cutting force of surface fixed is larger than the other fixed method, and the dispersion of the cutting force due to the fixture is small. The reason that the cutting force of the surface fixed is large, it is considered to be related to the contact area between the nipper and the fixture. In this study, we decided to use surface fixation that is stable and has small variations.

2.3. Evaluation of testing machine (change in cutting place).

As one of the evaluation items of the device, it was verified whether the cutting force varies depending on the cutting position of the wire. Table 3 shows the result of average cutting force and standard deviation. There are few differences between average force and standard deviation. It is considered that the value of cutting force is not significantly affected by the cutting position.

Table 3. Evaluation of cutting force with different fixed method.

a mm	17	22
b mm	80	
Average Force [kN]	0.376	0.374
Standard Deviation	0.0239	0.0247

Table 4. Evaluation of cutting force with different fixed method

	①	②	③	④	⑤	⑥	⑦	⑧	⑨	⑩	Total
Average Force [kN]	0.972	0.966	0.947	0.961	0.947	0.949	0.939	0.943	0.938	0.939	0.950
Error Rate	2.30	1.72	0.34	1.14	-0.31	0.10	1.20	0.78	1.28	1.14	-

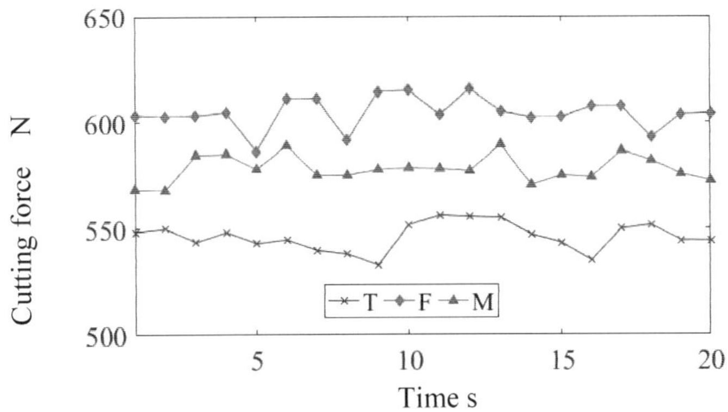

Figure 5. Comparison of cutting force by manufacturer.

2.4. Evaluation of testing machine (repeatability).

To decide the number of measurement, the tool was repositioned after one set (10 measurements) and these were repeated ten times (total 100 measurements). The experimental conditions are as follows: the distance from the pivot point to the wire a = 17 mm and to the rod b = 80 mm, the fixed method is the surface fixation, and the cutting material is copper wire.

Table 4 shows the result of comparing the average cutting force of 10 sets. As described above, one set is the average value of the 10 measurements, for example, in the fourth set, the average value of 10 measurements is 0.961 kN. The average cutting force of the ten sets was 0.950 kN, The difference between the average value of each set and total is at most 2.3%, and the error rate is small. Therefore, the number of measurements is shown to be valid with at least one set (10 measurements).

2.5. Cutting force per manufacturer

In order to examine the relationship between the cutting edge angle and the cutting force, we make the cutting experiment using the nipper of three manufacturer, T, M and F company. In experiment, after cutting the wire material 20 times, measured the cutting edge angle with a microscope.

The measured cutting force of each manufacturer show in figure 5. In this figure, although the measured values have dispersion, it is clear that there is a difference in cutting force for nipper of each manufacturer. Figure 6 is a schematic view of the cutting edge angle, and table 5 shows the cutting edge angle and average load of each company. The nipper with small edge angle has a small average cutting force, in case of large edge angle has a large average force. As a matter of course, the cutting ability can be improved by decreasing the edge angle.

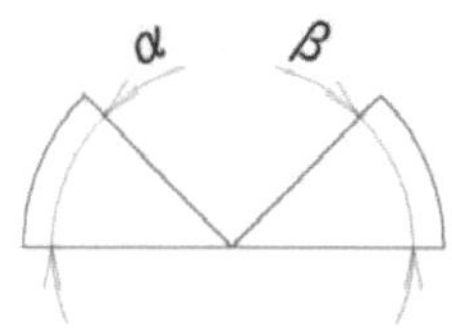

Figure 6. Angle of nipper edge.

Table 5. Comparison of cutting force and edge angle.

	T	F	M
Left edge /α °	45.0	49.1	49.1
Right edge /β °	39.5	45.0	44.8
Average force [N]	546	605	578

3. Summary

In this study, we developed and manufactured device to measure the cutting force. In order to improve the repeatability of the measurement, we first examined the fixed method of the hand work tool and showed that the surface fixation can be measured stably. In the number of measurements, it is shown to be valid with at least one set (10 measurements) as a result of 100 measurements. Moreover, in actual cutting test using the nipper of three manufacturer, the relationship between the edge angle and cutting force was clarified.

Reference

[1] JIS B 4625:1995. Slant edge cutting nippers.
[2] Kanazawa K, Li Y, In S and Koizumi C 1996 *J. Japan Soc. Prec. Eng.* **62** 1340-1344. (Japanese).
[3] Kanazawa K, Nakagawa A and Miura T, 1997 *J. Japan Soc. Prec. Eng.* **63** 70-75. (Japanese).
[4] Onuki T and Kanazawa K, *Cutting of aluminum alloy with a knife−edge cutter*, Materials and processing conference, (2006) 179-180. (Japanese).
[5] Marsot J, Claudon L and Jacqmin M 2007 *Appl. Ergonom.* **38** 83-89.
[6] Karltun J, Vogel K, Bergstrand M and Eklund J, 2016 *Appl. Ergonom.* **56** 92-100.